T
57
.95
.M35
1995

AGD-5854

DECISION MAKING AND FORECASTING
With Emphasis on Model Building and Policy Analysis

McGraw-Hill Series in Industrial Engineering and Management Science

Consulting Editors

Kenneth E. Case, *Department of Industrial Engineering and Management, Oklahoma State University*

Philip M. Wolfe, *Department of Industrial and Management Systems Engineering, Arizona State University*

DECISION MAKING AND FORECASTING
With Emphasis on Model Building and Policy Analysis

Kneale T. Marshall

U.S. Naval Postgraduate School

Robert M. Oliver

University of California at Berkeley

McGraw-Hill, Inc.

New York   St. Louis   San Francisco   Auckland   Bogotá   Caracas   Lisbon
London   Madrid   Mexico City   Milan   Montreal   New Delhi
San Juan   Singapore   Sydney   Tokyo   Toronto

This book was set in Times Roman by American Composition & Graphics, Inc.
The editors were Eric M. Munson and John M. Morriss;
the production supervisor was Denise L. Puryear.
The cover was designed by Ed Smith Design.
Project supervision was done by Tage Publishing Service, Inc.
R. R. Donnelley & Sons Company was printer and binder.

DECISION MAKING AND FORECASTING
With Emphasis on Model Building and Policy Analysis

This book is printed on acid-free paper.

1 2 3 4 5 6 7 8 9 0   DOC   DOC   9 0 9 8 7 6 5

ISBN 0-07-048027-3

Library of Congress Cataloging-in-Publication Data

Marshall, Kneale T.
  Decision making and forecasting: with emphasis on model building
and policy analysis / Kneale T. Marshall, Robert M. Oliver.
      p.   cm.
  Includes bibliographical references and index.
  ISBN 0-07-048027-3
  1. Decision-making—Mathematical models.   2. Operations research.
I. Oliver, Robert M.   II. Title
T57.95.M35   1995
658.4'033—dc20   95-869

ABOUT THE AUTHORS

Kneale T. Marshall is Professor of Operations Research at the Naval Postgraduate School in Monterey, California. He received his doctorate in Operations Research from the University of California at Berkeley in 1966. During his career he has held positions as Chairman of the Operations Research Department, Dean of Information and Policy Sciences, and Provost and Academic Dean. He served as senior science advisor to the Chief of Naval Personnel in the Office of the Chief of Naval Operations, and has been a technical advisor to both private companies and agencies within the Department of Defense. He served as associated editor for *Operations Research* for over 10 years and is author of over 40 papers, journal articles and books in education and manpower forecasting and modeling, and in stochastic processes.

Robert M. Oliver is Professor of Operations Research and Engineering Science at the University of California, Berkeley. He received his doctorate in Physics at the Massachusetts Institute of Technology in 1957 and is a recipient of the Fulbright Scholarship to the University of London and the Lanchester Prize of the Operations Research Society of America for his research in predicting and scheduling U.S. Post Office mail flow and sorting operations. He has held numerous appointments as technical advisor and consultant to corporations and U.S. government agencies and is the author of over 100 papers, journal articles, and books in forecasting, filtering, prediction, and decision-making models of operations research.

To the influence of
Morse and Kimball
and
To Donna and Sophie

CONTENTS

9 Advanced Concepts

PREFACE

The purpose of this book is to bring the scientific method to decision making and forecasting. Our objectives are best summarized from a passage in Morse and Kimball [1951]:

> Operations Research is a scientific method of providing executive departments with a quantitative basis for decisions regarding the operations under their control.

The emphasis throughout this book is to give the reader exposure and experience in the art and science of model-building, to show the close connection between forecasting and decision making, and to illustrate how decision trees and influence diagrams can be used as a common framework for discussion by both the mathematically trained analyst and the client or decision maker who may not have sophisticated quantitative skills.

For more years than we care to admit, both authors have taught courses in Operations Research and Management Science curricula that dealt with applied probability, forecasting, and decision-making models for undergraduate and graduate college students. These courses were taught separately without any attempt to link their common topics and we were always confronted with the inbuilt dilemma of presenting techniques to solve a mathematical problem once the model was clearly laid out before the student rather than explaining how to design models that represent real problems. This natural conflict was reinforced by the apparent lack of rigor in the art of model building when compared with the more formal mathematical foundations of the techniques themselves and by our own frustration in having to teach "standard" operations research and management science courses which seldom focussed on the central theme of making better decisions in real-world environments.

INTEGRATING FORECASTS IN DECISION MODELS

Several important developments have occurred during the past decade which offers all of us not only an opportunity to integrate forecasting and decision making but also to use graphical aids in the formulation and design of mathematical models and the presentation of results to decision makers. The natural language of influence diagrams and decision trees has found substantial application in a small number of industrial consulting companies but is only offered in the curricula of a very few universities. We hope that will change.

The essential ingredients of an integrated forecasting and decision-making formulation are

- development of a clear understanding of how uncertainty is to be handled and a logical framework in which to discuss and analyze rational courses of action.
- clear identification of the sources of uncertainty, identifying the effects of new sources of information and new forecasts and whether the decision maker and forecaster are one and the same or different people or institutions.
- careful and thoughtful problem formulation, including who makes decisions, which factors are known, which ones are uncertain, which ones directly influence decisions, and the timing of information and decisions.
- identification of objectives of the decision maker(s) and the decision-making environment. It is important to understand how objectives have different effects on strategies, and how objectives depend on the client(s).
- recognition of multiple attributes and the understanding of the extent to which there may be tradeoffs among them.
- emphasis on sensitivity analyses, recognizing that models are only crude approximations to real-world problems.
- communication of results and insights to decision makers.

Throughout, we repeatedly make use of four principles: those of coherence of probabilities, consistent units of measurement, optimality in sequential decision making, and parsimony in the construction of models. Perhaps even more important but not explicitly stated is the principle of common sense: if a model gives solutions that do not make sense it is probably the wrong model!

Beginning with single decision problems with only one uncertainty, we study a number of one-stage decision problems for perishable commodities where multiple uncertainties are present. This set of problems includes those known as yield management problems. The extension to sequential decisions, many periods of time, multiple forecasts, several attributes and criteria, and the use of nonlinear utilities are complications which are introduced as one attempts to solve real-world problems.

In this book we see how to design the most important features of realistic decision problems into mathematical models that reveal their structure and give insight. We emphasize model building rather than an axiomatic unifying theory, but references are given where the interested reader can pursue the mathematical theory if he or she chooses.

In this way we hope to bridge the gap between decision theorists and decision makers whose primary interest is to understand their particular problem but who have little time to pursue a unifying theory.

Because we place great emphasis on model formulation we rely on graphical techniques that use influence diagrams and decision trees. The analyses that follow the formulation are themselves often graphical. Such techniques have proved useful both in building models with client input as a prime factor and in analysis to explain results and give insights. We show how forecasting must be integrated with decision making in a coherent manner, recognizing the reality that forecasters and decision makers are almost always different individuals or organizations. We make frequent use of the economic value of forecasts and how these are related to discrimination and calibration of forecasts. The methods and models we describe can be developed on modern personal computer spreadsheet software although there are now several commercial packages linking influence diagrams and decision trees.

Many books on decision theory present a unified and consistent mathematical theory based on a given set of axioms. Emphasis is on the mathematical structure and its properties derived through theorems. The examples are often contrived to illustrate these properties; of necessity there is little regard for realism. Such texts can be thought of as technique oriented. The student studying them becomes well versed in the theory and its properties, but when faced with a real decision problem, often finds it hard to recognize how the theory applies or how one goes about the task of building a mathematical model to explore alternatives and different outcomes. Another approach is to base problem formulation on in-depth and detailed case studies of real problems. Real problems are almost always encumbered with complex details making it hard for the student to understand their structure or the sources of uncertainties. Our goal is to provide a text that lies between abstract theory and detailed case studies.

USE OF THE TEXT IN INSTRUCTION

The level of the text is aimed towards business and economics students entering a quantitative business program or a final year course for science or engineering students. Both of us have taught graduate and undergraduate courses using the material in the book. We have always found that there is more material than can be taught in a one-semester course, particularly with graduate students whose education is primarily mathematical and seem to lack the basic skills of model building.

For undergraduate students Chapter 1 and portions of Chapter 2 can be followed by Chapters 4 and 5 and portions of Chapters 6 and 7. A graduate course could require Chapters 1, 2, and 3 as preliminary reading with emphasis on Chapter 4 and material from 6, 7, 8, and 9 depending on which topics the instructor wanted to emphasize. The prerequisites include a knowledge of calculus, elementary linear algebra, and probability theory; we have found that even the better-prepared graduate students with backgrounds in probability and statistics do not feel comfortable dealing with more than two random variables. Conditional independence and its implications takes substantial time to sink in; working with formulations of real, not toy, problems may be the only way to develop this expertise as common sense can be used to check mathematical results. The text includes a body of core material, followed by more advanced sections that are clearly indicated as such, and which could be used with students possessing a more mature mathematical background. Elementary and advanced problems are included with each chapter. Some are thought to be difficult.

ACKNOWLEDGMENTS

This book had its origins in a seminar at Berkeley during the sabbatical visit of one of the authors and a series of eight lectures given in the fall of 1990 at the Institute of Statistics and Operations Research, Victoria University, New Zealand where the other author was on sabbatical leave. The focus in the lectures was on the application of decision trees and influence diagrams to forecasting and decision analysis. We assembled notes on several problems that either were simplified models of real-world decision problems or which illustrated how forecasting could be integrated with decision making and provide economic value to the endeavor.

We want to thank the faculty, staff, and students of the Institute of Statistics and Operations Research at Victoria University, Wellington, New Zealand for their original interest in the topics and for providing secretarial, computing, and other services for a sabbatical visitor. Corrections to the first set of notes were the result of many helpful suggestions and comments made by G. Anthony Vignaux. We are most appreciative of the support of the chairs of the departments of our host institutions, Professor Peter Purdue of the Operations Research Department at the U.S. Naval Postgraduate School, and Professor Shmuel Oren of the Industrial Engineering and Operations Research Department at the University of California, Berkeley, for affording us the support and encouragement in scheduling the original seminars based on early versions with incomplete notes.

Our many thanks to a large number of students who had to suffer through early offerings of the courses and problems; they are too numerous to name individually except for Zvi Covaliu, a former doctoral student at the University of California, and Teo Weng Lim of the Singapore Air Force, a former masters student at the Naval Postgraduate School. Dr. Covaliu taught an undergraduate version of the course in his final year and unselfishly shared many problem formulations to illustrate new ideas and suggest better ways of teaching them. Both Dr. Covaliu and Captain Lim continually contributed many constructive criticisms that led to correction of errors and greater clarification of our presentation. We are very grateful to both of them. Earlier versions of the book have been used in the classroom by four different teachers. Our reviewers included Zvi Covaliu, George Washington University; Jane Fraser, The Ohio State University; Simon French, University of Leeds; Craig Kirkwood, Arizona State University; Stephen Pollock, University of Michigan; and Robert Winkler, Duke University; all of whom made suggestions that resulted in very substantial changes and, in our opinion, improvements to the manuscript.

We take responsibility for any remaining errors and hope that interested readers who find them, or have suggestions for correcting and improving the text, will share their ideas with us.

Kneale T. Marshall
Robert M. Oliver

DECISION MAKING AND FORECASTING
With Emphasis on Model Building and Policy Analysis

CHAPTER
1

BASIC
CONCEPTS

1.1 INTRODUCTION

The purpose of this book is to present an approach to modeling, analyzing, and integrating forecasts into decision making in insightful ways. The reader might ask, "Why another book on decision making or forecasting?" Many books are available on these topics, a number of which we have drawn on and from which we quote. But most of them emphasize analysis of models rather than model building and treat decision making, information gathering, and forecasting as separate topics. We stress model building, the linkages between decision making and forecasting, graphical methods, and sensitivity analyses that examine the decisions and policy spaces. Throughout the book we are keenly interested in the economic value of information and forecasts derived from relevant data rather than the examination of forecasting techniques isolated from decision making. We have drawn on our own experiences in making decisions in management positions and in teaching the material in graduate and undergraduate courses.

The term *decision making* is very broad, encompassing a wide range of possible topics, from a simple choice between two alternatives, each with known certain outcomes, to an axiomatic mathematical logic found in texts on statistical decision theory. In this book the most important common characteristic of the problems we consider is that decisions must be made when one does not know with certainty their effect because of the randomness of future events. Some of our examples involve using past and current data to forecast and so reduce uncertainty about the future. Others depend heavily on individual subjective judgments about the future, and still others use forecasts that combine expert opinion with historical data. Some have simple outcome measures that are directly quantifiable, and others have conflicting outcome measures not all of which are directly quantifiable. Some involve making sequential decisions over time, so that in making today's decision we must consider what decisions we should make in the future.

It is usual to treat the subject of forecasting uncertain futures separately from decision making. Most of the literature in forecasting makes no more than passing reference to how it might be applied in making decisions. In this book, one of our aims is to demonstrate how forecasting and decision making are necessarily intertwined. Emphasis is always on demonstrating how various factors can be modeled in ways useful to a decision *maker*.

The word *maker* is emphasized to stress that our models are not ones that simply present an "optimal" solution. It may seem redundant to state that decision makers make decisions! But all too often the analyst, who has the technical expertise to develop and analyze models to assist such a person, forgets this fact and is disappointed when the decision maker does not accept his or her "optimal" solution. In fact, the decision maker is often more interested in obtaining insights and understanding how decisions change with varying conditions or assumptions and much less in a particular numerical value that may be a solution under a restrictive set of assumptions. A decision maker at top levels of management is more concerned with the appropriateness of the model than with the mathematical techniques used to solve a given model. That is why in this book we emphasize model building and analysis rather than the mathematical techniques themselves.

We emphasize graphical representations of results as these are often more easily understood than algebraic formulations. They are also useful in showing which decisions are best over ranges of possible parameter values rather than only offering point solutions. Using a word the reader will see repeatedly in this book, our models and the techniques we use to solve them are designed to give insight to the decision maker who has the ultimate responsibility for making the decision. Although we discuss numerical algorithms for making forecasts and solving decision problems, our primary interests are in developing concepts, formulating, and designing good models, rather than in numerically efficient computations.

The remainder of this chapter describes the essentials of the type of decision problems discussed in this book, sets up the mathematical notation used, reviews the material the reader will need in the area of probability theory, introduces the reader to the elements of influence diagrams and decision trees, and illustrates their use in model building and analysis through a number of examples. This chapter concludes with an overview of the remaining chapters in the book.

1.2 THE IMPORTANCE OF MODELS IN DECISION MAKING

A significant portion of this book deals with what is conventionally called *Decision Analysis*, including the use of influence diagrams and decision trees, suitable methods for measuring outcomes, and preferences among choices. Decision analysis has proved extremely useful for "executive" or "strategic" decision making. An article by Corner and Kirkwood in 1991 lists over eighty applications published in the open literature from 1970 through 1989. The rate of growth in applications is very great, and the fields where they have been applied include environmental, medical, management, legal, accounting, transportation, finance, and governance, to name only a few. To quote from their paper,

> Decision analysis provides tools for quantitatively analyzing decisions with uncertainty and/or multiple conflicting objectives. These tools are especially useful when there is limited directly relevant data so that expert judgment plays a significant role in the decision making process. Such situations include government policy making and regulation, strategic business decisions, and such risky personal decisions as selecting a treatment for a serious medical problem. (p. 206)

The importance of making quantitative formulations of decision problems cannot be overemphasized, even though they have not always been used by decision makers to solve problems. The increasing complexity of decision making and the availability of computers and large data bases in industry and government make it imperative that we understand the costs and benefits of the timely use of relevant information and how decisions taken at one point in time depend on the decisions that follow.

Insufficient attention has been given in the classroom to the formulation and study of decision-making models. We hope that modeling in decision making will become more important in the undergraduate and graduate education of students in business, science, and engineering. We know of only a few institutions of higher education where a serious effort is made to teach this important subject—even in the operations research and management science fields the primary focus is on the mathematical techniques rather than problem formulation and the derivation of useful insights.

1.3   THE NATURE OF A DECISION PROBLEM

There are basic common threads to the problems discussed in this book. In every problem a decision maker has a number of alternate choices of actions from which one or more must be chosen to achieve some desired objective. Often, the set of alternative actions consists of two elements, either do something or do nothing. At other times the set of alternatives consists of a small number of alternate choices. In some examples in Chapter 2 the reader will see that the decision maker can have infinitely many choices. We denote the set of decision alternatives by $\mathcal{D}$; it is usually obvious in our examples what this set of possible decisions is.

In multistage problems where decisions are made in sequence over time from different sets, we denote by $\mathcal{D}_i$ the set of decision alternatives at stage i. A generic decision is denoted d, an element of $\mathcal{D}$, or $d \in \mathcal{D}$ for short. It is important that lowercase d not be confused with upper case D, which we use to denote a generic decision before it is made. For example, a decision maker may have to decide whether or not to proceed with a certain course of action (e.g., submit a proposal, develop a product, repair a facility). In this case $\mathcal{D}$ would contain two elements, "take the action" and "do not take the action"; thus d must be one of these decisions. But before the decision maker has decided on a specific decision, the decision is referred to by D.

Whenever a decision has to be made, there is a certain amount of information available and known to the decision maker. This is denoted by I and is called the information set. Depending on the assumptions of the particular problem, I may be quite limited. In some cases we compare how "good" a decision we can make knowing a given set of information with how well we could do if we had perfect knowledge of the future. By this scheme we can have a measure of the value of getting better information, and we can often determine the maximum value that can be obtained from any forecast. A single-stage decision problem can be thought of as choosing a decision from a set $\mathcal{D}$ when knowing only a limited set of information I in order to achieve the best expected result. In making this statement we implicitly assume that the result can be measured and predicted. We consider cases where the result is measured directly in some well-understood unit such as money. We also consider cases where it is not obvious how to quantify the results and where the value placed on them depends significantly on the person making the decision. Emphasis is always on gaining insight into the problem and looking at sensitivity of decisions to values of the uncertain parameters. For this reason, graphical methods are used extensively, especially influence diagrams and decision trees. The elements of these are described in Section 1.7. More advanced concepts are introduced at appropriate times in later chapters.

We emphasize that our goal is to present models and analytic methods to solve multiperiod sequential decision problems. When there is only one decision to be made that is to be chosen from a small set of alternatives (i.e., there is only one set $\mathcal{D}$ that has only a few members), one could make a list of the expected result from each alternative, rank them, and choose the best. Some examples in Chapters 4 and 5 could be treated this way, as could multistage problems, at least conceptually, by listing all combinations of one alternative from each stage. This is not our approach. It is our belief that by modeling sequential problems with influence diagrams and decision trees we gain much greater insight into the structure and important interrelationships in a problem.

1.4 MODELING UNCERTAINTY WITH PROBABILITY

Uncertainty of event outcomes is treated in the usual way with probabilities.[1] There are at least three interpretations of probability to be found in the literature, namely, the frequency approach, the measure approach, and subjective probability. A detailed comparison of these is beyond the scope of this book, but we describe briefly the main points of these three approaches.

A reader who has completed an introductory or intermediate course in probability will almost certainly have been exposed to the *frequency interpretation of probability*. This is the one most commonly taught and possibly the most appropriate when proba-

---

[1] We use the word *alternative* when discussing decision options and *outcome* when referring to a particular random event that occurs. The reader is cautioned that some software packages do not distinguish in this way and refer to both as *states*.

bility is to be applied to statistical theory for analysis of historical data. To demonstrate what we mean by the frequency interpretation, consider rolling a die many times and counting the fraction of times you obtain the number 3. Because rolling the die has six possible outcomes, all equally likely in the case of a fair die, in a very large number of rolls you would expect the fraction that produce a 3 to be approximately one-sixth. This can be shown to be true by experiment, and we call this fraction the probability of the event that a 3 is obtained on a single roll. An equivalent interpretation is that of the odds of an event. When the die is rolled there are five possible outcomes that are *not* 3, and only one 3. Thus, we say that the odds of getting a 3 are one to five or the odds of not getting a 3 are five to one. We use both ideas in this book. In the financial field and in many real-world applications where decisions are directly influenced by betting or taking risks, clients are more comfortable in quoting the odds that a particular event will or will not occur than they are in calculating a conditional probability. For example, in evaluating the risk of an individual applying for insurance or a credit card, for a number of reasons the important estimate is the odds that the individual will turn out to be "good." The term "good" may refer to the on-time payment of bills, not defaulting on a mortgage, or not presenting a claim against an insurance policy. We use both styles of modeling uncertainty; in some cases formulation in terms of odds is more natural, but in other cases it is more natural to use conditional probabilities.

The *measure approach*, usually found in advanced courses in probability, "measures" the event of interest. In this interpretation the reader can think of the probability as indicating the "size" of an event on a scale where an event that is certain to occur measures 1. This interpretation is not used explicitly in this book.

A third interpretation is that probability encompasses the *subjective* views of a decision maker as to the chance of a future uncertain event happening. In many cases of decision making it may be neither useful nor possible to measure historical data, either because the conditions prevailing at the decision point are significantly different from those at previous times or because a decision is being made in new or different circumstances where historical data are not available or not relevant. Some illustrative examples follow.

In the next chapter we consider some decision problems associated with a choice of plays in a football game. In order to analyze them, we need to know both the probability of scoring for each decision a coach can make and the relative gain for winning versus tieing the game. In that problem we need to assess various scoring probabilities. It can be considered a certainty today that a team in either a major professional league or a national college conference keeps statistical data on their performance in many different situations. But it is important to think through how applicable are these data to estimating the current probability of scoring. Historically the team may have scored on a conversion kick 95% of the time and scored on a two-point try 60% of the time. But if the team has just scored, the game almost tied after being in a losing position, and the championship within reach, the coach may estimate the success probability on this particular two-point attempt to be 0.85. This is a subjective probability. It captures the judgment of the decision maker. When human behavior and/or competitive forces are involved that could distort historic averages, it is often appropriate to use such judgments.

Another example involves a medical decision (see Table 1.2 (p. 30)). Suppose a patient has a medical problem that, although not life-threatening, markedly reduces the person's ability to enjoy life. Surgery has the potential to remove many of the negative effects of the problem, but surgery also brings risks. If it is successful the patient will have a significantly improved life. If not, it could make life worse and even lead to death. For a patient to be able to make a carefully thought-out decision on whether or not to have surgery, a vital piece of information is the probability that the surgery will be successful. To be more specific, suppose the decision is whether or not to have heart bypass surgery to relieve angina. The patient might obtain national statistics for successful surgery for a person of the same age, sex, physical condition, and the like. Suppose this is 0.7, but the patient's doctor states the probability of success as 0.85. Now the patient must consider which to believe. The doctor may, either intentionally or unintentionally, have overestimated the success probability of surgery. On the other hand, if you have been a patient of this doctor for some time so that he or she is completely familiar with your medical history, then the 0.85 may not be an optimistic estimate but truly reflect the best judgment of the doctor for your success; it may be more accurate than a national average. In such a situation it is quite appropriate for the doctor and/or patient to use subjective probability judgments.[2]

It is by no means always appropriate to use subjective probabilities. Consider, for example, a lottery where the decision to be made is whether or not to play. Such gambling games (including virtually all those for which significant amounts of money can be won playing against the house in casinos) have calculable probabilities of winning on each play. In deciding whether or not to play, many factors should be considered, but your subjective judgment as to your probability of success has no place in making a rational decision. The probability of winning is simply a function of the rules of the game and is the same for all players.

1.5 A BRIEF REVIEW OF PROBABILITY

No matter which interpretation of probability one uses, the same laws, rules, and notation apply. Our readers are expected to be familiar with the elements of introductory probability theory. Extensive use is made of the concepts of expected value, conditional probability, conditional expectation, and conditional independence. In this section we briefly review the most important concepts of probability and describe the structure of the notation used in this book. When there are exceptions to our general rules for notation we always try to point these out. Although we have included this material more as a review for students who have completed a course in probability, we hope it is useful as an introduction to those who have never studied the subject. However, it is not possible to include a thorough treatment of the subject, and readers who have not completed such a course are strongly urged to do so. Several well-known texts are suggested in the Summary section of this chapter.

---

[2] For a detailed discussion of this type of problem see Behn and Vaupel (1982).

1.5.1 Outcomes, Events, and Probabilities

Two basic concepts in probability theory are a *random outcome* and *a random event*, usually referred to simply as outcome and event. An outcome is the specific result of an uncertain operation or experiment. Some examples are:

Operation	Outcome
1. Toss a single coin.	Obtain a head.
2. Roll a die twice.	Obtain a 3 on 1st roll, 5 on 2nd.
3. Schedule 2 P.M. meeting.	Start at 2:05 P.M.
4. Take a decision course.	Obtain a B grade.

The set of all possible outcomes for a given operation or experiment is usually denoted by S. In our first example $S = \{\text{head, tail}\}$, and in the second it is the set of ordered pairs of numbers where each element of the pair is an integer from 1 to 6. Mathematically, $S = \{(i, j): i = 1, 2, \ldots, 6, j = 1, 2, \ldots, 6,\}$ and has thirty-six elements. If the meeting in our third example is canceled if it does not start by 2:15, we might use $S = \{s: 2{:}00 \le s \le 2{:}15\}$, which has infinitely many elements. If we measure time only to the nearest minute, we could use $S = \{2{:}00, 2{:}01, 2{:}02, \ldots, 2{:}15\}$, which has sixteen elements. For the fourth example S is the set of all possible grades. Note that the outcomes are each elements of the respective set.

An event is a subset of the set of outcomes. It may be a single outcome, in which case event and outcome can be used interchangeably, or it may be a more complex subset, as the following examples illustrate:

Operation	Event
1. Toss a single coin.	Obtain a head or tail.
2. Roll a die twice.	Obtain a total of 8.
3. Schedule 2 P.M. meeting.	Start before 2:10 P.M.
4. Take a decision course.	Obtain a passing grade.

For our first example the event is S itself. For the second it is any ordered pair of numbers that add to 8, that is, $\{(6, 2), (5, 3), (4, 4), (3, 5), (2, 6)\}$. For the third example it is any starting time less than 2:10, and in the fourth any grade equal to or better than a passing grade.

Let A denote an event. Probability statements about events are denoted as follows:

$$Pr\{A\} = \text{Probability that event } A \text{ occurs,}$$

using { } to contain the statement of the event. Probabilities are *always* numbers that cannot be either negative or larger than 1. Mathematically, for any event A, $0 \le Pr\{A\} \le 1$. If $Pr\{A\} = 1$, then event A is certain to occur; note that $Pr\{S\} = 1$. If $Pr\{A\} = 0$, there is a zero chance that event A will occur. For most of the decision examples in this book the reader can safely interpret this to mean that event A cannot occur. But in general this is not necessarily true, as we point out later in this section.

In many problems one starts with simple outcomes that are equally likely to occur, but one is often required to calculate the probability of occurrence of more complex events that are no longer equally likely. The reader must be ever vigilant to this; an erroneous assumption of equally likely events is often a cause of error in calculating probabilities.

To be able to calculate the probability of occurrence of some combination of outcomes or events, we use some concepts from set theory. For example, suppose we roll a die once and wish to determine the probability of the event "obtain an even number." For this to happen the roll must result in a 2, 4, or 6. If any one of these three outcomes occurs, the event of interest occurs. The probability of each is 1/6, but how do we combine them in the proper way? The key to answering this question is to note that no two of the three outcomes can occur on the same roll. This may seem obvious, but it is this property that is important. Let A and B represent any two events. We say that A and B are *mutually exclusive* if they cannot occur simultaneously in the same performance of an experiment or operation. Using set notation, let C be the union of these events. This is denoted mathematically by $C = A \cup B$, which means that as long as at least one of the two events A and B occurs, then C occurs. If A and B are *mutually exclusive*, then

$$Pr\{A \cup B\} = Pr\{A\} + Pr\{B\}. \tag{1.1}$$

When events are equivalent, as C is equivalent to $A \cup B$, their probabilities are equal, so we can find $Pr\{C\} = Pr\{A \cup B\}$ by adding the probabilities of A and B. We call this result the additive law:

> Let A and B be two mutually exclusive events. If an event C occurs whenever at least one of the events A or B occurs, the probability that C occurs is obtained by summing the probabilities of A and B occurring.

One particular application of this rule is in finding the probability that a given event does not occur. For any event "A occurs," denoted simply by A, the complementary event "A does not occur" is denoted by A^C. Clearly, A and A^C are mutually exclusive, so $Pr\{A \cup A^C\} = Pr\{A\} + Pr\{A^C\}$. But $Pr\{A \cup A^C\}$ in words is "the probability that either A occurs or A^C occurs," which must by definition equal one. So

$$Pr\{A^C\} = 1 - Pr\{A\}. \tag{1.2}$$

The additive law extends to any number of mutually exclusive events. Returning to the roll of a die, let A_i be the event "the roll results in number i" for $i = 1, 2, \ldots, 6$. If C represents the event that an even is obtained, we see that

$$Pr\{C\} = Pr\{A_2 \cup A_4 \cup A_6\} = Pr\{A_2\} + Pr\{A_4\} + Pr\{A_6\} = 1/2.$$

Does this result agree with your intuition?

In general two events may not be mutually exclusive, but some event of interest may be equivalent to their union. For this more general case the probability of the union is given by the sum of the probabilities of each minus the probability of their intersection:

$$Pr\{A \cup B\} = Pr\{A\} + Pr\{B\} - Pr\{A \cap B\}. \tag{1.3}$$

The expression $C = A \cap B$ means that event C occurs only when both of the events A and B occur.

A common way to demonstrate Equations (1.1) and (1.3) graphically is with the use of a Venn diagram. Suppose we represent the set of all possible outcomes S that might occur by a rectangle as in Figure 1-1. Think of the area of any event shown in the rectangle as being equal to the probability that the event occurs. Thus the area of the rectangle is 1. The occurrence of events A and B are shown by two circles within the rectangle. In Figure 1-1(a) the two circles have no area in common—they do not overlap or intersect. This is the graphical equivalent of stating that A and B are mutually exclusive events; there are no outcomes common to both event A and event B. Note that the total area is the sum of the individual areas, illustrating Equation (1.1). In Figure 1-1(b) A and B are not mutually exclusive. Their intersection is clearly indicated and represents the subset of outcomes common to both A and B. When A and B intersect, if we simply add the areas of each we count the area in the intersection twice. To correct for this, after adding the two areas we then subtract the area of the intersection (Equation (1.3)).

As a simple numerical example consider the rolling of a die described earlier. Consider the two events "the number obtained is at most 5" and "the number obtained is at least 3." Call these events B and C, respectively, and let us construct a Venn diagram to illustrate their occurrence. The rectangle in Figure 1-2 represents the sample space S of all possible outcomes. This has been divided into six equal pieces whose areas represent the probability of occurrence of each of the outcomes or simple events A_i. The shaded area represents the intersection of events B and C. As $Pr\{B\} = 5/6$, $Pr\{C\} = 4/6$, and $Pr\{B \cap C\} = 3/6$, Equation (1.3) gives $Pr\{B \cup C\} = 1$. Does this agree with your intuition?

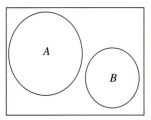 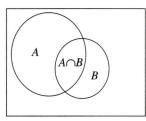

(a) A and B Mutually Exclusive   (b) A and B Intersect

FIGURE 1-1
Venn diagrams to depict events.

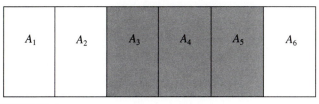

FIGURE 1-2
Venn diagram for the roll of a die.

1.5.2   Conditional Probability and Independence

In order to be able to use Equation (1.3) to calculate the probability of the union of two events that are not mutually exclusive we need to be able to calculate the probability of the intersection. We now introduce the very important concept of *conditional probability*. Suppose events A and B can both occur at the same time and that we wish to calculate the probability that B occurs, given that A has occurred. We write this mathematically as $Pr\{B \mid A\}$. A graphical depiction of how this is calculated can be obtained from the Venn diagram in Figure 1-1(b). Assuming that the area of the image representing an event is equal to the probability that the event occurs, the area of the rectangle is 1, the area of the large and small circles $Pr\{A\}$ and $Pr\{B\}$, respectively, and the area of the intersection of the two circles $Pr\{A \cap B\}$. This result is usually stated as the very important *multiplicative law* of probability:

Let A and B be two events. The probability that both A and B occur is obtained by

$$Pr\{A \cap B\} = Pr\{B \mid A\} Pr\{A\}. \tag{1.4}$$

Alternatively, the reader can see that the conditional probability that B occurs, given that A has occurred, is the area of the overlap of the two circles relative to area of the A circle. Thus, on dividing by a nonzero $Pr\{A\}$,

$$Pr\{B \mid A\} = Pr\{A \cap B\}/Pr\{A\}. \tag{1.5}$$

As an example, consider the problem of tossing a coin twice demonstrated in Figure 1-3. The square represents all possible outcomes. The area above the horizontal line represents the event that the first of the two tosses is a head, that below a tail. Similarly, the area to the left of the vertical line represent the event that the second of the two tosses is a head, that to the right a tail. Each of these two lines divides the total area in half, so if the total area of the square is 1, the area of each of the four inner squares is 1/4. The upper left square represents the event that both tosses are heads. The event "one head is obtained in two tosses" is represented by the combination of the lower left and upper

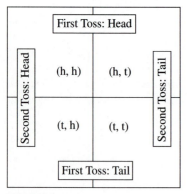

FIGURE 1-3
Venn diagram for two coin tosses.

right areas, whereas the event "at least one head is obtained in two tosses" is represented by the combination of the lower left, upper left, and upper right areas. Note that these last two events are not mutually exclusive; their intersection is the lower left and upper right areas, that is, the first of these events. This occurs because the first event is contained in, or is a subset of, the second event.

Suppose that B represents the event "both tosses are heads" and A "at least one toss is a head." From Figure 1-3 we see that A occurs if we obtain any one of the three mutually exclusive events (h, h), (h, t), or (t, h), where the elements of these vectors represent the outcomes of the first and second tosses respectively. Using the additive law, $Pr\{A\} = 3/4$. Note that for this example $B = A \cap B$ and that event B occurs only if (h, h) occurs, so $Pr\{A \cap B\} = 1/4$. Using Equation (1.5) we see that $Pr\{B \mid A\} = 1/3$. Does this result agree with your intuition? This problem is often stated in the following form: Given that a family with two children has at least one girl, what is the probability that both children are girls? Assuming that the chances are even that any child born is of either sex the answer is 1/3. Many students argue that since you know one child is a girl, and the probability that the other child is a girl is 1/2, then the answer should be 1/2. The fallacy of this argument is in an underlying implicit assumption that the three events {two girls}, {one girl and one boy} and {two boys} are equally likely to occur. If they were, because the event "at least one girl" occurs whenever either of the first two occur, the answer of 1/2 would be correct. But this is not true; the second event is twice as likely to occur as either the first or the third, as can be seen in the Venn diagram in Figure 1-3.

Suppose now that A and B are the events "1st toss is a head" and "2nd toss is a head," respectively, so $Pr\{A \cap B\}$ is the probability that both tosses are heads. Most people would agree that whether or not the first toss produces a head the probability of a head on the second choice is still 1/2. We say that the two events are *independent*.[3] In general we say that events A and B are *independent* if and only if

$$Pr\{A \cap B\} = Pr\{A\}Pr\{B\}. \tag{1.6}$$

For the coin tossing example we see that the probability of obtaining two heads is 1/4, as indicated by the upper left area in Figure 1-3. Note that if Equation (1.6) holds (A and B are independent), this together with (1.5) shows that $Pr\{B \mid A\} = Pr\{B\}$ as should be expected. This expression tells us that our assessment of the probability that B occurs is the same no matter what we know about the occurrence of A whenever A and B are independent.

The reader is cautioned not to confuse the independence and mutually exclusive properties of sets or events; this is a common mistake for newcomers to probability. *If two events that can occur with positive probability are mutually exclusive, then they must be dependent.* If you know that one of two mutually exclusive events has occurred,

---

[3] Strictly speaking we should say they are statistically independent. In many applied problems, especially those referring to complex mechanical or electrical equipment, it may be reasonable to assume that the failure events of two or more components are statistically independent even though the components are linked operationally.

you know that the other cannot have occurred, so the probability they both occur is zero and Equation (1.6) cannot hold. Suppose that we toss a coin once with A and B representing obtaining a head and tail, respectively. Clearly you cannot obtain both a head and a tail on the same toss and so $Pr\{A \cap B\} = 0$. But $Pr\{A\} = Pr\{B\} = 1/2$ and so Equation (1.6) does not hold.

1.5.3 Partitions and the Law of Total Probability

One of the most useful results that has widespread application in calculating probabilities is known as the law of total probability. It relies on the concepts of mutually exclusive events, the additive law, conditional probability, and what is called a partition of the set of possible outcomes.

Recall that the set of all possible outcomes to which probabilities can be assigned is S. A partition of S is any collection of mutually exclusive events in S whose union is S. Formally, let $\{A_i: i = 1, 2, \ldots, n\}$ be n subsets of S. This collection forms a partition of S if and only if

$$A_i \cap A_j = \emptyset \text{ (the empty set) when } i \neq j$$

and

$$\bigcup_{i=1}^{n} A_i = S.$$

A particularly simple partition is formed by any event A together with its complement A^C. Graphically, one can think of dividing the rectangle in a Venn diagram into a number of nonoverlapping pieces that completely cover the rectangle. A specific example is shown in Figure 1-3 where the four small squares form a partition of the overall square. Another is shown in Figure 1-2 by the six inner rectangles. In general, we assume that there are n mutually exclusive events that cover S, and some event B that can occur when some or possibly all of the partition events occur. Figure 1-4 shows a partition with $n = 4$ and an event B represented by an oval that intersects with all four events A_1, A_2, A_3, and A_4 in the partition. Notice that the area, and hence the probability, of event B is equal to the sum of the areas represented by the intersection of each A_i with B. We can write this mathematically as

$$Pr\{B\} = Pr\{A_1 \cap B\} + Pr\{A_2 \cap B\} + Pr\{A_3 \cap B\} + Pr\{A_4 \cap B\}.$$

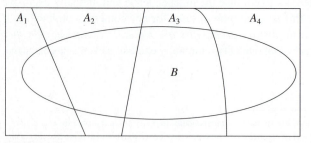

FIGURE 1-4
The law of total probability in graphical form.

In general, let the events $A_1, A_2, A_3, \ldots, A_n$ be a partition of S. For any event B in S the probability of B satisfies

$$Pr\{B\} = \sum_{i=1}^{n} Pr\{A_i \cap B\}, \tag{1.7}$$

and by using Equation (1.5)

$$Pr\{B\} = \sum_{i=1}^{n} Pr\{B|A_i\} Pr\{A_i\}. \tag{1.8}$$

Equation (1.8) is usually referred to as the *law of total probability*. It is often used to calculate the probability of some event that is difficult to determine directly, but where the conditional probabilities on the right-hand side, together with the partition probabilities, are easily found. It is used frequently in this book in solving and analyzing decision problems. The following simple example is used as an illustration.

Suppose that it is estimated that one in every 1,000 people in the age group 55 to 65 has a certain type of cancer. Of those that do, 40% have large cancers (say tumors larger than 2 centimeters), and 60% have small cancers (less than 2 centimeters). Let us assume that a diagnostic test has been developed for this disease but that it is not perfect. In clinical tests suppose that when the test is administered to people known to have the disease with large cancers, 99% show a positive result; when administered to people with small cancers, 65% show positive; on those known not to have the disease 10% show a positive result. If this test is administered widely on a regular basis to this age group, what fraction of tests indicate cancer is present? Let A_1 be the event "large cancer," A_2 the event "small cancer," and A_3 be the event "no cancer." These three events are mutually exclusive and form a partition of the sample space (the condition of the population being tested) with $Pr\{A_1\} = 0.0004$, $Pr\{A_2\} = 0.0006$, and $Pr\{A_3\} = 0.9990$. Let B be the event "a test result is positive." From the clinical test results

$$Pr\{B|A_1\} = 0.99, Pr\{B|A_2\} = 0.65, Pr\{B|A_3\} = 0.10.$$

Using Equation (1.8), we find that

$$Pr\{B\} = 0.99 \times 0.0004 + 0.65 \times 0.0006 + 0.10 \times 0.999 = 0.100686.$$

It is extremely important in all modeling to be as precise as possible when defining variables and other factors that are to be used in a mathematical representation of a problem. This is especially true when defining uncertain events. Note that B in the example is the event that the cancer test shows positive. This is not the same as the event "the person has cancer." Denote this by C, and note that $Pr\{C\} = 1 - Pr\{A_3\}$. It is easy to see that $Pr\{C\} = 0.001$, but how does one find the probability that a person has cancer, given that a test of that person gives a positive result, that is, $Pr\{C|B\}$? To answer this question we need the following result.

1.5.4 Bayes' Rule

Let B and C be two uncertain events where B is a forecast of the occurrence of event C. If $Pr\{B \cap C\} = Pr\{B\}Pr\{C\}$, the events B and C are said to be independent. Because B

is a forecast of C, the last thing we want is for B to be independent of C! We show later in the book that the more dependent they are, the more useful is the forecast. By definition, the conditional probability that B occurs given that C has occurred is given by

$$Pr\{B|C\} = \frac{Pr\{B \cap C\}}{Pr\{C\}}, \tag{1.9}$$

and by interchanging B and C

$$Pr\{C|B\} = \frac{Pr\{B \cap C\}}{Pr\{B\}}. \tag{1.10}$$

Suppose we have data that enable us to estimate $Pr\{B|C\}$, the probability that B will occur given that C has occurred, but that in a given decision problem we require $Pr\{C \mid B\}$. Using Equations (1.9) and (1.10) it is straightforward to show that

$$Pr\{C|B\} = \frac{Pr\{B|C\} Pr\{C\}}{Pr\{B\}}. \tag{1.11}$$

Equation (1.11) is known as Bayes' Rule and is used extensively throughout the book.

Returning to the cancer example, the test for the disease can be thought of as a forecast of the disease itself, and $Pr\{C \mid B\}$ is the probability of having cancer given a positive test result. Of the three terms on the right side of Equation (1.11) $Pr\{C\} = 0.001$ and $Pr\{B\} = 0.100686$. To find the $Pr\{B \mid C\}$ term, we proceed as follows. Note that $C = A_1 \cup A_2$, so

$$Pr\{B \mid C\} = Pr\{B \mid A_1 \cup A_2\} = Pr\{B \cap (A_1 \cup A_2)\}/Pr\{A_1 \cup A_2\}.$$

But $B \cap (A_1 \cup A_2) = (B \cap A_1) \cup (B \cap A_2)$ (see Problem 1.2), and the events in parentheses are mutually exclusive (see Problem 1.2). Thus,

$$Pr\{B \mid C\} = \left[Pr\{B \mid A_1\}Pr\{A_1\} + Pr\{B \mid A_2\}Pr\{A_2\}\right]/Pr\{C\}$$
$$= [0.99 \times 0.0.0004 + 0.65 \times 0.0006]/0.001 = 0.768.$$

Substituting these results into Equation (1.11) gives

$$Pr\{C \mid B\} = 0.007806.$$

Notice that fewer than 1% of those who test positive with this test will actually have cancer.

1.5.5   Random Variables and Distributions

Many uncertain phenomena can be measured on a numerical scale. Examples include (1) the number of empty parking spaces you find in your favorite parking lot on arriving at work tomorrow, (2) the opening price of IBM stock tomorrow on the New York Stock Exchange, (3) your flight time today from San Francisco to London, (4) total revenues for your company in the next quarter, and so on. None of these quantities is known with certainty before it occurs, yet we would like to denote it with a mathematical symbol. In probability theory such a quantity is referred to as a *random variable*, and in this book these are denoted by uppercase letters such as X, Y. The value they take on is denoted by lowercase letters such as x, y. For Example (1), suppose X is the number of empty parking places you find on arrival at work where the lot has a total of five places; X could take on any value x from the set $\{0, 1, 2, 3, 4, \text{or } 5\}$. The set of possible outcomes of the random variable is denoted with script notation, for example, $\mathcal{X}$ for X and $\mathcal{Y}$ for Y. For the most part in this text we deal with uncertain events that can take on only a few values and so $\mathcal{X}$ or $\mathcal{Y}$ would usually contain only a few elements.

For the parking place example, suppose when you arrived today you found two empty places. Then X took on the value 2. The concepts of an outcome and an event already introduced are often statements about the numerical value or numerical range taken on by the random variable. If you found two empty places, the outcome $\{X = 2\}$ occurred. An event of particular interest is $\{X > 0\}$, that is, you have somewhere to park.

In most texts on probability two categories of random variables are discussed, namely, *discrete* and *continuous*. The parking place random variable is an example that can take on only discrete values. Stock prices, such as those of Example (2) are usually quoted in dollars to the nearest one-eighth so they can also be treated as discrete random variables. In Example (3), you may decide that the flight time is equally likely to be any value between 10.5 and 11.25 hours. This would be an example of a continuous random variable. Example (4) would also usually be treated as continuous. The mechanics of mathematical computations are quite different for each type. For a discrete random variable we define what is referred to as a *probability mass function* (pmf for short) $p(x)$, where

$$p(x) = Pr\{X = x\} = Pr\{X \text{ takes on the value } x\}.$$

It is understood that x is one of the set of possible values that X can take on; we denote this set $\mathcal{X}$. Every pmf must satisfy

$$\sum_{x \in X} p(x) = 1, \quad p(x) \geq 0. \tag{1.12}$$

To illustrate, suppose you are to roll a die once. Clearly, you cannot predict the outcome with certainty before it occurs, but the set of possible outcomes is $\mathcal{X} = \{1, 2, 3, 4, 5, 6\}$. The pmf for this example (if the die is fair) is

$$p(x) = 1/6, \quad x \in \mathcal{X}.$$

The expression $x \in \mathcal{X}$ is a shorthand way of stating that x is a member of the set $\mathcal{X}$, or in this example, x is any of the integers 1 through 6. A second example is the parking place described. Here $\mathcal{X} = \{0, 1, 2, 3, 4, 5\}$, but the probabilities may not be equal. They could be any positive numbers that add to 1. An example is shown in the second column of Table 1.1.

Most of the random variables found in this book are discrete. Continuous random variables are found only in the models of yield management in Sections 2.5 through 2.9 in Chapter 2. Calculations involving continuous random variables require the use of both differential and integral calculus. The equivalent to the pmf is what is called the *probability density function* (pdf). Unlike a discrete random variable, a continuous ran-

TABLE 1.1
Parking place probabilities

Empty parking places x	Probability mass function (pmf) $p(x)$	Cumulative distribution function (cdf) $P(x)$
0	0.10	0.10
1	0.30	0.40
2	0.40	0.80
3	0.10	0.90
4	0.06	0.96
5	0.04	1.00

dom variable takes on any particular value with probability zero. This is true for *every* value it can take on, and yet whenever it is observed it always takes on one of these values. The pdf at a given value $x \in X$ can be thought of as the propensity of the random variable to take on the value x. If the pdf is denoted $p(x)$, the equivalent to Equation (1.12) for continuous random variables is

$$\int_{x \in X} p(x)\, dx = 1, \quad p(x) \geq 0. \qquad (1.13)$$

As an example, consider the flight time example from San Francisco to London where it was equally likely to be any number between 10.5 and 11.25. Thus, X is the set of numbers between these limits, or $X = \{x;\ 10.5 \leq x \leq 11.25\}$. The equally likely assumption implies that

$$p(x) = 4/3 \qquad \text{if } 10.5 \leq x \leq 11.25,$$

$$= 0 \qquad \text{otherwise.}$$

The reader should check that Equation (1.13) holds. The reader should be careful not to confuse a pdf with a probability. As this example shows, pdf's can exceed 1, whereas probabilities cannot.

Both discrete and continuous random variables have associated with them a *cumulative distribution function* (cdf). For any random variable X its cdf gives the probability that X does not exceed a given value x. In this book, the pmf and pdf are both denoted $p(x)$, and the corresponding cdf is denoted $P(x)$. From its definition $P(x) = Pr\{X \leq x\}$. For discrete random variables it is found by summing the terms of the pmf for all outcomes not exceeding a given value. If $X = \{x_1, x_2, x_3, \ldots, x_n\}$,

$$P(x) = \sum_{x_i \leq x} p(x_i), \quad x \geq 0. \qquad (1.14)$$

Column 3 in Table 1.1 shows the cdf for the parking example, and this is shown plotted in Figure 1-5. Note that it is a discontinuous function and that at each integer x the value of the function is indicated (the upper of the two points in each case).

For continuous random variables the cdf is found by integrating the pdf over all outcomes not exceeding x. If $X = \{x; x \geq 0\}$,

$$P(x) = \int_0^x p(u)\,du, \quad u \geq 0. \tag{1.15}$$

Thus, in the flight time example the reader should check that

$$Pr\{X \leq x\} = \int_{10.5}^x \frac{4}{3}\,du = \frac{4(x-10.5)}{3}$$

for $10.5 \leq x \leq 11.25$ and plot this result. You should find that the probability that the flight time does not exceed 11 hours is 2/3.

Suppose that one is interested in finding the probability that a random outcome falls in a certain range, say, larger than a but no larger than b. This can easily be found from the cdf as

$$Pr\{a < X \leq b\} = P(b) - P(a). \tag{1.16}$$

This follows from Equation (1.3) when event $A = \{X > a\}$ and $B = \{X \leq b\}$ (see Problem 1.5).

Most texts on probability introduce and study standard distributions of random variables. The most common ones include: for the discrete case Bernoulli trials, the binomial, negative binomial, geometric, Poisson, and hypergeometric; for the continuous case the uniform (the flight time above is an example), exponential, gamma, beta, normal, and a collection of distributions that play key roles in statistical analyses. None of these are found in this text, with the exception of Bernoulli trials, the uniform and the

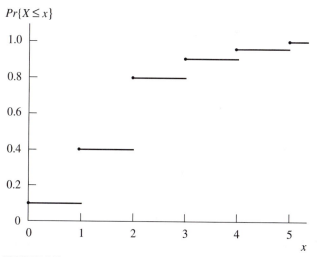

FIGURE 1-5
Cumulative distribution function for parking example.

exponential. A Bernoulli trial is a discrete random variable that can take on the value 0 or 1. It is commonly used in models for indicating whether an event of interest occurs or not. For a Bernoulli trial it is common to define $p(1) = Pr\{X = 1\}$ to be p, in which case $Pr\{X = 0\} = (1 - p)$ because X cannot take on any other values.

When more than one random variable is present in a problem, confusion can occur as to which density, mass function, or cdf is associated with which random variable. When this occurs, we use a subscript to indicate the appropriate association. For example $P_X(x) = Pr\{X \le x\}$, $P_Y(y) = Pr\{Y \le y\}$ are the cdf's of X and Y while $p_Z(z)$ is the density or mass function of Z.

1.5.6   Expected Values

When faced with an uncertain phenomenon, it is natural to ask what outcome value to "expect." It is not at all obvious how one can answer such a question. One might think of choosing the value that has the highest probability of occurring. But think of the simple case of rolling a die. Since all six numbers are equally likely to occur, there is no single value answer. If you are interested in the most likely outcome of the sum of two rolls of a die, the answer is seven. When such a number does exist, it is called the mode of the distribution of the random variable. A second way to answer the question is find that value for which the random variable is equally likely to be larger or smaller. For the flight time example this is 10 hours 52 minutes and 30 seconds. This measure is called the *median* of the distribution. A third way to answer the question is to multiply every possible value that can occur with the probability it occurs and find this weighted average. This is called the expected value, and it is found throughout this text. One justification for its use is that if many outcomes of a random variable are observed, as the number of observations increases, the probability that their average differs from the expected value approaches zero. This result is known as the law of large numbers.

The expected value of a random variable X is denoted $E[X]$. For a Bernoulli trial,

$$E[X] = 1 \times Pr\{X = 1\} + 0 \times Pr\{X = 0\} = p.$$

In general, if a random variable X can take on discrete value x_i with probability $p(x_i)$, where x_i is one of n possible values, its expected value is given by

$$E[X] = \sum_{i=1}^{n} x_i p(x_i) . \tag{1.17}$$

When the random variable is continuous with X the set of all nonnegative numbers, its expected value is given by

$$E[X] = \int_0^\infty x p(x) \, dx . \tag{1.18}$$

For the flight time example, Equation (1.18) gives 10 hours 52 minutes and 30 seconds, the same as the median.[4] For the single roll of a fair die Equation (1.17) gives $E[X] = 3.5$. The reader may find it strange that what is referred to as the expected value cannot in fact be obtained on any random outcome! This is just an unfortunate semantic consequence of the terminology. One should not infer from this that the concept is flawed; expected value is probably the most widely accepted measure used to summarize the possible outcomes of a random event. A common, but not the only interpretation is that if the average is taken of many observations of the random event it will be close to the expected value. We discuss this further in Section 1.9 when we consider a number of ways to measure the result of a decision problem.

When two random variables X and Y are related by a deterministic relationship, we say that Y is a function of X or that $Y = f(X)$. The expected value of Y is found by adding the various values that Y can take on weighted by the probability of the X that produces the given Y value. When X is a continuous random variable, this is

$$E[Y] = \int y p_Y(y)\, dy = E[f(X)] = \int f(x)\, p_X(x)\, dx. \qquad (1.19)$$

The integrals would be replaced with summations if X and Y were discrete random variables; the integration or summation is over the whole range of possible outcomes of X or Y.

One function of a random variable is called the *variance* and has found widespread use in measuring the variability of a random variable. It is used to summarize the entire distribution of a random variable by a single number rather than by a pmf or pdf. It measures the variability of the random variable about its expected value and is defined to be

$$Var[X] = E[(X - E[X])^2].$$

It can also be written in the form

$$Var[X] = E[X^2] - E[X]^2.$$

In the first of these expressions note that if X does not vary, then it is always equal to $E[X]$ and the variance is zero. In the second, if X does not vary, then $E[X^2] = E[X]^2$, and again we see that the variance is zero. The more X can differ from its expected value with positive probability, the higher the variance. Using Equation (1.19), for a discrete random variable,

$$Var[X] = \sum_{x \in X} x^2 p(x) - \left(\sum_{x \in X} x p(x) \right)^2,$$

and for a continuous random variable,

$$Var[X] = \int_{x \in X} x^2 p(x)\, dx - \left(\int_{x \in X} x p(x)\, dx \right)^2.$$

---

[4] For this to be true the random variable must have a symmetric pdf, i.e., one whose shape to the right of the median is a mirror image of its shape to the left of the median.

The (positive) square root of the variance is called the *standard deviation* and is often quoted along with an average or mean to summarize a set of data. For the single roll of a die the expected value is 3.5, the variance is approximately 2.92, and standard deviation 1.71. For the flight time example, the expected value is 10.875, the variance is approximately 0.047 hrs^2, and the standard deviation 13 minutes.

1.5.7   Marginal, Conditional, and Joint Probabilities

We are often concerned with the description of probabilities with two or more random variables. For the sake of illustration, let us denote these two uncertain quantities by X and Y. Assume that the outcomes of X are $x \in X$ and of Y are $y \in \mathcal{Y}$. The random variable X has a (marginal) pmf of

$$p_X(x) = Pr\{X = x\}, \quad x \in X,$$

Similarly, the random variable Y has a pmf of

$$p_Y(y) = Pr\{Y = y\}, \quad y \in \mathcal{Y}.$$

These pmf's tell us how frequently each category or outcome is likely to occur. How these are estimated is the subject of a great amount of statistical literature and can indeed be a difficult problem. As our emphasis in this book is on decision problems we usually assume that these are known.

Consistent with our earlier notation using events, when it is necessary to denote probabilities conditional on a given state of knowledge, the standard notation of a vertical line is used. Thus $p_X(x \mid I)$, or more simply $p(x \mid I)$, indicates the pmf or pdf of X, given that the information in the set I is known; $p_{X|Y}(x \mid y)$ indicates the pmf or pdf of X, given the random variable Y has taken on the value y.

Examples of conditional probabilities for X given *a known value of Y* express the effect that different values of $Y = y$ will have on the uncertainty of X. It cannot always be assumed that these are known or estimated from data that result from the observation of many outcomes that actually occur. It may be that what is known is the joint distribution of X and Y or the conditional distribution of X given Y. It is important to understand the differences between them and how to calculate one from the other using the laws of probability. The notation that we use is the following:

1. The joint probability of X and Y is denoted by

$$p_{X,Y}(x, y) = Pr\{X = x, Y = y\}, \quad x \in X \text{ and } y \in \mathcal{Y}.$$

2. The conditional probability of Y, given X, is denoted by

$$p_{Y|X}(y \mid x) = Pr\{Y = y \mid X = x\}, \quad x \in X, y \in \mathcal{Y}.$$

3. The conditional probability of X, given Y, is denoted by

$$p_{X|Y}(x \mid y) = Pr\{X = x \mid Y = y\}, \quad x \in X, y \in \mathcal{Y}.$$

The laws of probability apply, so that each of these distributions is nonnegative and the conditional probabilities over the outcomes of the random variable of interest (not the conditioning event) sum to 1.

The sets X and Y depend on the particular problem, the quantity and quality of data from which estimates must be made or the underlying description of the problem. If we know the joint distribution, we can obtain whatever marginal and/or conditional distributions are needed.

With few exceptions, bold lowercase letters (such as **f, x, p**) are used to denote vectors, and bold uppercase letters (such as **F, X, P**) are used to denote arrays or matrices. For example, if a random variable X measuring the financial risk of an individual has three outcomes, "good," "bad," and "acceptable," or $X = \{g, b, a\}$ for short, the probability of these different outcomes being $p(g) = 0.3, p(b) = 0.5, p(a) = 0.2$, we could also denote the pmf of X in vector form by $\mathbf{p}_X = (0.3, 0.5, 0.2)$; the ordering of the probabilities in this vector coincides with the ordering in the set X. On the other hand, if we knew that the conditional probabilities of these three outcomes depended on whether the individual had large or small investments, denoted by $Y = \{large, small\}$ or $\{l, s\}$ for short, we would use a matrix $\mathbf{P} = [p_{X|Y}(x|y)]$ with rows of the matrix corresponding to l and s, respectively, and with columns corresponding to g, b, and a.

For any two random variables X and Y,

$$p_{X,Y}(x,y) = p_{Y|X}(y \mid x)p_X(x) = p_{X|Y}(x \mid y)p_Y(y). \tag{1.20}$$

Notice that from the two expressions to the right of the left most equality sign, one can divide both sides by $p_X(x)$ (assuming it is not zero) to obtain Bayes' rule in a slightly different form:

$$p_{Y|X}(y|x) = \frac{p_{X|Y}(x|y)\,p_Y(y)}{p_X(x)}. \tag{1.21}$$

Another way of understanding this equation is to note that it can be written and expressed as the conditional probability of $Y = y$ given $X = x$, which is equal to the ratio of two factors times the marginal probability of Y. Sometimes the marginal probability of Y is called the *prior* and the conditional probability of Y, given X, is called the *posterior* of Y in the light of the new information or evidence provided by knowing $X = x$. It is the assessments of the distribution of random quantities such as these that give us forecasts of uncertainty (with or without conditions) in our decision models as the reader will see in Chapter 3.

When two random variables have a joint distribution, one measure of their interaction known as the *covariance* is

$$Cov[X, Y] = E[XY] - E[X]E[Y].$$

The first term is found using Equation (1.19); for discrete random variables it is

$$E[XY] = \sum_{x \in X} \sum_{y \in Y} xy p_{X,Y}(x, y) \ ,$$

and for continuous random variables

$$E[XY] = \int_{x \in X} \int_{y \in \mathcal{Y}} xy p_{X,Y}(x, y) .$$

A related measure is known as the *correlation coefficient*, usually denoted by $\rho_{X,Y}$; it is defined as

$$\rho_{X,Y} = \frac{Cov[X, Y]}{\sqrt{Var[X] Var[Y]}} .$$

The covariance and correlation coefficient have important uses later in the book when we investigate properties of forecasts.

1.5.8 Probabilities and Odds

Probabilities are not the only way to describe uncertainty. A common method often used in gambling is what is called odds. The concept is used in this book for some of our examples when it is more appropriate in those cases than does using probability directly.

Probability and odds are related. First, we point out that there are two definitions of odds in common use; one is the odds of winning, the other of losing. In horse racing for example, if the odds offered on a horse are 3 to 1, the person giving the odds believes that the chance that it will lose is three times as large as the chance it will win, that is that the probability of winning is 1/4. Thus the 3-to-1 odds is the odds of losing. If we denote by p the probability that the event of interest (winning) occurs, the odds O in this case are given by

Odds of losing: $O = (1 - p)/p.$

The reason for this appears to be that you can easily calculate your winnings if you bet on the horse; you will receive $(1 + O)$ times the amount bet if it wins. In this book we use the

Odds of winning: $O = p/(1 - p),$

which for the horse race example would be 1 to 3, the reciprocal of the odds of losing. Given the winning odds it is easy to see that the probability of winning is found by

$$p = O/(1 + O).$$

Whereas probabilities of all possible outcomes add to 1, this is not true with odds. Suppose n outcomes can occur, and let $O(i)$ be the odds that outcome i occurs, $i = 1, 2, \ldots, n$. To check that this set of odds is consistent, they must satisfy

$$\sum_{i=1}^{n} \frac{O(i)}{(1 + O(i))} = 1 .$$

In general, calculating items of interest is much more difficult with odds than it is with probability.

1.5.9 Conditional Independence

The concept of independence of events was introduced in Section 1.5.2. This idea extends to random variables. Let X and Y be any two random variables with pmf's or pdf's $p_X(x)$ and $p_Y(y)$, and outcome sets X and Y respectively. X and Y are said to be independent random variables if and only if $p_{X,Y}(x,y) = p_X(x)p_Y(y)$ for every $x \in X$ and $y \in Y$. This last statement is very important. The factorization must hold for every combination of possible outcomes.

The models of many real decision problems contain more than two random variables. Let us suppose we have three, the X and Y just mentioned together with Z with pmf or pdf $p_Z(z)$ and outcome set Z. All three are independent if and only if we can factor the joint distribution as

$$p_{X,Y,Z}(x,y,z) = p_X(x)p_Y(y)p_Z(z).$$

But there are many ways in which dependencies can be present. For example, our earlier results can be used to write the preceding equation in the form

$$p_{X,Y,Z}(x,y,z) = p_{X,Y|Z}(x,y \mid z)p_Z(z).$$

It is quite possible for X and Y to be dependent random variables, and also to be independent if the value of Z is known. If this is true, we say that X and Y are *conditionally independent*, given Z, and indicate this mathematically by

$$p_{X,Y|Z}(x,y \mid z) = p_{X|Z}(x \mid z)\, p_{Y|Z}(y \mid z). \tag{1.22}$$

This must hold for every combination of x in X, y in Y, and z in Z.

When taking expectations of two or more random variables, to avoid confusion, we sometimes use the same subscripting notation as in Equations (1.20), (1.21), and (1.22). For example, suppose we have three random variables X, Y, and Z and wish to find the expected value of a function $f(X, Y)$ conditional on Z. We denote this by

$$E_{X,Y|Z}[f(X, Y) \mid Z = z] = \sum_x \sum_y f(x, y)\, p_{X,Y|Z}(x, y|z) \qquad . \tag{1.23}$$

Conditional independence arises naturally in forecasting. Suppose that you observe the results of two financial services that predict what the Dow Jones industrial average will be at some future time. Let X and Y be their respective forecasts of the same outcome, which we call Z. Assuming there is no collusion between the services, it is probably reasonable to assume that Equation (1.22) holds. You would not on the other hand expect X or Y to be independent of Z; it is not desirable for a forecast to be independent of the phenomena being forecast! You expect X and Z to be dependent and Y and Z to be dependent. If they are, then X and Y are dependent, but this is not inconsistent with X and Y being conditionally independent, given Z. Conditional dependencies can become quite complex. A simple graphical way to portray these conditional dependencies and independencies uses what are referred to as *influence diagrams*. The reader will find an introduction to these diagrams in Section 1.7; they are used extensively throughout the book.

1.5.10   Coherence

The reader will find that we frequently refer to and use the concept of *coherence*. We use this concept in two ways, one of which is to refer to the laws of probability, the other refers to the rational behavior of a decision maker. In both cases the notion of *consistency* is paramount. That is to say, the models and their analysis must be internally consistent. The use of the laws of probability will ensure consistency in the mathematical analysis. For example, (1) probabilities must be between 0 and 1, (2) they must add to 1 over all possible outcomes, and (3) joint, conditional, and marginal distributions must be consistent. Similarly, the application of decision analysis in model building must include the requirement that the decision maker act rationally. The reader will find that we return to this principle of coherence throughout the text.

We end this section with an example of a set of probabilities that are not coherent. Suppose that in your morning newspaper you find the following weather forecast: there is a 40% chance of rain today, a 30% chance of rain tomorrow, and a 20% chance it will not rain on either day. To check whether these satisfy the laws of probability, let A be the event "it rains today" and B be the event "it rains tomorrow." We are told that $Pr\{A\} = 0.4$, $Pr\{B\} = 0.3$, and $1 - Pr\{A \cup B\} = 0.2$. Substituting these into Equation (1.3) (p. 8) gives $Pr\{A \cap B\} = Pr\{\text{it rains on both days}\} = -0.1$. Clearly Item (1) in the above paragraph is violated. The reader should be able to show that the chance of no rain on either day must be at least 30%.

As the number of random variables increases, the task of checking for consistency increases rapidly. Even for three random variables the dependencies among them can be quite complex as we just pointed out; they may all three be independent, two of them may be conditionally independent given the third, or there may be no conditional independence. Depending on the situation, using conditional probability correctly can be quite a formidable task. The reader will shortly be introduced to influence diagrams, which play a central role in modeling decision problems and ensuring coherence among the conditional and marginal distributions of a number of dependent random variables.

1.6   THE ROLE OF FORECASTING

Perhaps the earliest recorded attempt to synthesize forecasting models in decision analysis was made by Bunn (1978). He states in his introduction,

> It is recognized that in the general context of policy formulation, the decision maker must effect a coherent synthesis of all the available evidence including, particularly, forecast models.

This claim was in direct contradiction of an apparent belief summarized in an earlier publication by Granger and Newbold (1975) that stated:

One thing that we would certainly argue is that the forecasting task can be completely sep-arated from that of the decision maker. If we pursue this quest for a scientific method further, principles are clearly now required to enable us to select out the best criterion of forecast evaluation.

One of the important contributions of this book is the emphasis placed on the synthesis of forecasting and decision making. The reader will not find recipes for specific forecasting formulas, but several chapters are devoted to forecasting concepts that pertain to making better decisions.

In many areas, particularly in the physical sciences, the primary focus of forecasting is on postulating and testing models that provide a better understanding of a natural process. In such cases it may be credible to focus on single criteria of forecast performance to decide how experimental results fit a postulated (deterministic) model. Although the theory of statistical tests of fitting error, particularly those of squared error, is well developed, it is not particularly useful in decision analysis where there are usually several fundamental sources of uncertainty with different conditional dependencies that influence the objectives of the decision maker in different ways.

Although forecasting is a fascinating and important theoretical and practical topic in its own right, we view the role of forecasting as secondary to decision making. One utilizes relevant information and forecasts in order to make better decisions and not as ends in themselves. For this reason one should not commit to a forecasting methodology such as regression, Box-Jenkins time-series methods, exponential smoothing, Kalman filtering, to name a few, until one is convinced that the underlying prediction problem is meaningful in the context of the decision problem. Notwithstanding their considerable popularity, many measures of forecast performance based on mean squared error and point predictors evaluate inappropriate aspects of the forecasting problem and are quite often irrelevant to decision making. In situations where rare events or the tails of distributions are important to the decision maker, measures such as mean squared error (MSE) and mean absolute deviation (MAD) are virtually useless.

Although we do not enter into a lengthy treatise on the design of forecasting systems or the analysis and use of specific forecasting formulas, one can address many issues of forecast quality independently of much of the detail that goes into a particular forecast or the full details of the decision problem where the forecasts are used. Corporations and government have invested billions of dollars in large and comprehensive management information systems whose purpose is to add value through better forecasting and ultimately, better decisions. But this objective is elusive because investment in relevant data bases and forecasts has often been judged on technical considerations that are not linked with decision making.

From our point of view, the most important requirement is that *data gathering, forecasting, and decision making be a coherent enterprise with consistent objectives. The designer(s) of the data bases and the forecasting model(s) and the decision maker(s) should be of one mind; even when this is not the case, as can easily occur when data and forecasts are obtained from different external sources, the decision maker must have a good understanding of the forecasting model outputs and of their quality; the important idea is that forecasts must be coherent with the needs of the decision model.*

1.7   ELEMENTS OF INFLUENCE DIAGRAMS
AND DECISION TREES

Decision trees have been the stock-in-trade of decision analysis since the 1960s. Notation has not been consistent in the literature as the analysis of influence diagrams has developed. The reader should take care when reading the references or related material to note any differences between the notation there and in this book, which are as follows.

Three distinct types of nodes are used in both influence diagrams and decision trees:

Square nodes represent decisions, and each one has associated with it a decision set $\mathcal{D}$.

Circular nodes represent random quantities or events, and each one has associated with it a random outcome set such as X or Y.

Diamond nodes represent results of the decision process.[5] The set of possible results is denoted $\mathcal{R}$ with typical element r.

Any particular result will usually depend on the decision(s) taken and random outcome(s) that occur. Before these are known, we refer to the result as R. In later parts of the book, particularly in Chapter 9, we need to distinguish between the name of the result node and the mathematical function that may depend on chance quantities and decisions used to calculate the result. At that time we refer to the result or value node but retain our convention for showing how the result is a function of one or more variables of interest.

We list and order nodes in both influence diagrams and decision trees from left to right to coincide with the progression of time. Although there are some disadvantages to this convention, as we shall see when we cover topics of arc reversal in chance influence diagrams in later chapters, it helps stress the importance of the timing of events relative to one another; in our experience it is a significant aid in using influence diagrams in model building, even when the assessment of the conditional probabilities used in problem formulation may not agree with this time ordering.

Both influence diagrams and decision trees use the basic elements of networks or graphs, namely, nodes, branches, and directed arcs. Their purpose is to describe a given decision problem, and they do so in quite different ways. *It is very important that the reader clearly differentiate the uses of directed arcs in influence diagrams and branches in decision trees.* Even though the branches are traversed from left to right (i.e., downstream in the decision tree), we avoid using an arrow in order to emphasize the important distinction between the branch in the tree and the arc in the influence diagram.

---

[5] Much of the literature and software in decision analysis refers to what we call result nodes as value nodes. We prefer the term result nodes as this is more general. It includes cases where results are measured by losses rather than values, and cases where it is not obvious how results should be quantified.

1.7.1 Directed Arcs in Influence Diagrams

Directed arcs are used to denote possible conditional dependence in influence diagrams. If a directed arc joins two nodes, it indicates some type of dependence between the nodes. Figure 1-6 is used to illustrate six basic node/directed arc combinations that can occur. In (a) a directed arc joins two chance nodes that represent the random events X and Y. This indicates two facts about the random events; (1) they may be statistically dependent, and (2) the outcome of the random event X will be known when the probability distribution of Y is assessed. In (b) the directed arc that joins chance node X to decision node D indicates that (1) the value of the uncertain event X is observed and known to the decision maker before the decision D is made, and (2) the value of X may influence the decision D. In (c) the directed arc that joins decision node D to chance node X indicates that (1) the decision D is made before the random event X occurs, and (2) the probability distribution of X may depend on the particular decision made. In (d) the directed arc that connects two decision nodes D_1 and D_2 indicates that (1) the decision selected for D_1 is known to the decision maker before D_2 is made, and (2) the decision taken for D_1 may influence the decision D_2. In (e) the directed arc indicates that the result R of the decision process depends on the outcome of the random event X, and in (f) that R depends on the decision alternative chosen. The reader will find many examples of influence diagrams throughout the book.

1.7.2 Branches in Decision Trees

Decision trees display the sequences of decisions and random events that occur in all possible scenarios of the decision problem. Branches are used to denote possible outcomes of random events or possible alternative decisions. From any given decision node there will be as many branches as there are possible decisions (i.e., elements of the associated set $\mathcal{D}$). From any given random event node there will be as many branches as there are outcomes that the random event can take on (i.e., elements of the associated set X).

In the next two sections we formulate the first of many examples that are used to illustrate the model-building and analysis ideas and methods developed throughout the book. Although the procedures necessary to solve and analyze more complex problems are not covered until later chapters, the reader should find that the examples we describe are simple to follow using the basic ideas without a formal mathematical structure. In all

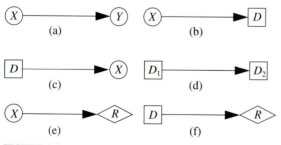

FIGURE 1-6
Directed arcs in influence diagram.

our examples we proceed in a deliberate sequence of steps. We (1) describe the problem in words, (2) identify the random events, decisions, and results that can occur, (3) describe the logic and interdependencies of the decision problem with an influence diagram, (4) use this logic together with the decision sets and random outcome sets to build a decision tree, (5) determine the optimal decisions in terms of the parameters, and (6) analyze the optimal policies as functions of the model parameters in a sensitivity analysis. By using numerous examples we hope to be able to give the reader an ability to build models as well as analyze their implications.

1.8 THE DECISION SAPLING

A simple model that demonstrates the essence of decision making under uncertainty is one where a decision D must be made either to continue in the current mode of operation for which there is little or no uncertainty, or move in a new direction that has an uncertain outcome X but has potentially both larger and smaller payoffs. The influence diagram for this problem is shown in Figure 1-7. The set $\mathcal{D}$ consists of two decision choices, (1) a riskless alternative RA and (2) a risky venture RV; the set X has elements 0 and 1, where if the risky venture is chosen, X is 1 if it is a success and 0 if not. The decision tree for this problem is shown in Figure 1-8. Because of its simplicity and the fact that we use and refer to it in a number of different ways throughout the book we call it the *decision sapling*. Some examples follow that illustrate its widespread application from the personal to the strategic management level. Later in the book the reader will see how it can be used as a tool to elicit information from decision makers that helps quantify results that have no obvious numeric measurement scale.

1. *Choosing an investment account.* You are considering placing your savings in a savings and loan account which is insured up to $100,000. It pays a modest interest rate and is guaranteed against loss of capital. An alternative is to place your savings in a stock mutual fund that has potential for substantially higher returns but could result in a loss of capital if the market declines. The savings and loan decision is the riskless alternative, and the stock mutual fund the risky venture. Which would you choose?
2. *Playing a lottery.* Consider a person's decision whether or not to participate in a state or national lottery where the object is to choose 6 integers from the set 1 through 49. For a very modest amount (in the USA it is usually $1), you have a chance to gain great wealth, although the chance of success is extremely small (approximately 1 in 14,000,000 for

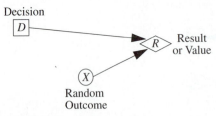

FIGURE 1-7
Influence diagram for the decision sapling.

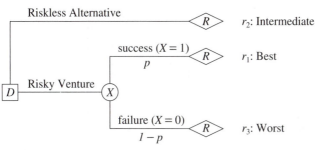

Riskless Alternative — R r_2: Intermediate

success ($X = 1$) — R r_1: Best
p

D Risky Venture X

failure ($X = 0$) — R r_3: Worst
$1 - p$

FIGURE 1-8
The decision sapling.

this example). If you choose to participate, you take the risky venture. If not, you take the riskless alternative. Would you play?

3. *Changing jobs.* Suppose you have been working for a company for a number of years; your job is no longer as personally satisfying as it was, but it gives you financial security and a well-known work environment. A new job offer comes along that initially pays the same, but appears to have greater potential for higher rewards. Do you risk taking on the new job or stay in your current position?

4. *Elective surgery.* A more serious personal decision problem arises in the case of an adverse medical condition. Suppose your health has deteriorated because of age and poor diet and that you now suffer from angina. The disease is not an immediate threat to life but causes crippling chest pains. These can be controlled with medication, but they seriously limit your ability to participate in some of your favorite physical activities. Your doctor suggests bypass heart surgery to improve your condition. Of course the surgery is risky, but if successful you could again take up some activities you used to enjoy. If not successful, it could leave you even more debilitated or may even be fatal. Do you agree to surgery?

5. *A civil lawsuit decision.* You have taken a professional to court to recover losses you claim are due to malpractice and are asking for $2,000,000. The case has cost you $50,000, has been heard, and the jury is deliberating. The attorney for the defendant offers immediate settlement of $700,000. Do you settle (the riskless alternative) or wait to hear the jury's decision?

6. *Product development.* Your research and development branch has been working on improvements to one of your products that, if successful, would result in substantially larger profits. Not all the problems have been eliminated from the product-improvement techniques, so there is risk of actually reducing the product quality (introduction of new car models is a well-known example). Should you go into production of the new "improved" product or stay with the current one?

7. *Nuclear power plant construction.* A consortium of large corporations who build conventional nuclear power plants using well-understood technology has been conducting research and development into an advanced type of reactor. If successful, it would be more efficient and would greatly reduce the nuclear waste problem. There are still many unknowns in the future technology, and a failure of such a plant might be more catastrophic than the conventional type. The consortium is at a point where it needs to decide whether or not to continue with the conventional technology or switch to the advanced design.

TABLE 1.2
Examples of decision sapling problems

Example	Best	Intermediate (riskless)	Worst	Risky venture probabilities
		Results		
Investment account	Stock fund high return and capital gain	Savings and loan interest	Capital loss and/or no return	Estimated in financial literature
Lottery	Independent wealth	0	Lose ticket cost	Determined by game rules
Changing jobs	Increased job satisfaction and income	Same job, known environment, job security	New job is worse or less secure	Subjective, state of economy, industry type
Elective surgery	Improved life	Live with condition	Worse condition or death	Subjective expert opinions available
Civil lawsuit	$1,950,000	$650,000	-$50,000	Subjective
Product development	Improved profits, market share	Reliable product	Poor product, warrantee and recall costs, market share loss	R&D results, marketing test results
Nuclear power plant	Unit cost reduction, nuclear waste reduction	Known technology and costs	Less safe facility, higher costs	R&D results, test results

The fundamentals of these example problems are summarized in Table 1.2. Although these problems differ radically in scope and technical details, they all have at their core the decision problem illustrated in Figures 1-7 and 1-8. Figure 1-7 depicts the problem *structure*. There are three important events; a decision D is made before the observation of an uncertain event X, which in turn is observed before the result R of the decision problem is measured. The two directed arcs, from D to R and X to R, show that both the decision and random event can influence the result. The fact that no arc is drawn from D to X indicates that the probability distribution assigned to X does not depend on the decision. Having structured the problem, we next determine the set of possible decisions or outcomes at each node. At the decision node the possible choices are RA and RV, so we denote a set $\mathcal{D} = \{RA,RV\}$. The possible outcomes of the random event X are success or failure of the risky venture; we could define the set X to be {success, failure}, or $\{1,0\}$ if we set X equal to 1 when success occurs and 0 when it does not. This latter notation is used frequently throughout the book, including the remainder of the section. Let p be the probability of success if the risky venture is chosen, and let the payoffs of the three possible results be:

r_1 if the risky venture is chosen and is a success,

r_2 if the riskless alternative is chosen,

r_3 if the risky venture is chosen and is a failure,

so $\mathcal{R} = \{r_1, r_2, r_3\}$. In some problems the elements of $\mathcal{R}$ are numeric, in others they are not. For example, in Table 1.2 the civil lawsuit has three numeric results; in the elective surgery they have not been quantified. When they are numeric, we assume that $r_1 > r_2 > r_3$. When they are not, there is assumed to be an ordering determined by the preference of the decision maker such that r_1 is preferred to r_2, which is preferred to r_3. The reader should note that this ordering implies that one of the decisions does not dominate. It typifies the essential ingredient of most real decision problems—the decision maker must weigh the potential for higher payoff or efficiency against increased risk. In the remainder of this chapter the results are assumed to be numeric.

Because the result depends on both the decision and the random outcome, we define a result function $R(x, d)$, where

$$R(0, RA) = R(1, RA) = r_2,$$

$$R(0, RV) = r_3,$$

$$R(1, RV) = r_1.$$

Using Equation (1.19) (p. 19) with $Pr\{X = 1\} = p$, we see that the expected return when the riskless alternative RA is chosen is r_2 and when the risky venture is RV is chosen is $pr_1 + (1 - p)r_3$. Using expected payoff as the criterion for evaluating the result, the optimal policy is:[6]

---

[6] Section 1.9 contains a discussion on the choice of a criterion.

$$\text{Choose RA if } r_2 > pr_1 + (1 - p)r_3,$$

$$\text{Choose RV if } r_2 < pr_1 + (1 - p)r_3,$$

which means that we follow one policy when the top inequality is satisfied, the other policy with the bottom inequality. We are indifferent to the policy when there is equality. The value of p that satisfies $r_2 = pr_1 + (1 - p)r_3$ is therefore called the *indifference probability* and is denoted by $\bar{p}$, the solution of

$$\bar{p} = (r_2 - r_3)/(r_1 - r_3).$$

Figure 1-9 illustrates the important point that for fixed values of r_1, r_2, and r_3 there is a whole range of p values for which each of the two decision choices is optimal. This is a simple example of a policy space analysis. The reader will see more complex examples of this idea as he or she proceeds through the book, especially in Chapter 5. In order to choose the best decision it may not be necessary to know p with great precision. In some problems p can be calculated precisely (such as in the lottery). In others it can be estimated from past data. But in many problems, especially those more strategic in nature, one often has to rely on expert or subjective judgment that cannot be corroborated by careful measurement. It is comforting to know that the same policy is often optimal for a whole range of probability values rather than for a single number.

1.8.1 The Value of Perfect Information

Suppose you could buy information before your decision is made that predicted perfectly the outcome of the risky venture. If success of the "risky" venture were guaranteed, you would take the risky venture and get r_1, whereas if failure were guaranteed, you would take the riskless alternative and get r_2; thus, you would always be able to avoid the worst outcome r_3. Because the risky venture results in a win a fraction p of the time, your expected return with perfect information would be $pr_1 + (1 - p)r_2$. Without it you must make your decision using only p, r_1, r_2, and r_3. If p were less than $(r_2 - r_3)/(r_1 - r_3)$, you would take the riskless alternative, your payoff would be r_2, and the expected increase in expected payoff due to perfect information, say Δ, would be $p(r_1 - r_2)$. If p were greater than $(r_2 - r_3)/(r_1 - r_3)$, you would take the risky venture, your payoff would be $pr_1 + (1 - p)r_3$, and Δ would be $(1 - p)(r_2 - r_3)$. Figure 1-10 shows a plot of Δ as a function of p. Notice that perfect information has a maximum value when $p = \bar{p}$, the indifference probability where it did not matter which policy you used.

Before leaving this section it is appropriate to say a few words about the payoffs. In the examples we gave and summarized in Table 1.2 the reader can see that it is not always obvious how payoffs should be measured. For the medical problem one would have to put measures on "living with angina," "improved life" due to successful surgery, and perhaps "death" for unsuccessful surgery. Even when measures seem to be quantifiable, as in the lottery problem, using expected monetary payoffs may not be the appropriate decision criterion. Lotteries run by a state or a large commercial enterprise usually offer very high prizes with very low probability of winning, where the expected payoff is negative $(pr_1 + (1 - p)r_3 < r_2 = 0)$. Because the expected payoff is negative, why do

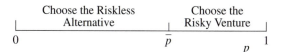

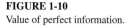

FIGURE 1-9
Policy space for the decision sapling.

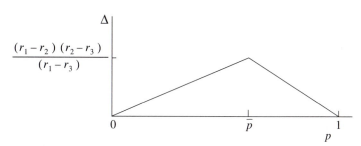

FIGURE 1-10
Value of perfect information.

millions of people play these highly popular lotteries? One explanation is that they view their payoffs not directly in monetary terms, but as r_1—independent wealth, r_2—lifestyle and wealth unchanged, r_3—lose \$1. Chapter 6 discusses the problems of deciding on appropriate result measures under risk and how to quantify them.

1.9 CRITERIA FOR COMPARING RESULTS

Over the years a number of ways have been suggested of measuring the best course of action when making a decision whose result depends on future uncertain events. Although they are not widely accepted or used today, we feel that it is important to review the main ones as they illustrate the essential problem of risk and dealing with uncertainty. But first we discuss the important concept of *dominance*.

There are times when a particular decision, say d_1, will always lead to at least as good a result as that for any in a subset of alternative decisions, say $\mathcal{D}_1$, for every possible outcome of an uncertain event. When this is true, we say that d_1 *dominates* $\mathcal{D}_1$. If our objective is to maximize a result $R(X, d)$ that depends on the decision d and the random variable X, then d_1 *dominates* $\mathcal{D}_1$ if and only if

$$R(x, d_1) \geq R(x, d) \text{ for all } d \in \mathcal{D}_1 \text{ and } x \in X.$$

Whenever a dominating decision is found the other decisions in $\mathcal{D}_1$ can be eliminated from the problem because none of them can ever improve on the result for any outcome of X. Note that in Table 1.3, which contains the result function in units of \$1,000 for the civil lawsuit shown in Table 1.2, neither of the two decisions dominates the other, so neither can be eliminated.

Dealing with uncertainty of future outcomes is not easy for most decision makers. Readers should try hard to imagine themselves in such a position as they study each example; think of the decision problem, and not simply the analysis. If you were involved

TABLE 1.3
Result function for civil lawsuit example

	Trial outcome (units of $1,000)	
Plaintiff's decision	$X=1$	$X=0$
$d = \mathrm{RA}$	650	650
$d = \mathrm{RV}$	1,950	−50

in this lawsuit and estimated your chances of winning the case are only 50%, would you behave differently if the units were 10 cents instead of $1,000? For many individuals the answer is probably yes. Only a very rich person might consider giving up a certainty of $650,000 for a 50% chance of winning $2,000,000. Many more may risk losing $5 when they could win $195 rather than accept $65. For most people one's propensity to accept risk depends on the size of the possible losses and gains relative to one's wealth. We return to this important concept in later chapters. In the remainder of this section we discuss several criteria that can be used in decision making, including some that do not require any estimates of the probabilities of underlying uncertain events.

One way to resolve a decision dilemma is to first note the worst possible result for each decision alternative and then choose the alternative that gives the best of these worst results. This is the most pessimistic way to solve a decision problem under uncertainty and is referred to as the *maximin* criterion. Using the data in Table 1.2 for the civil lawsuit problem, the worst possible outcomes for RA and RV are $650,000 and −$50,000, respectively. From these we see that using the maximin criterion would lead to the decision to take the riskless alternative—settle the case.

The opposite extreme to maximin is *maximax*. Here one finds the best possible outcome for each decision alternative and chooses the alternative that gives the largest. Again using the civil lawsuit example, for RA and RV they are $650,000 and $1,950,000 respectively, so using the maximax criterion would lead to the decision to wait for the jury. Such a criterion might be suitable for the most optimistic gambler.

A third criterion, referred to as the *minimax regret*, is a little more complex. The idea is to look at the loss, or regret, for *not* choosing the best alternative for each possible random outcome, find the maximum regret for each decision alternative, then choose the decision that minimizes these maximums. The procedure is as follows: For each outcome of the random event (1) find the best possible return over all decision alternatives, and (2) for each decision alternative find the loss of return (regret) by choosing that alternative. For each decision alternative find the maximum of these regrets. Finally, find the minimum of these maximums to determine the desired decision alternative. For our lawsuit example in Table 1.2, if the jury awards the $2,000,000 the best outcome occurs when the plaintiff waits for this jury decision for a net return of $1,950,000. If the jury does not award the $2,000,000 the best outcome occurs when the plaintiff settles for $700,000. There is no regret for choosing RV if the jury finds for the plaintiff, but a regret of $700,000 if they do not. There is a regret of $1,300,000 for choosing RA if the jury finds for the plaintiff, but no regret if they do not. Thus the maximum regret for choosing RV is $700,000, and for RA it is $1,300,000. The minimum of $700,000 is obtained by choosing RV and not accepting the settlement offer.

One advantage that all three criteria have is that they can be determined without any knowledge of, or attempt to estimate, the distribution of the random event. Two of the three are conservative, whereas the third can carry a high risk. In reality a decision maker always has some knowledge of the distribution of the underlying random variable(s), from either past data, experience, judgment, or a combination of these and therefore should be able to do better. We always assume some such knowledge and choose a criterion that makes use of it.

The criterion that we apply in most parts of this book is expected value. The term comes from the frequency interpretation of probability and can be misleading. For problems that are solved repeatedly the term is appropriate; the law of large numbers tells us that if the results of many replications of a decision problem are averaged, it is very unlikely that this average will differ significantly from the expected value. However, the appropriate use of expected value as a criterion does not depend on this frequency interpretation. Many probabilities used by decision makers are subjective, as we shall often see in this book. If one interprets such a probability as a measure on a scale where 1.0 means you are certain an event will occur and 0 means you are certain it will not, expected value is simply a weighted average of possible results that reflects one's knowledge and judgment. For each decision alternative (or vector of alternatives in sequential decision problems) we usually find the expected result $E[R(X, d)]$ and the decision, denoted with an asterisk, d^*, that maximizes this over all d in $\mathcal{D}$. When it is not appropriate to use a simple linear return function, we use utility, which can be thought of as a nonlinear function of R and maximize expected utility. The reader will find these concepts in Chapter 6.

1.10 CLARIFYING TERMINOLOGY

Throughout this book there are frequent references to the terms *objectives, characteristics, attributes, criteria, trade-offs,* and *constraints*. It is important to understand clearly what we mean when we use these terms. By *objectives* we mean the purpose or end toward which our efforts in problem-solving are directed. For example, our objective may be to outperform our competitors in a business venture. A more limited objective would be to reliably detect cancer when it is present. The objective of a football coach may be to win the championship or maintain a certain minimum position in the league. By *characteristic* we mean a feature or quality that pertains to, is distinctive of, or that indicates the character of an object. For example, a person's age or the performance of a car would represent characteristics. By *attribute* we mean a well-defined measurable value of a characteristic. For example, if the characteristic is the performance of a car, we may measure this by the time it takes for the car to reach a certain speed from a standing start; or we could use the time it takes to accelerate from 40 to 60 mph. The *criterion* is the way we measure the level to which the objectives are met by the different decision choices using the chosen attribute(s). For example, we may maximize the minimum profit or market share achieved using a maximin criterion, or we may maximize the expected value of profit or market share. A *trade-off* is an exchange of one or more attributes within a criterion to achieve a benefit or advantage. The terms criteria, attributes, trade-offs, and objectives are frequently confused. We always attempt to state clearly what we intend when using these words.

1.11 BOOK OVERVIEW

The timing of events as well as knowledge of information and possible alternatives are critical to good decision making. In many programmed automated systems such as guidance control systems on a rocket, space vehicle, industrial robot, and so on, decisions are made rapidly so as to seem almost continuous. Such systems are often modeled in continuous time. In this book we treat problems where decisions are made relatively infrequently (perhaps only once), so that the time between sequential decisions is significant and becomes an assumed integral part of the models. Thus, underlying all our models is the concept of a "stage" or "time period" that is not always explicitly identified. We think of a stage as a point in time when a decision has to be made. The times between successive stages are not necessarily equal in length. For example, suppose you are at a point in time where you start to observe some phenomenon to gather information before making a decision. You might start gathering information on new cars from local dealers to decide on whether or not to buy a new car, and if so, what make and model. After studying all the information for a week you must decide whether or not to buy a given car, abandon the project, or continue gathering more information. If you decide ahead of time that you will either buy a given car or abandon the project without undertaking a search for more information, then you undertook a one-stage decision problem. If after the first week of information gathering you consider getting further information, then your problem has at least two stages. It need not take a week to complete the second search for data, in which case the stages are not of the same length. It should be clear to the reader what we mean by a stage in the various examples throughout the book.

A key objective of the book is to demonstrate the interaction between forecasting uncertain outcomes and using the results of such a forecast in a decision model. By a forecast we mean the use of information and experts who develop models to reduce the uncertainty of some future event, using information that may or may not be available to a decision maker. Of course a decision maker always has *some* information in the form of historical records, knowledge of operations, or subjective feelings about uncertain event outcomes. We refer to such limited information as a *baseline forecast*. Chapter 2 describes and analyzes some simple decision problems that only use baseline forecasts. It starts with examples where the number of both decision choices and possible random outcomes is quite small. We discuss in some detail several models that can be used to describe and quantify the effects of a last-minute decision that must be taken by a coach of a football team when a choice must be made between two plays. Even though the specific problem has actually arisen on several notable occasions between college teams, the importance of the formulation lies in the fact that the problem can be cast in several different ways depending on who the decision maker is. There follows a number of variants of the well-known "newsboy" problem to demonstrate examples of decision problems with infinitely many decision alternatives and infinitely many possible random outcomes. Although in many real problems both of these factors are finite, modeling and analysis are greatly simplified by assuming continuous probability distributions and continuous decision functions. Real-life examples include yield management in airline fare discounting and hotel room rate discounting. The concept of "the value of perfect information" is illustrated as well as the concept of correlating the random demand with the decisions that are made. Optimal decisions are found for the various problems using marginal analysis and graphical methods. For those readers comfortable with using simple calculus, analytic methods are also included.

Chapter 3 treats the important role of forecasting in decision making. It discusses and illustrates two basic forecast types, probability forecasts and categorical forecasts. Examples of probability forecasts are "the chance of rain tomorrow is 30%" and "there is an 80% chance our proposal will be accepted and funded." Typical categorical forecast examples are "the maximum temperature tomorrow will be in the seventies" and "the market for our product will be between 200 and 250 next month." We introduce the fundamental concepts of forecast likelihood and decision probability. Both are conditional probabilities, and both play important roles in decision modeling. The first gives the probability of forecasting a certain outcome, given whether or not an important event occurred, and the second gives the reverse, the probability that an important event will occur, given a certain outcome is forecast to occur. Forecast likelihoods and decision probabilities are related by the laws of probability using Bayes' rule. It is very important to understand how they differ, what they mean, and how one can be obtained from the other.

Chapter 4 is devoted to model building. Little emphasis is given to this crucial area in the literature, where emphasis is usually on details of the mathematical analyses of models that are already assumed to be well formulated. We attempt to show the reader how, starting with a clean sheet of paper, one goes about structuring the important aspects of a model. The primary tool we use for understanding the problem and building a model is the influence diagram. Once the modeler and decision maker feel comfortable with the model, we show how one can construct and solve its associated decision tree. We cite several examples where the "wrong" model has been formulated to help the reader understand what constitutes a "good" model. Parsimony is an important feature of good model building.

Chapter 5 demonstrates how one carries out detailed sensitivity analyses on a decision problem once an "optimal" solution has been found. Not only do these analyses show the importance and the effects of assumptions on critical parameter values, but also they give considerable insight into problem structure. Examples are used to introduce the concepts of sequential decision making with and without forecasts. Specific problems include a revisit to the football example from Chapter 2, an agricultural example of protecting crops from the possibility of freezing weather, mechanical part testing versus replacement in a mechanical maintenance problem, and a problem that includes the possibility of two forecasts. In most of these problems the first decision is usually whether to buy additional information or commit resources to improve forecasts; thus, we are interested in the economic value that forecasting brings to the solution of the overall problem.

Chapter 6 introduces examples where it is either not appropriate to measure results with simple linear measures or where results are not easily quantifiable such as win, lose, or tie a football game. Risk may be a significant factor. Often, we need to be able to extract from decision makers their subjective judgments as to the likelihood of uncertain events happening, as well as the relative importance of possible outcomes. To do this we introduce subjective probability, utility theory and the ideas behind nonlinear payoff functions, and show how these concepts can be used to model the decision maker's attitude toward risk.

Chapter 7 extends the results of earlier chapters to problems where we must consider several rather than one attribute and results are multidimensional. For example, in purchasing a car, cost may be only one of many attributes that are important; others may include measures of performance such as reliability, safety, quality, or characteristic such as color or manufacturer. We show how to trade off conflicting measures and extend the utility theory results of Chapter 6 to the multiattribute case. We show how the traditional models of examining cost-benefit analyses can be easily extended to those situations where uncertainty plays a major role in determining which alternatives ought to be selected. Finally, we include a discussion and treatment of the analytic hierarchy process (AHP) which is a well-known and popular method for ranking and choosing among alternatives. We show that its assumptions of and reliance on unitless measures and a priori normalizations are inconsistent with our principle of consistent measurement units.

In Chapter 8 the reader is introduced to the problem of measuring the quality of a forecast. The concepts of forecast calibration, discrimination, and correlation are studied and can be related to the better-known concepts of least-squares error and Brier scores. We also show how one maintains coherent forecasts if one chooses to convert a probability forecast into a categorical forecast or aggregates detailed categories into those useful for a particular decision problem. Finally, we pay special attention to the quality and economic value of forecasts as well as the situations that arise when forecasts and decisions are made by different people or organizations.

Finally, Chapter 9 gives a more in-depth analysis of large-scale influence diagrams and decision trees. In real-world problems the decision tree often becomes extremely large and bushy. Dependencies between random variables are often difficult, if not impossible, to discern. For this reason influence diagrams become very important with their ability to illustrate graphically the important connections between decisions, random events, and outcomes and the timing of information that is relevant to decision making. Although we are not primarily concerned with computational efficiency, we review the general structure of algorithms based on the principle of optimality that explicitly recognize the history of events and decisions that influence costs, utilities, and conditional probabilities. Our models attempt to capture the essential and relevant data required to solve such decision problems in a format that is useful in computing algorithms. The structure of such algorithms is demonstrated in formulating and solving a nuclear reactor decision problem. The chapter also includes a deeper study of influence diagrams and graphical procedures for reducing their complexity before one formulates the appropriate decision tree. We describe arc reversal in chance influence diagrams (without decision nodes) that illustrate the dependencies between random variables and how graphical procedures can be properly used to make inferences and lead to simplification of the influence diagram. Finally, in influence diagrams that include decision nodes we use the principles of optimality and coherence to obtain graphical reduction procedures that allow us to simplify the influence diagram.

Each chapter, including this one, has a "Summary and Insights" section that includes what the authors believe are the most important results and conclusions that can be drawn from the chapter as well as a brief summary of the literature sources and their contribution to the topics contained in each chapter. Many of these can be used by the reader interested in pursuing special topics in greater detail. Articles and books that either we refer to in the text or in our opinion are useful for supplementary reading are also included at the end of each chapter. We hope that we have been fair and comprehensive in our selection of other authors who have contributed to this important and growing field; if not, we apologize in advance for any omissions.

Also included with each chapter is a set of problems that amplify and extend results in the text. More difficult and challenging sections of chapters are denoted by an asterisk, as are the more difficult problems.

1.12 SUMMARY AND INSIGHTS

Toward the end of each chapter we summarize the important ideas and insights gained from earlier sections. Although the examples in this introductory chapter are simple they serve to illustrate some important concepts.

The essence of decision making under uncertainty is captured by the simple decision sapling shown in Figure 1-8 (p. 29). The choice of decisions usually includes ones with low risk but "average" payoff, together with high-risk choices that can result in high or low payoffs. The question of how much larger the payoff must be to be commensurate with the increase in risk, or how large the success probability of the risky venture for a given payoff, is answered in Figure 1-9 (p. 33) in terms of the payoffs and the probability of success in the risky venture. Even with this simple model we find that there is not just one unique value but rather there are ranges of values for payoffs and the success probability for which it is optimal to make the same decision.

Decisions and choices are usually made before event outcomes are observed, a fact that makes decision making and forecasting under uncertainty such an important and fascinating subject. However, it is useful to inquire into the structure and value of optimal decisions if one could foresee the future by having perfect information on uncertain outcomes before decisions are made.

The importance of a coherent synthesis of all relevant information and forecasts for the decision maker cannot be overemphasized. Timing and content of relevant data are critical components of good decision making.

One of the critical issues in making decisions is how to deal with uncertainty; we rely on the theory of probability. There are numerous books and professional journals dealing with probability and statistics models. However, for the purposes of our book, one might refer to Ross (1990), Larson (1982, first five chapters), or Mendenhall, Wackerly, and Scheaffer (1990, first six chapters) for a frequency approach, and Lindley (1985), Lee (1989), or O'Hagan (1988) for a subjective approach.

The application of decision analysis in model building must include the requirement that the decision maker act rationally. For an excellent discussion on coherence and rational behavior, see Lindley (1985), Bunn (1984), French (1993), and Smith (1988a). An early description of decision trees and the basic treatment of rollback and the calculation of optimal decisions can be found in Raiffa (1968). For a simplified mathematical treatment of many decision problem formulations, see Behn and Vaupel (1982), Bunn (1984), and Clemen (1990).

Influence diagrams are a more recent development that greatly aid in problem formulation and in the practical application of model design that can be used to visualize the structure and interpret the results of decision problems. See Howard and Matheson (1984), Shachter (1986), and many of the papers in Oliver and Smith (1990) not only for their individual contributions but also for the many references to the literature which they provide.

A recent survey of the decision analysis literature has been made by Corner and Kirkwood (1991), and a recent book by Clemen (1990) introduces the use of influence diagrams and decision trees to the art and science of making decisions.

PROBLEMS

1.1 For the parking space problem pmf in Table 1.1 (p. 16), find
a. The probability of finding a place to park.
b. The expected number, variance, and standard deviation of the number of places you find.

1.2 Let A_1 and A_2 be two mutually exclusive events and B a third event. Using a Venn diagram, show that
a. $B \cap (A_1 \cup A_2) = (B \cap A_1) \cup (B \cap A_2)$.
b. $(B \cap A_1)$ and $(B \cap A_2)$ are mutually exclusive.

c. $Pr\{B \mid (A_1 \cup A_2)\} = \dfrac{Pr\{B \mid A_1\} Pr\{A_1\} + Pr\{B \mid A_2\} Pr\{A_2\}}{Pr\{A_1\} + Pr\{A_2\}}$.

1.3 For the cancer example in Section 1.5.3, find $Pr\{A_1 \mid B\}$ and $Pr\{A_2 \mid B\}$, and interpret your result.

1.4 Two (fair) die are rolled. Let X_1 and X_2 be their outcomes. Let Y be the total obtained on a given roll, so that $Y = X_1 + X_2$.
a. Find the set of possible outcomes $\mathcal{Y}$.
b. Find the pmf of Y.

1.5 Verify Equation (1.16) (p. 17) using events $A = \{X > a\}$ and $B = \{X \leq b\}$ together with Equation (1.3) (p. 8).

1.6 Let X and Y be two discrete random variables with the joint pmf $p_{X,Y}(x,y)$ shown in the following table:

		Random variable Y		
Joint probability mass function		1	2	3
Random variable X	1	0.13	0.05	0.03
	2	0.08	0.14	0.13
	3	0.02	0.15	0.12
	4	0.05	0.06	0.04

a. Find $p_X(x)$ and $p_Y(y)$.

b. Are X and Y independent?

c. Find $E[X]$, $E[Y]$, $Var[X]$, $Var[Y]$, and $Cov[X, Y]$.

1.7 The following odds were published on May 6, 1994, on fifteen horses who were to run in the Kentucky Derby on May 7, 1994: 20-1, 12-1, 30-1, 8-5, 30-1, 30-1, 8-1, 15-1, 6-1, 3-1, 30-1, 15-1, 15-1, 30-1, and 30-1. Show that these odds are not coherent and hence cannot be the true odds. Which law of probability do they violate?

1.8 Suppose that you are to decide whether or not to change restaurants for lunch. Your current restaurant gives you a satisfaction rating of 0.7 on a scale from 0 (the worst) to 1.0 (the best possible). From friends you think there are even chances of a new restaurant giving you satisfaction ratings of 0.85 and 0.6.

a. Should you try the new restaurant?

b. Find your indifference probability between the current and alternative restaurants. Interpret the result.

1.9 A patient suffering from angina is advised by his personal physician that major improvement in daily living could be achieved through bypass surgery. The physician estimates that this patient has an 80% chance of successful surgery. What factors should the patient take into consideration in making a decision as to whether or not to undergo the surgery? What further information is needed? Comment on how it might be obtained.

1.10 Let X_i be a random variable that represents the change in the price of a common stock on day i, and let Y_i be the corresponding trading volume. Explain in words the meaning of the following:

a. $E[X_i] = 0$.

b. $Cov[Y_i, Y_{i+1}] > 0$.

c. $Cov[X_i, Y_i] > 0$.

d. $E[X_i] > 0$.

e. $Pr\{Y_{i+1} = y | X_i = x\} = Pr\{Y_{i+1} = y\}$.

1.11 *Let X be a nonnegative random variable and c be some positive constant. Show that

a. $E[X] = \int_0^\infty [1 - P_X(x)]\, dx,$

b. If $Y = Min\{X, c\}$, $E[Y] = \int_0^c [1 - P_X(x)]\, dx,$

c. If $Y = Max\{X, c\}$, $E[Y] = c + \int_c^\infty [1 - P_X(x)]\, dx.$

1.12 As an investment banker suppose you have to choose between investing $1 million in one of the following two alternatives, each of which can be a success or a failure:

• Project A, where, if the venture is successful, the investment will triple in value in five years, and, if not, the investment will be lost,

• Project B where, if the venture is successful, the investment will increase 50% in value in five years, and, if not, it will decrease 50% in value.

a. Does either alternative dominate?

b. Draw the influence diagram and decision tree for this problem.

c. Find the best decision using the maximin, maximax, and minimax regret criteria.

d. Simplify the decision tree if the probability that Project B fails is zero (it may be a guaranteed investment certificate making 8.45% tax-free interest compounded annually).

e. Using expected value of the investment as the criterion together with the assumption in (d), what is the minimum probability of success for Project A you would require for it to be preferred to project B?

f. Comment on the choice of expected value of the investment as a criterion if the total assets of the institution making the decision were (1) $2 million, (2) $200 million.

CHAPTER
2

USING
BASELINE
FORECASTS

2.1 INTRODUCTION

Decision making under uncertainty requires that one or more decisions be made before one or more uncertain quantities are observed. If one had perfect information about the uncertain quantities, one would only need to make lists of the alternatives and their consequences, and by some criterion select the best or most desirable from the list. Of course in real problems perfect information is almost never available. The decision maker must estimate probabilities of random event outcomes or obtain (usually at a price) a forecast of the outcomes from an expert source.

The decision problems considered in this chapter are simple enough in structure that formulating and building the model are not considered difficult tasks. In these problems the decision maker uses whatever experience and knowledge are at hand to estimate the probabilities of event outcomes without obtaining expert advice in the form of a forecast. These event probabilities are known as *baseline forecasts* and are based on an information set I that includes beliefs or estimates made from historical data about the random quantities of interest. These baseline forecasts differ from the more specialized forecasts we discuss in later chapters in that they do not depend on information in addition to the basic information set I. The reader familiar with Bayesian terminology can think of a baseline forecast as a starting prior.

In Section 2.2 we study a crop protection problem where the baseline forecast describes the probability or odds that freezing temperatures will occur during a critical period before harvesting. This information is all we have to work with when we are asked whether to protect or not protect a crop against freezing. In Section 2.3 we formulate a football coach's decision problem and a related bet between two people watching the same college football game; in neither example do we have available an explicit formal forecast of events or uncertain quantities other than a belief that the coach is more or less likely to select a two-point conversion play rather than the more usual single-point conversion attempt. For the betting problem we study two cases, one where the coach's decision is known before the bet is made, the other where it is not. Thus, in one case, we have perfect information on the coach's decision, whereas in the other case the coach's decision is explicitly treated as a random quantity; in the latter case we need to state our belief that the coach is likely to select one or the other alternative.

Section 2.4 poses a problem that is often faced by a wholesale or retail distributor of products; in this case the commodity is an expensive special-purpose industrial battery. The problem formulation quickly demonstrates how decision trees can grow rapidly even for problems of simple structure; it serves to motivate the use of "continuous" models where we consider infinitely many decision choices. The newsboy problem described in Section 2.5 is probably the best-known single-stage decision problem in the management science and operations research literature where there are infinitely many decision choices. It illustrates trade-offs between having too much and too little inventory of a perishable commodity. In recent years this simple model and its variants have been used in pricing and allocation of airline seats, rental autos, and hotel rooms; these are usually known as yield management problems because they undertake the formulation and solution of risky decisions to make marginal improvements under existing operating conditions when outcomes and rewards are uncertain. The value of perfect information is illustrated in Section 2.6, and examples are included in Sections 2.7 and 2.8. The one characteristic that these problems have in common is that they are all concerned with perishable services or goods in which the decision maker must decide between accepting a discounted price that is available immediately or reject that offer in anticipation of an uncertain future higher price. A newspaper is usually worthless the day after it is published; an empty airline seat is worthless once the plane has left the gate on a given flight; revenue lost by an empty hotel room on a given night can never be recovered. In each case a guaranteed lower price might be more favorable than planning on a higher price which might not be realized. Section 2.9 shows how to make use of additional information related the unknown demand in the newsboy problem. Such information can be thought of as a forecast, and is intended to introduce the reader to material in the following chapters.

The problems in this chapter involve formulations where our objective is to find a decision that maximizes expected return (minimizes expected cost) over all decision alternatives $d \in \mathcal{D}$. For simplicity we assume that there is only one uncertain quantity X and that the decision maker's estimate (belief) that it will assume the value x, given his or her information set I, is $p_X(x \mid I)$, the probability mass function (pmf). If x does occur and decision d is made, the result (payoff or loss) will be $R(x, d)$. If we assume that the expected result is to be maximized and denote this maximum by r^*, when X is a discrete random variable our problem can be stated mathematically as

$$r^* = Max_d E[R(X, d) | I] = Max_d \sum_x R(x, d) p_X(x | I) ,$$ (2.1)

where it is understood that the summation is over all $x \in X$. Alternatively, if X is a continuous random variable, we specify the uncertainty in terms of a probability density function (pdf) and write

$$r^* = Max_d E[R(X, d) | I] = Max_d \int R(x, d) p_X(x | I) dx .$$ (2.2)

If the expected result is to be minimized, replace *Max* in these equations with *Min*, and denote the minimum result (loss) by l^*. A decision d^* that maximizes the expected return (minimizes the expected loss) is called a *Bayes decision*. We refer to the formulations in Equations (2.1) and (2.2) as the decision problem *with a baseline forecast* $p_X(x | I)$.

Although we have not shown the dependence explicitly in those equations, the reader should keep in mind that the probability distribution of X may be conditioned on the decision d as well as the information set I. In the future this dependence will be explicitly shown whenever the occasion warrants. The reader must understand that the assessment of $p_X(x | I)$ or $p_X(x | d, I)$ is, indeed, a form of forecast because it expresses a belief about the probability of a future event occurring. It may or may not be based on past data.

Almost any problem that can be represented with a baseline forecast can benefit from the use of one or more forecasts, opinions, or beliefs that are conditioned on special information or knowledge of factors that influence the uncertain quantities or events occurring after the decision(s) has been made. As we shall see in Chapter 3, the role of forecasting is to use any or all relevant information that is available to sharpen our knowledge and reduce our uncertainties on possible outcomes of quantities that affect the decisions selected and the final results. Information may be in the form of historical observations of the quantities of interest or of some additional variables that influence, or are influenced by, the quantities we use in our decision problem. For example, cloudy skies may presage rain, a forthcoming dividend may lead to a stock price increase, the presence of an antigen in the blood may increase the likelihood of prostate cancer, or a photograph of military trucks of a certain type may lead the observer to increase his belief of the presence of a nuclear weapon or a mobile missile. There are many examples where the observation of a quantity or event different from the one being used in the decision problem has relevance to the uncertainties and may therefore directly affect the decision one makes. A simple example for the newsboy problem with additional information is shown in Section 2.9, but we defer a detailed discussion of such problems and concentrate on the formulation of single stage-decision problems.

2.2   A CROP PROTECTION DECISION

Many commercial crops are subject to damage when temperatures drop below a certain level; for our example we assume that damage occurs whenever temperatures fall to freezing or below. Methods are available in some cases to protect the crop (or its value) from such damage, but all cost money. Citrus fruit orchards can be protected by smudge pots, certain fruits such as strawberries by spraying and freezing. One can also think of insurance as a protective device as it protects the *value* of the crop, not the crop itself.

A decision maker (farmer, producer, insurer) must make a decision D whether or not to protect a crop. Let

$$D = \text{NP} \quad \text{if the decision is to Not Protect,}$$

$$D = \text{P} \quad \text{if the decision is to Protect.}$$

Thus $\mathcal{D} = \{\text{P, NP}\}$. Let the random variable X represent the uncertain weather with

$$X = 1 \text{ if it freezes,}$$

$$X = 0 \text{ if it does not,}$$

so $X = \{1, 0\}$. Let r_1 be the loss incurred if freezing weather occurs and the crop was not protected, and r_2 the cost of protecting the crop, where $r_1 > r_2$. There is no cost if the crop is not protected and it does not freeze. This problem has the structure of a decision sapling shown in Figure 1-8 (p. 29) where P is the riskless alternative and NP is the risky venture; it is shown below in Figure 2-1.

Assume that the farmer uses a baseline forecast that it will freeze; that is, one that can be made simply by someone without special knowledge or expertise. This might correspond to a quick estimate that the probability of a freeze equals the fraction of time that freezing temperatures were recorded last year or simply to a best guess. Obviously, this forecast might be much better if the effect of time of the year, the location of nearby storms, and satellite photographs of weather patterns or meteorological expertise were available. As greater expertise is brought to bear, higher-quality and more valuable forecasts can be presented to the decision maker. We defer the important contribution of forecasting to the next chapter and content ourselves meanwhile with using what is known as a baseline forecast.

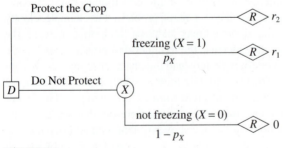

FIGURE 2-1
The crop protection decision problem.

TABLE 2.1
Loss function for the crop protection problem

	Weather condition	
Protect decision (D)	not freezing ($X = 0$)	freezing ($X = 1$)
Do Not Protect (NP)	0	r_1
Protect (P)	r_2	r_2

The loss clearly depends on both the decision D and the weather condition X, as is shown in Table 2.1. Another way to think of the formulation is that you pay an insurance premium r_2 in order to insure yourself against the very large loss r_1. It seems intuitively obvious that the answer to this problem should depend on the cost/loss ratio r_2/r_1. In real crop protection problems the set of decisions $\mathcal{D}$ might include a large number of alternatives and a very large number of weather outcomes in X in which there are many categories besides freezing.

When the decision maker has very limited information concerning the weather, such as newspaper reports or long-term forecasts provided by the weather service, he or she assesses the probability that freezing weather will occur, that is the number[1] $Pr\{X = 1\} = p_X$. We compare the expected loss at the X node in Figure 2-1, $r_1 p_X$, to the certain loss from the Protect decision, r_2, and choose that decision which leads to the minimum expected loss. Thus, the optimal decision, d^*, is

$$d^* = \text{P (Protect)} \qquad \text{when } r_2/r_1 < p_X < 1.$$

$$d^* = \text{NP (Do Not Protect)} \quad \text{when} \quad 0 < p_X < r_2/r_1.$$

Suppose our baseline forecast of p_X is 0.372.[2] We should protect the crop if the cost of doing so is no more than 37.2% of the loss that would occur if we did not and the freezing temperatures occurred. If we let $l_B{}^*$ denote the minimum expected loss incurred, where the subscript B denotes that a baseline forecast is used, then

$$l_B{}^* = r_2 \qquad \text{when } 0 < r_2/r_1 < 0.372,$$

$$= 0.372 r_1 \quad \text{when } 0.372 < r_2/r_1 < 1.$$

The uppermost curve (thin line) in Figure 3-17 (p. 120) shows a plot of this minimal expected loss for the case $r_1 = 1$ (setting r_1 to 1 is equivalent to measuring losses in terms of units of r_1).

---

[1] The explicit dependence on the information set I has been suppressed. This will often be the case throughout the book, but the reader should keep in mind that an underlying information set is always present any time a probability is assessed.

[2] The reason for choosing this number will become clearer in Chapter 3; it is consistent with the data shown in Table 3.9 (p. 119).

We revisit this problem in Chapters and when we discuss the economic value of more discriminatory forecasts and sensitivity of the optimal solutions.

2.3 A DECISION IN FOOTBALL

We now turn to a decision problem that can arise in the American game of college football. Suppose you are the coach of a college team. In a critical game for the championship your team has just scored with less than a minute remaining, so that the score is 16 to 17 against you. You must decide whether to try for the kick conversion that, if successful, would tie the game with no time left to win, or attempt a two-point conversion and possibly win the game.[3] We first structure the problem, identify the decision sets, random outcome sets, and the result in a graphical formulation of the problem.

2.3.1 The Coach's Decision

The influence diagram for this problem is shown in Figure 2-2. First, note that there is one square decision node, one round chance node, and one diamond result or value node. The sequence of nodes from left to right reflects the actual timing of events: (1) you decide on the play, (2) the play is performed, (3) the result of the game is known. Because you believe that scoring from a one-point kick conversion has a much higher probability of occurring than it does when a two-point conversion is attempted, the coach's decision influences the probability distribution associated with the chance node. Thus, we must draw a directed arc from the node D to the node X. The result of the game, denoted by node R, is influenced by both the decision and the outcome of the play. If you make the kick conversion, the result can only be "tie" or "lose," but if you choose the two-point conversion, the result can be "win" or "lose." Which of these will be the actual result will depend on X, the outcome of the play. Thus, we must draw directed arcs from both nodes D and X to node R. This diagram contains all the influences between the coach's decision, the team's play, and the result of the game. Suppose for a moment that the team is just as likely to score on a kick conversion as to score two points (not a realistic assumption). Then, the arc from the square decision node to the circular chance node could be eliminated, because the decision choice would not affect the probability of scoring. Of course, if this were the case, the reader can come up with the "optimal" decision immediately. But we would expect the probability of scoring to be much higher for the kick than the two-point conversion.

In this simple problem there is only one decision set $\mathcal{D}$, which consists of two elements: (1) the kick conversion, which we denote by K, or (2) the two-point conversion, which we denote by T; thus, $\mathcal{D} = \{K,T\}$. Whichever decision is taken, the outcome is

---

[3] A similar situation arises in the British game of rugby. Suppose your team is three points behind with only minutes left in the game and it has been awarded a penalty near your opponent's line. A successful penalty kick with high probability of success will tie the game. A more risky decision is to attempt a try for four points and, if successful, win the game.

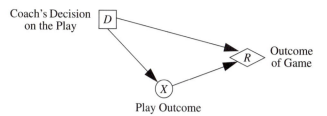

FIGURE 2-2
Influence diagram for the football problem.

uncertain. Let X be a random variable that takes on the value 1 if the team scores and 0 if it does not. Thus, we have a random outcome set $X = \{1, 0\}$. The result set $\mathcal{R}$ is {win, lose, tie}.

The decision tree for this problem is shown in Figure 2-3. Note the same ordering of nodes from left to right as in the influence diagram. Because $\mathcal{D} = \{K, T\}$ has two elements, two branches leave the square decision node D, each representing one of the two alternative decisions. These branches terminate at chance nodes. Note that both are labeled X as these nodes represent the outcome of the play. To each of these are attached two branches that represent the elements of the set $X = \{1, 0\}$. These branches are joined to diamond nodes that indicate the result of the game. Note that there is a path from D to a terminal node for every possible combination of decision and play outcome.

The reader can see that the influence diagram has a very different structure from the decision tree; a directed arc in an influence diagram represents an "influence" of one event or decision on another event or decision, whereas a branch in a decision tree represents a possible decision (action) or random event outcome.

In order to solve the coach's decision problem, he or she needs to estimate the probabilities of scoring for each decision as well as assign relative values to the three

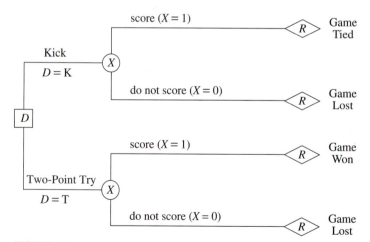

FIGURE 2-3
Decision tree for the football example.

possible results of the game. This last point, the quantification of results such as "win," "lose," or "draw," requires more advanced concepts than we have yet developed, and so we postpone trying to solve the coach's decision problem until later in the book.[4] The next section illustrates a related but different decision problem associated with the same game and shows how an analysis can be carried out when the scoring probabilities are not well estimated.

2.3.2   Betting on the Football Game

Two friends, Nancy and Ron, are in the stands watching the football game described and contemplate betting one against the other on the outcome that will result after the coach has made his decision to have the team kick or attempt the two-point conversion. Nancy offers the following bets to Ron, and he agrees:

1. If the coach decides on the two-point conversion and the team scores, Ron pays Nancy $2. If they do not score, Nancy pays Ron $1.
2. If the coach decides on the kick and the team scores, Ron pays Nancy $0.25. If they do not score, Nancy pays Ron $1.

We want to compare the two bets and see whether each person might prefer one to the other or be indifferent to either one.

Recall that we defined the random variable X to be 1 if the team scores and 0 if it does not. If the coach has the team kick for one point, we denoted this by K; if the two-point conversion is chosen, we denoted it by T. The result R was the outcome of the game.

The first point we want to make is that the timing of Nancy's bet is critical. Either she makes the bet not knowing the coach's decision or she is able to make her bet just after she knows the coach's decision but before the conversion play is attempted. We consider the latter case first, leaving aside how she received or how much she paid to get the information about the coach's decision.

The influence diagram for this problem is shown in Figure 2-4. Note the differences between this and Figure 2-2 (p. 49). The first major difference is that from the point of view of a spectator *the coach's decision must be treated as an uncertain event and is displayed as a round node*; in contrast with Figure 2-2, we now show it as a random node Y with outcome set $\mathcal{Y} = \{k, t\}$. Notice that we have changed our uppercase notation from K and T to k and t to indicate that outcomes are now considered to be random events rather than decision alternatives.[5] The decision node D now refers to Nancy's decision; if we let B represent "Bet" and NB represent "Do Not Bet," then $\mathcal{D} = \{B, NB\}$. The set X is unchanged. The result R is now the payoff to Nancy. It depends on the

---

[4] Problems 2.1 and 2.2 at the end of the chapter give the reader a hint as to how one can quantify such results.

[5] In much of the literature on influence diagrams the outcome and alternative sets are described in both cases as "states". We distinguish between them.

coach's decision Y, Nancy's decision D, and the outcome of the play X; $R(Y, X, D)$ is shown in Table 2.2. The entries in this table constitute the four elements of the set $\mathcal{R} = \{2, 0.25, 0, -1\}$.

The directed arcs in Figure 2-4 show that (1) the coach's decision is known to Nancy before she decides whether or not to bet, and that decision will probably influence Nancy's decision, (2) the coach's decision precedes the play outcome and will affect the probability distribution of that outcome, and (3) the coach's decision, Nancy's decision, and the play outcome all affect the result.

Let us assume that

$$Pr\{X = 1 \mid Y = k\} = p_k > Pr\{X = 1 \mid Y = t\} = p_t.$$

Almost everyone would agree that p_t is less than p_k. Because the chance of kicking an extra point is thought to be higher than the chance of scoring two points, it seems reasonable to place a smaller bet on the scoring outcome with the higher odds (or probability of success), and this is shown in Table 2.2. Nancy's problem is to decide whether or not to bet with Ron, given the known payoffs and whatever information or guesses she may have on the values of p_k and p_t. Before making her decision she must choose an appropriate criterion by which to judge the results; we assume the criterion used is expected return on the bet.

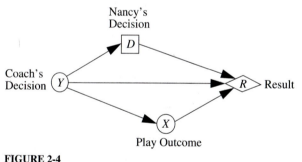

FIGURE 2-4
Nancy's influence diagram, coach's decision known.

TABLE 2.2
The payoff to Nancy

Nancy's decision	Random outcomes (coach's decision, play result)			
	$(Y, X) = (k, 1)$	$(Y, X) = (k, 0)$	$(Y, X) = (t, 1)$	$(Y, X) = (t, 0)$
$D = NB$	0	0	0	0
$D = B$	0.25	−1.00	2.00	−1.00

2.3.3   Analysis of the Football Betting Problem

The decision tree for Nancy's problem is shown in Figure 2-5. Note that it consists of two decision saplings connected to the Y node. Using expected value to measure the result and applying Equation (1.19) (p. 19), if the coach's decision results in a kick and

1. Nancy bets $(D = B)$, the expected return is

$$E[R(Y,X,B)|Y=k] = 0.25p_k + (-1)(1 - p_k) = (1.25)p_k - 1.$$

2. Nancy does not bet $(D = NB)$, then

$$E[R(Y,X,NB)|Y=k] = 0.$$

If the coach selects the two-point conversion and

1. Nancy bets,

$$E[R(Y,X,B)|Y=t] = 2p_t + (-1)(1 - p_t) = 3p_t - 1.$$

2. Nancy does not bet,

$$E[R(Y,X,NB)|Y=t] = 0.$$

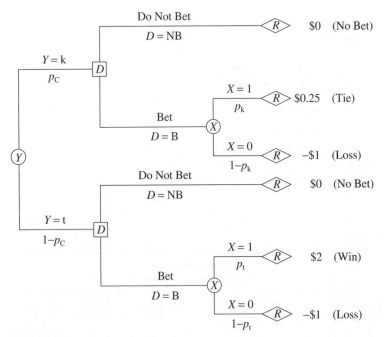

FIGURE 2-5
Nancy's decision tree, coach's decision known.

Nancy believes that $p_t < p_k$ but will bet with Ron only if she believes she can obtain a nonnegative expected value for the particular bet. Because she has agreed on the payoff amounts, she must believe that

$$(1.25)p_k - 1 > 0 \text{ and } 3p_t - 1 > 0, \text{ or equivalently, } p_k > 4/5 \text{ and } p_t > 1/3.$$

But Ron has also agreed to the bets, and because he wins whatever Nancy loses, he must believe that $p_k < 4/5$ and $p_t < 1/3$. There is nothing in this particular problem that requires them to agree on the same values of the success probabilities. For both of them their estimates are at least in part subjective. This situation frequently arises in real decision problems.

Suppose that Nancy believes that there is a 90% chance of the team scoring from a kick whereas Ron believes there is only a 75% chance. If the coach decides on a kick, they will bet, because they both expect positive returns, Nancy 0.125 and Ron 0.0625. Both Nancy and Ron will be indifferent to a "kick" or "two-point conversion" choice by the coach provided the expected gain of each bet is equal. This is true when $p_t = 5p_k/12$. The reader can check that Nancy prefers the coach to choose the two-point conversion when she believes that $p_t > 5p_k/12$; the reverse is true for Ron.

The *coach* is not supposed to think of placing bets on the outcomes of the point conversion; his decision is based on criteria such as win-loss records, love and pride of his team members, and the effect the win or tie may have on his chances of going to a bowl game or perhaps even his future employment as a football coach. However, if by chance he is interested in a "mental" rather than an "actual" gamble and if his payoff were identical to that of Nancy, he would pick the alternative that maximizes expected gain.

We now come to a more difficult and interesting betting problem for Nancy, which is what she should do if she does *not* know the coach's decision before placing the bet. The influence diagram for this problem is shown in Figure 2-6. Note two important differences between Figures 2-4 and 2-6. First, the node D for Nancy's decision is shown to the left of the coach's decision Y (again treated as a random variable), because Nancy must make her decision *before* she knows what the coach's decision will be. Second, there is no arc between D and Y because Nancy's decision as to whether or not to bet has no influence over the coach's decision. We now build the decision tree for this problem. Note the left to right ordering of nodes in Figure 2-6. This order will be preserved in the decision tree, and so we start with node D in Figure 2-7; the decision set is the same as in

Nancy's
Decision

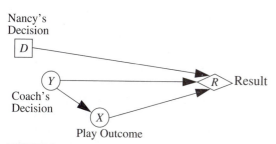

FIGURE 2-6
Nancy's influence diagram, coach's decision unknown.

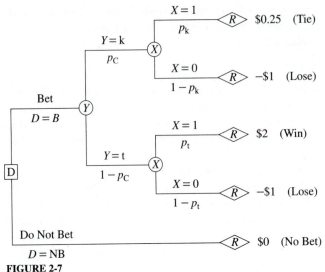

FIGURE 2-7
Nancy's decision tree, coach's decision unknown.

the previous problem, $\mathcal{D} = \{B, NB\}$, so two branches leave node D. The branch corresponding to NB (do not bet) ends in a result node R. The branch corresponding to B ends in node Y. The set of possible outcomes of Y has two elements, $\mathcal{Y} = \{k, t\}$ with two branches leaving node Y. Each one of these ends in node X, the outcome of the play, and because $X = \{1, 0\}$ represents "score" and "do not score," respectively, there are two branches leaving each node X. All branches leaving X nodes end in result nodes R.

Unlike the previous case, if Nancy bets before she knows the coach's decision she must assess the probability that the coach will select decision alternative k or t. We assume that her decision will be to bet only if the expected return is nonnegative. Let p_C be her assessment of the probability that the coach will select k, so $p_C = Pr\{Y = k\}$. Then

$$E[R(Y,X,B)] = p_C(1.25p_k - 1) + (1 - p_C)(3p_t - 1),$$

$$E[R(Y, X, NB)] = 0. \qquad (2.3)$$

Nancy would bet only if the first of these expressions is greater than zero; of course, the decision she selects depends on p_C, p_k, and p_t.

The need for estimating the value of perfect information appears repeatedly in our examination of different decision and forecasting problems. We illustrate using Nancy's betting problems. Nancy's two decision trees in Figures 2-5 and 2-7 can be used to calculate the expected value of knowing the coach's decision before a bet is placed. Let us compare the optimal expected payoff for selected values of p_k, p_t, and p_C.

First, let $p_k > 4/5$ and $p_t > 1/3$. From Figure 2-5 (the case where Nancy knows the coach's decision before she places a bet) her optimal decision is to bet no matter what the coach decides. Let the top branch occur (the coach chooses the kick) with probability p_C and the bottom branch with probability $1 - p_C$ so that the expected return (not knowing the coach's decision but knowing the decision Nancy would take *after* knowing his decision) is

$$E_Y[Max_D E_X[R(Y, X)|Y, D]] = p_C(1.25p_k - 1) + (1 - p_C)(3p_t - 1). \qquad (2.4)$$

From Figure 2-7 (the case where Nancy places a bet before she knows the coach's decision) the expected payoff from a bet is positive and equal to the right-hand term in Equation (2.4), so it is always optimal for Nancy to bet independent of the coach's play probability p_C. We see that for this example the optimal expected payoff is the same whether or not Nancy knows the coach's decision before placing the bet. When $p_k > 4/5$ and $p_t > 1/3$, *there is no value to her having information on the coach's decision before she places her bet.*

Suppose now that $p_k = 0.95$ and $p_t = 0.2$; the team is very good at kicking, but very unlikely to score with a two-point conversion. From Figure 2-5 Nancy's optimal decision is to bet if the coach decides to kick, but not for the two-point conversion, and her expected payoff is

$$E_Y[Max_D E_X[R(Y, X)|Y, D]] = p_C[1.25p_k - 1] = 0.1875p_C. \qquad (2.5)$$

From Figure 2-7 we see that her optimal decision is to bet only if she believes that the probability that the coach will choose the kick is greater than 0.68, in which case her optimal expected payoff is $0.5875p_C - 0.4$. This expression is always less than that in Equation (2.5) when $p_C < 1$. If she assumes that there is only a 95% chance the coach will choose the kick, her expected payoff is 0.158, whereas Equation (2.5) gives 0.178. The difference of 0.020 is the value to Nancy of knowing the coach's decision before placing the bet (her 2-cents' worth!). If she assumes that there is an even chance that the coach will choose the kick, she would not bet, and the value of knowing the coach's decision would increase to 0.094.

This simple example illustrates that the value of information depends on both the structure of a decision model and the value of its parameters. Often there is no value to added information, and hence we should not be willing to pay anything for it. At other times it can be quite valuable. We return to this football example in Chapter 5 where we study whole regions of parameter values in which the same decision policy is optimal.

An even more extreme case occurs if Nancy could hire a clairvoyant who could predict with certainty what the outcome of the play will be. What is the value of such perfect information and how much should she be willing to pay to obtain this information? Does she need to know the coach's decision as well or is all the information contained in knowing the play outcome? (See Problem 2.4).

2.3.4 An Alternate Modeling of Outcomes: Point Scores

In formulating mathematical models of decision problems it is essential to emphasize clarity in the definition of decisions, decision alternatives, events and event outcomes. We have defined X as the event that measures success or failure of the play, so $X = \{1, 0\}$. We could have considered the point scores themselves, in which case the set of possible outcomes would have been $X = \{0, 1, 2\}$. The elements of this set map directly into the set of game results $R = \{$lose, tie, win$\}$. If we know which particular outcome (element of X) occurs, we know which element of R occurs. Thus, the game result using this definition of X would be *conditionally independent* of the coach's decision, and the influence diagram representation would have no directed arc from the decision node to the result node.

A tree and influence diagram can also be drawn for Nancy's betting problem when the uncertain quantity is the point score rather than the success or failure of the play; it is very important to be precise about the timing of Nancy's bet relative to what information she does or does not know at the time she makes the bet. As before, the square decision node for the coach is now replaced by a circular chance node; Nancy's formulation must include the influence of Y on X as well as the influence each of these random quantities may have on her decision to bet (see Problem 2.6). Her payoff as a function of the conversion score and of her decision to bet or not is shown in Table 2.3. Clearly, her bet is influenced by both the coach's decision and the points resulting from the conversion.

2.4 A LIMITED LIFE INVENTORY PROBLEM

Suppose that you are a wholesale supplier of expensive special-purpose batteries to industry and plan on purchasing up to three at a cost of r_C each. The sale price of each battery is r_S, and each one has a shelf life of one year. Any unsold batteries remaining at the end of the year are worthless (their salvage value is negligible compared to the cost of the battery). Demand for the battery is random but never more than 3. Let the random variable X denote the number of batteries demanded in a year, and let $p_i = Pr\{X = i\}$, $i = 0, 1, 2, 3$ (the pmf of X). Let $P(i) = Pr\{X \leq i\}$ be the cumulative distribution function (cdf). The problem is to decide the number of batteries D to purchase at the beginning of the year to maximize expected annual profit.

The influence diagram for this problem has the same structure as the one for the decision sapling shown in Figure 1-7 (p. 28). The order size D must be decided on before the random demand X is observed. Because both the number of batteries ordered and the demand for them will influence the resulting profit, both nodes D and X are connected to R by directed arcs. The three sets corresponding to the three nodes are $D = \{0, 1, 2, 3\}$, $X = \{0, 1, 2, 3\}$, $R = \{-r_C d + r_S Min\{x, d\}; x \in X, d \in D\}$. The elements of D are the possible numbers of batteries to purchase; the elements of X are the possible demands. An explanation of the elements of the set R is as follows: if more batteries are

TABLE 2.3
Nancy's payoff and point scores

Nancy's decision	Point score		
	$X = 0$	$X = 1$	$X = 2$
$D = $ B	−1	0.25	2
$D = $ NB	0	0	0

purchased than are demanded (i.e., $d > x$), the resulting profit will be $r_S x - r_C d$, whereas if more batteries are demanded than are purchased ($d < x$), the resulting profit will be $r_S d - r_C d$.

The decision tree for this problem is shown in Figure 2-8. Note how much more complex it is than the decision sapling in Figure 1-8 (p. 29), even though both are obtained from the same influence diagram in Figure 1-7. This is because the influence diagram shows problem structure without detailed outcomes, whereas the decision tree must contain a path from D to R for every possible sequence of decisions and random outcomes.

Let $R(X, d)$ be the profit if the demand is X and decision d is made. To solve our problem we find the expected profit for each decision d. These are

$$E[R(X, 0)] = 0,$$

$$E[R(X, 1)] = (r_S - r_C) - p_0 r_S,$$

$$E[R(X, 2)] = 2(r_S - r_C) - 2p_0 r_S - p_1 r_S,$$

$$E[R(X, 3)] = 3(r_S - r_C) - 3p_0 r_S - 2p_1 r_S - p_2 r_S.$$

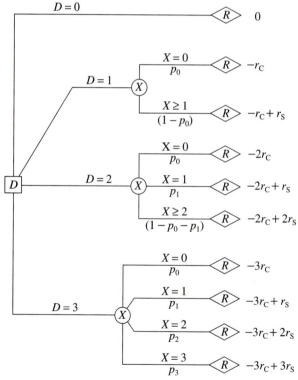

FIGURE 2-8
Decision tree for the battery purchase problem.

2.4.1   Marginal Analysis

Our problem is to choose the value of d that maximizes the expected profit. What if the demand could exceed 3? Perhaps it could be as high as 10, 20, or even in the hundreds. With numbers as high as these, the decision tree will become extremely large and bushy and drawing it will become impossible. In this section we show how one can exploit the special structure of this problem so that it can be represented and solved with a small decision tree.

The reader will probably have noticed the similarity in the above four expressions for $E[R(X, d)]$. It is left as an exercise (see Problem 2.7) to show that if the maximum demand is n, the expected profit when d are purchased is given by

$$E[R(X, d)] = d(r_S - r_C) - r_S \sum_{i=0}^{d-1} P(i), \quad d = 0, 1, \ldots, n, \tag{2.6}$$

where $P(i)$ is the cdf of the demand X. As d increases, the first term of this expression increases and the second term decreases. It can be shown that $E[R(X, d)]$ increases with d to a maximum value before starting to decrease. From Equation (2.6) it follows that

$$E[R(X, d+1)] - E[R(X, d)] = (r_S - r_C) - r_S P(d). \tag{2.7}$$

Suppose we consider a decision tree where we must choose between ordering $(d-1)$, d, or $(d+1)$, and we measure the payoff of each decision *relative to what it would be for choosing d*. This tree is shown in Figure 2-9. To illustrate how the payoffs

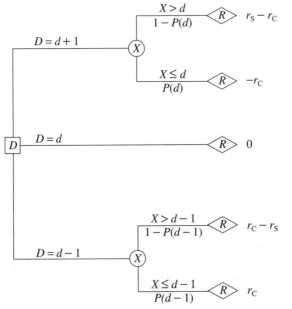

FIGURE 2-9
Decision tree for a marginal analysis.

are calculated, consider the top R node. To reach this node we order $(d + 1)$ and the demand exceeds d. Therefore, we sold the extra unit for r_S, and it cost us r_C for a net gain in payoff of $(r_S - r_C)$. If demand is d or less (second R node from the top), the unit is not sold and causes a net loss of $-r_C$. The reader can check the payoffs at the remaining R nodes. At the top X node, the net increase in expected payoff is $(r_S - r_C) - r_S P(d)$, and at the lower one $-(r_S - r_C) + r_S P(d - 1)$. These are precisely the expressions one would get using Equation (2.7). If d^* is the value of d that maximizes $E[R(X, d)]$, it must satisfy

$$1 - P(d^*) \le r_C/r_S \text{ and } 1 - P(d^*-1) \ge r_C/r_S. \tag{2.8}$$

What we have done is to consider the marginal effect of changing d^* by 1.

In order to find d^*, we must check all the inequalities in Equation (2.8); there are one fewer inequalities than there are elements of $\mathcal{D}$. For large $\mathcal{D}$ sets this can be a considerable task. Even though items must be purchased and are demanded in integer quantities, our analysis is greatly simplified by assuming a continuous random variable for X and a continuous decision set $\mathcal{D}$. This is illustrated in the next section.

2.5   THE NEWSBOY PROBLEM

Every student of Operations Research or Management Science has been introduced to the now classical optimization under uncertainty problem that is called the Newsboy or Christmas Tree problem. It is a simple example of a one-period inventory problem with uncertain demand.

The classic formulation of the problem is as follows. Suppose that a newsboy purchases papers at 15 cents each and sells them the same day for 25 cents each. The net profit for each paper sold is 10 cents. If he orders too few newspapers, there is demand that he cannot satisfy; he is unable to make as much profit as he could have had he ordered more newspapers. If he orders too many, he is left with worthless papers he cannot sell. The reader should notice that the battery example of the previous section has the same structure. That example was used to illustrate analysis for integer values of the demand and order quantity. Because of the large numbers involved and the remarks following the inequalities in Equation (2.8), it is usual to treat the newsboy problem as though the demand and order quantities are continuous.

The simplest model assumes that there is no salvage value for leftover newspapers, no back-ordering allowed, and that there are no hidden penalties such as opportunity costs or loss of future customers. Using notation consistent with that of Section 2.4, assume that the unit cost of a paper is r_C, and the sale price is $r_S > r_C$. Because there is no obvious upper limit to the number of newspapers that will be demanded or that we should order, the sets X and $\mathcal{D}$ are assumed to be the sets of all nonnegative numbers. The (assumed known) probability density function (pdf) for the uncertain demand, X, is $p(x)$ with cdf $P(x)$. The timing of the decision on how many papers to order is crucial: the decision on the number to order, $D = d$, must be made *before* the actual demand $X = x$ occurs. If too few papers are ordered, we shall be unable to meet all the demand and thus lose potential profit; if too many are ordered, we shall be left with worthless (day-old) newspapers that cannot be sold, in which case there may be a net loss. A more general

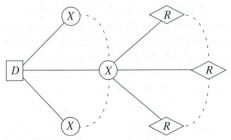

FIGURE 2-10
Decision tree for the newsboy problem.

formulation may relax the back-ordering rule, may include a salvage value for leftover papers or may even allow the purchase of papers at a higher cost once the demand is known. There are many variations.

The influence diagram for this problem is the same as the one for the decision sapling and the battery problem shown in Figure 1-7 (p. 28). It has one square decision node D, one circular random demand node X, and one diamond result node R that are shown from left to right in order of occurrence. There are only two arcs; both the order size and the demand affect the result, but the order size has no effect on the demand.

A decision tree for this problem would contain infinitely many random demand nodes and result nodes, and thus infinitely many branches between the nodes D and X, and X and R. In order to represent this graphically, we can draw a fan containing a typical branch as is shown in Figure 2-10. However, this approach gives little insight into either the structure of the problem or how to solve it. Greater insight is gained using the marginal analysis approach introduced in the preceding section.

Suppose we order d and the demand is X. The payoffs are given in Table 2.4. Figure 2-11 shows the decision tree to decide whether to order d or $(d + 1)$ by looking at the marginal payoff for ordering one additional unit more than d. We would be indifferent to $D = d$ and $D = d + 1$ when the expected payoff from the risky alternative is zero. This is true when d takes on the value d^*, where d^* satisfies

$$1 - P(d^*) = r_C/r_S. \tag{2.9}$$

The solution is well-known and easily stated: d^* is a number such that the stockout probability at this value is equal to the ratio of unit cost to unit revenue. A graphical method of solution is shown in Figure 2-12. The relationship between forecasts and decisions becomes apparent if we look at this graphical solution.

TABLE 2.4
Payoff matrix for newsboy problem

Demand	Payoff
$X \le d$	$r_S X - r_C d$
$X > d$	$r_S d - r_C d$

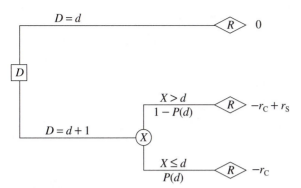

FIGURE 2-11
Decision tree for newsboy marginal analysis.

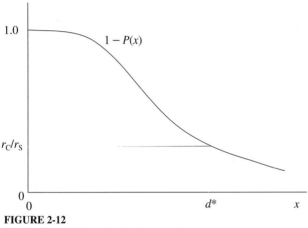

FIGURE 2-12
Solution for newsboy optimal order quantity.

If selling price is a 100% markup of unit cost (a common practice in many retail stores), $r_C/r_S = 0.5$ and the reorder quantity that maximizes expected profit is the *median* (not the expected value!) of the demand distribution. With high profit margins the r_C/r_S ratio is small, the stockout probability is small, and we are primarily interested in forecasting the behavior of the tail of the demand distribution. If the ratio is close to 1, profit margins are low, and we are most interested in predicting the probability of low demands and frequent stockouts because we do not want to be left with unsold inventory. The better we are able to estimate the demand distribution *in the vicinity of d^**, the better our ability to make good decisions. This leads us to consider the extreme case where we have perfect forecasts or perfect information about demand.

Let $R(X, d)$ be the profit when demand is X and d are ordered, so that

$$R(X, d) = r_S Min\{X, d\} - r_C d.$$

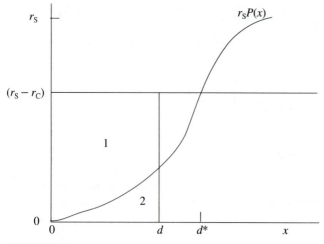

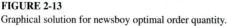

FIGURE 2-13
Graphical solution for newsboy optimal order quantity.

Since X is random, this is a random variable with expected value[6]

$$E\left[R\left(X, d\right)\right] \;=\; \int_0^\infty \left(r_S Min\left\{x, d\right\} - r_C d\right) p\left(x\right) dx. \tag{2.10}$$

For those readers familiar with calculus one can show that this expression simplifies, and using the results of Problem 1.11 of Chapter 1 can be written as[7]

$$E\left[R\left(X, d\right)\right] \;=\; \left(r_S - r_C\right) d - r_S \int_0^d P\left(x\right) dx. \tag{2.11}$$

Figure 2-13 shows a graphical representation of the terms in this equation. Suppose we choose some point d to the left of the point on the x axis where the horizontal line at $(r_S - r_C)$ crosses the function $r_S P(x)$. The first term in Equation (2.11) is represented by the rectangle made up of areas 1 and 2. The second term is the area under the $r_S P(x)$ curve to the left of d, that is, area 2. Thus, the expected profit is given by area 1. Clearly, this area increases as d moves to the right until it reaches the curve crossing point. If one starts with a point d to the right of the crossing point, one can show in a similar manner that the expected profit increases as this point is moved leftward toward the crossing point. This shows that the point at which they cross gives the optimal order

---

[6] Equation (2.10) is a particular form of Equation (2.2). We only use $p_X(x)$ in place of $p(x)$ when it is important to differentiate the particular random variable X under consideration from others in the problem formulation.

[7] Equation (2.11) is the continuous equivalent of Equation (2.6).

size d^* that maximizes expected profit. It is left to the reader to show that this is precisely the same d^* obtained via the marginal analysis and Equation (2.9).

Those readers familiar with calculus can find the optimal order size by differentiating Equation (2.11) and setting the derivative to zero,

$$\frac{\partial}{\partial d} E[R(X,d)] = (r_S - r_C) - r_S P(d) = 0.$$

The reader should check that the value of d that satisfies this is the same value d^* that satisfies Equation (2.9). The reader should also check that the second derivative is negative, thus ensuring that a maximum point has been found.

By substituting d^* from Equation (2.9) into Equation (2.11), it can be shown that (see Problem 2.8) the maximum expected total profit is equal to the product of marginal profit on each newspaper and of the conditional expected demand (the expected number sold):

$$r^* = E[R(X,d^*)] = (r_S - r_C) \int_0^{d^*} \frac{xp(x)\,dx}{P(d^*)}$$

$$= (r_S - r_C)\, E[X|X \le d^*].$$

(2.12)

It is easy to incorporate a number of other features into the newsboy problem: for example, the inclusion of a salvage value for leftover newspapers. If the newsboy buys his papers at $r_C = 15$ cents, sells them for $r_S = 25$ cents, he also may be able to sell leftover newspapers at a much lower price, say, $r_L = 10$ cents. Typically, the salvage value is less than the ordering price so that $r_L < r_C < r_S$. How does this new feature influence the structure of the optimal solutions, and will it increase or decrease the optimal order size?

Nothing about salvage value influences our assessment of the demand distribution. Only the payoff function is changed. We now include a salvage value of r_L per paper for any unsold inventory. The payoff becomes $r_S X + r_L(d - X) - r_C d$ when $X \le d$ and is unchanged at $r_S d - r_C d$ when $X > d$. Using these in Figure 2-11 it is easy to show that the optimal order quantity now satisfies

$$1 - P(d^*) = \frac{r_C - r_L}{r_S - r_L}.$$

When $r_L = 0$, we get the same solution shown in Equation (2.9); for any positive $r_L < r_C$, the ratio on the right is smaller than r_C/r_S. Thus, the new solution for d^* is larger than it would be without salvage value. Although we have deleted explicit notational dependence on the information set I, the reader should remember that the probability distribution is always based on an information set appropriate to the problem at hand.

2.6 THE VALUE OF PERFECT INFORMATION

Suppose we have *perfect information* on the demand and know that $X = x$ *before* we decide on the order size $D = d$. This situation is sometimes referred to as the *clairvoyant* problem. The influence of information and the timing of decisions are illustrated in Figure 2-14.

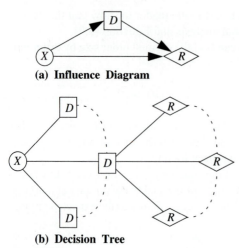

(a) **Influence Diagram**

(b) **Decision Tree**

FIGURE 2-14

The newsboy problem with perfect information.

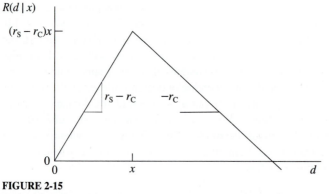

FIGURE 2-15

Profit with perfect information.

In Figure 2-14(a) the directed arc from X points into D. This indicates that perfect information about X (possibly through a perfect forecast of X) is obtained *before* the ordering decision is made; thus, it influences our decision choice, although it may not change the set of available alternatives. The diagram has changed from that in Figure 1-7 (p. 28) in three important ways: D is dependent on X, X precedes D, and the value that X takes on is known to the decision maker. Figure 2-14(b) shows a typical path in the decision tree. Because x is assumed known before d is made, we emphasize that the payoff is a function of d for a fixed x by denoting the profit by $R(d \mid x)$ and plot it in Figure 2-15.

Clearly the profit function $R(d \mid x)$ has a maximum at $d = x$; thus, the decision problem with perfect information has a particularly simple optimal solution denoted by $d(x)^{**} = x$: order the known demand and sell it to receive the *certain* profit $(r_S - r_C)x$.

Even though the ordering rule is known ahead of time, *a particular outcome X = x is uncertain* so that the expected profit *under perfect information* is

$$E[R(d^{**}|X)] = E[(r_S - r_C)X] = (r_S - r_C)E[X].$$

If we define the gain in expected profit from perfect information to be Δ, it is obtained by subtracting the expected optimal profit when demand is uncertain from the expected return with perfect information to obtain

$$\Delta = E[R(d^{**}|X) - R(X,d^*)] = (r_S - r_C)(E[X] - E[X \mid X \le d^*]) \ge 0.$$

It will never pay to expend more than this Δ on a forecast, not even one that is always perfect. When r_S is very much larger than r_C, profit margins are high and stockout probabilities are low. Thus, the conditional expectation $E[X \mid X \le d^*]$ is close to the unconditional expectation of the demand, and the term on the right may be close to zero. To the dismay of many forecasters, it may simply not be cost-effective to allocate resources to obtain perfect forecasts; crude estimates may suffice. There is an upper limit on how many computers, data-collection, storage, and analysis resources we should devote to improving forecasts; that is why simple estimates of the value of partial or perfect information should be considered before one recommends forecasting methods or expensive data analyses for a client.

2.7   AIRLINE SEAT ALLOCATION BASED ON PRICE AND DEMAND

In recent years, transportation companies have studied ways to increase the return from fare-paying passengers by considering optimal policies for allocating seats in first and second class, offering discount tickets in second class with cash payment well in advance of departure times, penalties for refunds on unused tickets, and overbooking policies. It has been frequently claimed in the press that this new form of "yield management" for United and American Airlines has led to the demise of the all-discount carriers such as World Airlines and Peoples Express.

With an airline industry that as a whole has excess capacity and with new designs that allow airline companies to make frequent changes in the space allocations between first, business, economy, and cargo classes, the interactions between space allocations, overbooking policies, and the prediction of customer demand for different fare classes becomes a most important aspect of the management of airline operations.

In what follows we discuss the structure of several single-stage models that are used to make decisions and design new policies; we pay particular attention to the way in which the timing of information plays a role in influencing decisions.

2.7.1   Space Allocation for Two Passenger Classes

The first problem we study is how to partition the total seating capacity, C, of an airplane so as to accommodate passengers in both first class and second class, the former paying a higher fare than the latter. Once the allocation is made, passengers in one class *cannot* use the seats of the other class even when, on departure, there is slack capacity in one class. The problem is somewhat artificial because the space allocation would not be made on the basis of demand for a single trip, but the model and method of analysis lead us directly into more realistic discount pricing models. The objective is to maximize expected return from passenger fares. Let

r_F = Ticket price for first class,

r_D = Ticket price for second class, $r_D < r_F$,

C = Total seating capacity of the airplane,

X = Random demand for fares in second class with (baseline) cdf $P_X(x)$,

Y = Random demand for fares in first class with (baseline) cdf $P_Y(y)$.

Assume that C, $P_X(x)$, $P_Y(y)$, r_F and r_D are known and that adding one more seat in second class results in one fewer seats in first class. We also assume that:

A1: All passengers show up and use their tickets,

A2: X and Y are independent random demands,

A3: All operating and aircraft purchase costs are fixed.

Assumption A2 may or may not be realistic depending on the market and pricing structure that create demand for first and second class. Later in the section we show how to relax this assumption.

Our problem is to decide how many seats, considered to be a continuous variable, to allocate to second class, so our decision set $\mathcal{D} = \{d\ ;\ 0 \le d \le C\}$. The random demand sets are $X = \{x\ ;\ x \ge 0\}$, and $\mathcal{Y} = \{y\ ;\ y \ge 0\}$. An influence diagram for the decision problem is shown in Figure 2-16. This is slightly more complicated than the newsboy problem because total revenue depends on *two* uncertain demands; note that we can interchange the timing of the X and Y nodes without affecting the resulting solution. The important point is that the decision D precedes *both* X and Y.

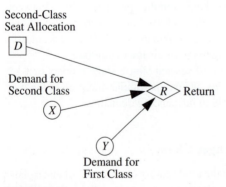

FIGURE 2-16
Influence diagram for the seat class allocation problem.

Let $R(X, Y, d)$ be the revenue when the demands for seats in second and first class are X and Y, respectively, and d of the C seats are allocated to second class. Expressions for $R(X, Y, d)$ are given in Table 2.5 for the different combinations of X and Y relative to C and d.

Total expected revenue can be obtained by multiplying these revenues by the appropriate probabilities and integrating (or summing) over x and y. We again use a marginal analysis.

Figure 2-17 is the appropriate decision tree for our analysis. Consider the path leading from node X to the bottom R node. When demand for second class does not exceed assigned capacity $(X \le d)$ and there is at least one empty seat in first class $(Y < C - d)$, changing a seat in first class to a seat in second class would not result in any increase or decrease in revenue; hence the payoff of 0 assigned to the bottom R node. If $X \le d$ and $Y \ge C - d$, changing a seat in first class to one in second would result in losing a fare r_F in first class (next higher R node). If demand for second class exceeds d and first class falls short of $C - d$, changing a seat results in an increase in revenue of a second class fare r_D. Finally, if second class demand exceeds d and first class demand is at least $C - d$, changing a seat results in a revenue change of $r_D - r_F$. Seat allocation is op-

TABLE 2.5
The revenue matrix for passenger demand

Demand for second class	Demand for first class	
	$Y < C - d$	$Y \ge C - d$
$X \le d$	$r_D X + r_F Y$	$r_D X + r_F(C - d)$
$X > d$	$r_D d + r_F Y$	$r_D d + r_F(C - d)$

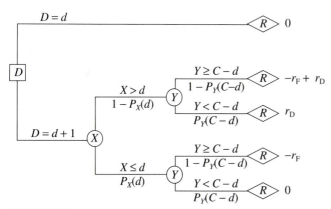

FIGURE 2-17
Marginal analysis decision tree.

timal or "balanced" when the net expected change in revenue is zero, so that the optimal seat allocation in second class must satisfy

$$[1 - P_X(d)][r_D P_Y(C - d) + [r_D - r_F][1 - P_Y(C - d)]] - r_F P_X(d)[1 - P_Y(C - d)] = 0.$$

This expression reduces to

$$r_D[1 - P_X(d)] = r_F[1 - P_Y(C - d)]. \tag{2.13}$$

By plotting $r_D[1 - P_X(d)]$ (left axis in Figure 2-18) and $r_F[1 - P_Y(C - d)]$ (right axis) as functions of d for $0 \le d \le C$, one can see that

1. When $r_D > r_F[1 - P_Y(C)]$, the curves cross and an interior solution for d^* is found.
2. When $r_D < r_F$, it is never optimal to allocate the total seating capacity C to the passengers in second class, that is $d^* < C$.
3. When $r_D < r_F[1 - P_Y(C)]$, the curves do not cross and the optimal allocation is $d^* = 0$; no seats are allocated to second class, and the entire space in the airplane is for passengers in first class. This situation is illustrated in the right-hand drawing of Figure 2-18.

Calculus can be used to solve this problem in the usual way. If we let $r(d) = E[R(X,Y,d)]$, taking expectations over the expressions for $R(X,Y,d)$ from Table 2.5, we get

$$r(d) = r_D \int_0^d [1 - P_X(x)]\, \partial x + r_F \int_0^{C-d} [1 - P_Y(y)]\, \partial y.$$

The derivative of this total expected return is

$$\frac{\partial}{\partial d} r(d) = r_D[1 - P_X(d)] - r_F[1 - P_Y(C - d)].$$

Setting this equal to zero gives Equation (2.13). Again, the reader should check that the second derivative confirms the concavity of $r(d)$.

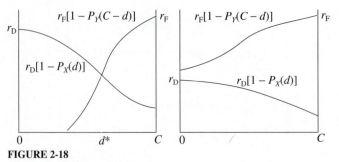

FIGURE 2-18
Graphical solution for space allocation.

We end this section with a numerical example. Assume that X and Y are independent and exponentially distributed with means $E[X] = 200$ and $E[Y] = 100$ respectively. Solving Equation (2.13) for the optimal allocation to second class yields[8]

$$d^* = \frac{2}{3}C + \frac{200}{3}\log\frac{r_D}{r_F}.$$

When $C = 300$ and $r_D/r_F = 1/2$, we find that $d^* = 153$, which means that the allocation to second class is just more than one-half the capacity. If $r_D/r_F = 0.8$, d^* increases to 185. Under these assumptions d^* increases linearly with the capacity C and the log of the price ratio of second class to first class.

2.7.2 Two-Class Space Allocation with Perfect Information

It is interesting to see how our models change if we have perfect information about the demand for seats in second class at price r_D. Assume that X is known to be x *before* the seat allocation decision is made but that demand for first class Y remains uncertain.

The influence diagram in Figure 2-19 shows that we have perfect information on X but not on Y and that revenues are derived from both seat types. Of course, the X node precedes the decision node D, and the Y node comes after D. The reader should be able to draw the typical paths in a decision tree that correspond to the influence diagram in Figure 2-19.

Even though the demand for second class is known, it is not at all obvious that one should sell tickets in second class to all who want them because this may consume space that could be sold later to more profitable passengers in first class, whose number is now uncertain. Let us assume that the actual demand for second class is larger than the seat allocation in second class ($x \geq d$). Of course, we shall collect ticket fares only from those

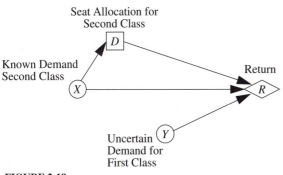

FIGURE 2-19
Known second class, uncertain demand for first class.

---

[8] The notation *log* is used for all logarithms in this book, including natural logarithms.

passengers in second class to whom we offer space. If we select d passengers in second class, the return is

$$R(x,Y,d) = r_D d + r_F Min\{Y; C - d\}$$

with expectation

$$r(d) = E[R(x, Y, d)] = r_D d + r_F \int_0^{C-d} [1 - P(y)] \partial y \tag{2.14}$$

independent of the distribution of X! Figure 2-20 shows a graphical representation of Equation (2.14) where we have assumed that the horizontal line at r_D crosses the curve $r_F[1-P_Y(y)]$ at a point $C - d'$ between 0 and C. Areas 1 and 2 represent the first and second terms in Equation (2.14), respectively. This total area will be a maximum when the point $C - d$ moves to the right to the value $C - d'$. Thus, d' satisfies

$$r_F[1 - P_Y(C-d')] = r_D.$$

Finally, the optimal allocation of seats in second class d^* is the smaller of d' and x, the demand for discount seats. Is it surprising that the optimal allocation to passengers in second class no longer depends on an assessment of demand for passengers in second class?

By substitution similar to that done with Equations (2.9) and (2.12) for the news-boy problem one can show that the expected value of the optimal allocation policy when d^* is positive is

$$r^* = r_D d^* + r_F P_Y(C - d^*)E[Y \mid Y < C - d^*] + r_F(C - d^*)(r_D/r_F)$$

$$= r_D C + (r_F - r_D)E[Y \mid Y < C-d^*].$$

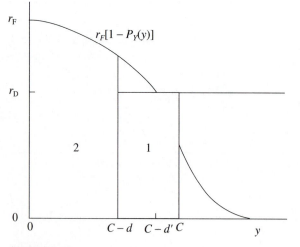

FIGURE 2-20
Graphical solution of seat allocation.

One can show that the expected return is always greater than when both X and Y are uncertain.

At this point we should ask ourselves what would happen if we relaxed the independence assumption on X and Y. A moment's reflection should convince the reader that we would add a directed arc from the X node to the Y node in the influence diagram of Figure 2-19 and that the algebraic derivations would require the conditional distribution function $P_{Y|X}(y \mid x)$ rather than the marginal distribution $P_Y(y)$. Nothing else would change in the formulation and derivation of the optimal solution.

2.7.3   Discount-Seat Allocations in One Passenger Class

In this section we consider the decision rules that can be used to allocate passenger seats at a discount fare in a class that has a fixed capacity. For example, we consider how many discount passenger tickets in second class we should sell given a known seating capacity in second class. The model used is similar to but different from the models of the preceding sections. Discount fares are usually purchased well in advance of the time of departure of the airplane. From the point of view of the airline company the purpose of discount seat allocation is to balance the reduced risk and economic benefit derived from early purchase of discount-fare passengers against the riskier but more profitable demand for full-fare seats. The optimal allocation of discount seats within second class is usually determined long after the space allocations to first-class and second-class sections have been determined; thus, we can assume that the space allocation problem discussed has already been solved. The structure of the discount-fare problem has one very important difference: *unused discount seats may be used by full-fare passengers.* Thus, in addition to the three assumptions we made in Section 2.7.1, we are able to state the very important additional assumption that:

A4: Full-fare passengers can be assigned unused discount seats.

Our attention is now focused exclusively on the demand for, and space allocated to, passengers in second class; this one passenger class is now subdivided into two fare groups: those using discount fares and those using full fares. Let

r_D = Discount ticket price in second class,
r_F = Full-fare ticket price in second class,
C = Seating capacity of the airplane in second class,
X = Random demand for discount fares with (baseline) cdf $P_X(x)$,
Y = Random demand for full fares with (baseline) cdf $P_Y(y)$.

Note that the capacity in second class C used in this problem corresponds to a solution for d^* in the space-allocation problem described in Section 2.7.1.

Our problem is to decide how many of the C seats should be allocated to discount fares. As before, the set $\mathcal{D}$ is $\{d; 0 \leq d \leq C\}$, X is $\{x; x \geq 0\}$, and $\mathcal{Y}$ is $\{y; y \geq 0\}$. We shall

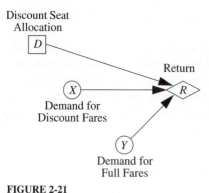

Discount Seat
Allocation

Return

Demand for
Discount Fares

Demand for
Full Fares

FIGURE 2-21
Influence diagram for discount fare problem.

see how A4 modifies the structure of the model and the assessment of the conditional probabilities and the solutions.

In Figure 2-21 it appears that the structure of the influence diagram is identical to that of Figure 2-16 except that the names of the random quantities and decision variable have changed.

Again, the problem is solved using a marginal analysis that compares the *certain return* of the passenger willing to purchase a discount-fare seat at r_D with the risky venture whose *expected return* is based on an uncertain future demand for a full-fare seat at price $r_F > r_D$. The payoff matrix for discount fares is given in Table 2.6. On comparing these entries with those in Table 2.5 we see that only the cell in the northeast corner has changed. This corresponds to the situation where discount allocations are underutilized by the discount passengers and there is excess demand for full fares. Assumption A4 now allows us to use discount-fare seats to satisfy demand from full fare passengers and gives us a flexibility in yield management that we did not have before. When the discount-fare demand does not exceed the discount-seat allocation, the marginal revenue for an additional discount seat is zero! When the demand for discount-fare seats exceeds the allocation, the addition of an extra discount-fare seat either returns r_D or results in a net loss of $r_D - r_F$. These marginal returns are shown in Figure 2-22. Note that the riskless alternative has been left out of the tree. It should be obvious to the reader by now

TABLE 2.6
The revenue matrix with discount fares

	Demand for full fare	
Demand for discount fare	$Y < C - d$	$Y \geq C - d$
$X \leq d$	$r_D X + r_F Y$	$r_D X + r_F \, Min\{Y, (C - X)\}$
$X > d$	$r_D d + r_F Y$	$r_D d + r_F (C - d)$

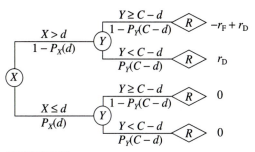

FIGURE 2-22
Marginal analysis event tree.

that all we need to do is to find the expected payoff for the risky venture and equate it to zero. Doing this shows us that the optimal allocation $d*$ is the solution of a "newsboy-like" problem:

$$1 - P_Y(C - d*) = \frac{r_D}{r_F}.$$

The optimal solution does not require an assessment of the demand for discount fares, but the assessment of full-fare demand is conditional on knowing how many discount tickets have been sold! As $d*$ is independent of X, there is no economic value to be gained from improved forecasts of passenger demand for discounted seats; the assumption that X and Y are independent random variables means that knowing the actual demand for discount tickets tells us nothing about the passenger demand for full-fare tickets. This formula for the optimal solution shows that there is no economic benefit in lowering the discount fare below $r_F[1 - P_Y(C)]$, because at this price the entire seating capacity would be already allocated to passengers paying full fare.

Recall that our assumption A2 was that X and Y are independent demands. We may want to assume that the demands for discount fares and full fares, X and Y, respectively, are dependent, in which case we must insert a directed arc between nodes X and Y in Figure 2-21. The reader can show that if Y depends on X (i.e., knowing the demand for discount fares affects our estimation of demand for full fares), the optimal decision $d*$ satisfies

$$1 - P_{Y|X}(C - d*|x) = \frac{r_D}{r_F},$$

where

$$P_{Y|X}(y|x) = Pr\{Y \le y|X = x\}.$$

2.7.4 A Dynamic Decision Model

The model in the preceding subsection could be used to determine an initial discount-seat allocation for a flight long before departure. Inherent in the model is an assumption that the demand for each type of fare does not change over time. As the flight time ap-

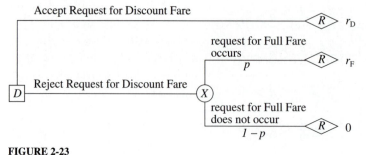

FIGURE 2-23
The dynamic discount fare problem.

proaches, a seat reserved for full fare may remain empty if the discount seats are filled early and a potential passenger who is not willing to pay full fare is denied a ticket. The airline might increase revenues if a ticket sales' person had the authority to fill a full-fare seat with a passenger for a discount seat, but when should this be allowed?

Today's airline reservation systems contain data that can be used dynamically to calculate and monitor changes in demand for various price fares. As the time for the flight approaches (for example, within 24 hours of gate time), suppose a request arrives for a discount seat and only full-fare seats are available. Suppose also that discount seats are not refundable. The airline could immediately fill an empty seat and receive r_D or deny the request in the hope of filling the seat with a full-fare passenger. The first alternative is riskless and the second risky. We see that as gate time approaches, the ticket selling decision can be represented by the decision sapling in Figure 2-23 with the probability of success p (filling an empty seat at full fare) changing dynamically over time. If the objective is to maximize expected revenue for a flight, when the discount-fare request arrives, the system should recompute p and sell an empty seat at a discount fare if and only if $r_D > pr_F$.

2.8 PRICING AND MARKETING OF HOTEL ROOMS

A large hotel has C rooms each with a regular rate of r_F per night. The demand for these rooms at this rate is a random variable Y with cdf $P_Y(y)$. Customers who make up this demand are primarily businesspeople who contact the hotel directly and pay this full rate. The hotel also sells rooms to a "consolidator" at a discount rate of r_D per night and receives payment for these rooms even if the consolidator does not succeed in filling all of them. The consolidator's customers are primarily tourists and tour groups who pay r_P per night. We assume that $r_D < r_P < r_F$. The consolidator's demand for rooms at their reduced price r_P is a random variable X with cdf $P_X(x)$. We assume that $E[X] > E[Y]$.

Let us first consider the decision problem faced by the hotel. They must make a decision D_h on how many rooms they should sell to the consolidator. Let $\mathcal{D}_h = \{d_h; 0 \le d_h \le C\}$, and $\mathcal{Y} = \{y; 0 \le y\}$. The appropriate influence diagram is structurally identical to that of Figure 1-7 (p. 28) for the "Newsboy Problem"; the node names and the random quantities differ. We do not need a node for the demand X that influences the consolidator's decision because the return to the hotel depends only on its own direct demand Y and the number of rooms it decides to sell to the consolidator. It should be clear to the reader that the consolidator has a separate and distinct decision problem, namely, how many rooms it should purchase from the hotel. This and other related problems are discussed later in the section.

The return to the hotel is given by

$$R_h(Y, d_h) = r_D d_h + r_F Min\{Y, C - d_h\}.$$

Taking expected values we obtain

$$E[R_h(Y, d_h)] = r_D d_h + r_F \int_0^{(C-d_h)} [1 - P_Y(y)] \, dy.$$

The value d_h* that maximizes the expected return to the hotel can be found graphically (as we did in Figure 2-20) by plotting r_D and $r_F[1 - P_Y(C - d)]$ versus d. The point where these lines cross is d_h*, which satisfies

$$1 - P_Y(C - d_h*) = r_D/r_F.$$

We now turn to the consolidator's decision problem D_c, which is to decide how many discounted rooms to purchase from the hotel. The reader can easily show that the influence diagram has the same structure but with D_c replacing D_h and X replacing Y. If the consolidator buys d_c rooms, their return is given by

$$R_c(X, d_c) = r_P Min\{X, d_c\} - r_D d_c$$

with expected values equal to

$$E[R_c(X, d_c)] = r_P \int_0^{d_c} [1 - P_X(x)] \, dx - r_D d_c.$$

The value d_c* is found by plotting $r_P[1 - P_X(d)]$ and r_D versus d. It satisfies

$$1 - P_X(d_c*) = r_D/r_P,$$

which again is a simple variant of the solution of a newsboy problem.

The following numerical example illustrates the results. Let $C = 1000$ rooms, $r_F = \$150$, $r_D = \$75$, and Y an exponential random variable with mean 500. Then $d_h* = 653$. This gives an expected rate of return to the hotel of $\$86,507$/night. The reader can show that without using the consolidator, the optimal expected nightly return would be $\$64,850$. Now suppose that the consolidator charges $\$100$ for each room (r_P) and at this price with their market, X is an exponential random variable with mean 2,000. Then, $d_c* = 575$. This gives an optimal expected return to the consolidator of $\$6,849$/night. The optimal values d_h* and d_c* are clearly different. It is easy to show that if the consolidator had to take the 653 rooms that the hotel would like to sell, this would reduce the consolidator's expected return by $\$113$. If the hotel agreed to sell only 575 rooms to the consolidator, the hotel's expected return would be reduced by $\$438$. In a real situation

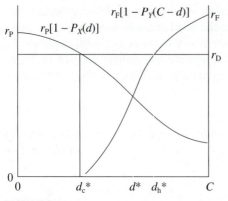

FIGURE 2-24
Optimal hotel, consolidator, and equilibrium room numbers.

the hotel and consolidator would probably negotiate an agreement between the 653 and 575 rooms. Can we help in such negotiations?

Suppose we combine the plots used to obtain d_h^* and d_c^* on a single graph. This is shown in Figure 2-24. Let d^* be the point[9] where the curves $r_F[1 - P_Y(C - d)]$ and $r_P[1 - P_X(d)]$ cross. Note that, depending on the amount of discount offered, the horizontal line for r_D could be above or below this cross-over point. For the case shown in Figure 2-24 $d_c^* < d^* < d_h^*$; if it lies below, then $d_h^* < d^* < d_c^*$. All three points would be equal when the discount price is set at

$$r_D = r_F[1 - P_Y(C - d^*)] = r_P[1 - P_X(d^*)].$$

In a negotiation the hotel would try to increase r_D and the consolidator would try to decrease it. Agreement would be reached when both agree on the equilibrium price that satisfies this equation.

Returning to the above numerical example, note that the consolidator would like to buy fewer rooms from the hotel than the hotel would like to sell. This indicates that the hotel discount price r_D is too high. The equilibrium solution gives a discount price r_D of \$72.70 per room and a discount room allocation of $d^* = 638$.

There are many variations of this problem one can study. As in the discount-seat airline problem in Section 2.7.3, one could assume that if the hotel runs its own discount operation and sets a deadline on the sale of discounted rooms, any unsold discounted rooms could be used to fill any unfilled demand at the regular rate that occurs after this deadline. See the problems at the end of the chapter for further analysis.

---

[9] It is possible that the curves do not cross as was demonstrated in Figure 2-18. We assume here that they do cross and that d^* is a unique point between 0 and C. Compare this problem with the airline seat problem of Section 2.7.1.

2.9   THE NEWSBOY PROBLEM WITH ADDITIONAL INFORMATION

We now show how to incorporate new information in the form of a forecast into the newsboy problem. Let $r_C = \$0.15$, $r_S = \$0.25$, and from past demand data assume that we have assessed the baseline pdf $p(x)$ of the uncertain next-day demand X to be

$$p(x) = (x - 500)/20{,}000 \quad \text{if } 500 \leq x < 550,$$
$$= 1/400 \qquad\qquad \text{if } 550 \leq x < 900,$$
$$= (950 - x)/20{,}000 \quad \text{if } 900 \leq x < 950.$$

Without further knowledge of demand the optimal order quantity d^* is found using Equation (2.9) to be 685; this gives a daily expected profit of $60.43.

   Suppose experience has shown that demand for this newspaper is related to that of a second paper, Y. If the demand for it were known to be y, the demand X is equally likely to be any number between $y + 100$ and $y + 150$. Thus, the conditional pdf of X, given Y, is

$$p(x|y) = 1/50 \quad \text{if } y + 100 < x < y + 150.$$

Using this in Equation (2.9) gives the optimal order quantity as a function of y, $d^*(y) = y + 120$; whatever the next day demand is determined to be for the second paper, we should order that number plus 120 of the first paper. To calculate the unconditional expected profit we also need to know the density of Y. In order for the pdf's $p(x|y)$, $p(x)$, and $p(y)$ to be coherent, Y must be equally likely to be between 400 and 800 (see Problem 2.22). This gives an expected daily profit using the optimal decision $d^*(y)$ of $71.00.

   The major contribution in knowing $\{Y = y\}$ is to reduce the uncertainty about X. What this simple example illustrates is that perfect information on a quantity Y can help to predict another random quantity X that, along with the decision D, explicitly influences profit. In our example we have discovered that by knowing Y we can increase our expected daily profit by $10.57. Clearly, this is an upper bound on what knowing Y is worth. Recall from Section 2.6 that if X itself is known, the expected profit is $(r_S - r_C)E[X]$, which for this problem is $72.50. Although knowing Y does not guarantee that we can predict X with certainty, it increases the daily expected profit substantially. Any further improvement in the prediction of X can only increase it further by at most $1.50.

   Evidence on some "related" random variable may be very useful in helping us predict the outcome of a random variable X, the uncertain quantity that matters in our decision problem. Figure 3-1 (p. 85) shows a typical influence diagram with a random event X together with a forecast F related to this event. To this point we have assumed that we know the probability density or mass function $p_X(x|I)$, where I is an appropriate information set. The next chapter discusses how different types of forecasts can be incorporated into a decision model.

2.10   SUMMARY AND INSIGHTS

The simple decision sapling model introduced in Chapter 1 is used to illustrate a decision problem with a baseline forecast for the probability of success of a risky venture. The minimal expected loss function for this problem is seen (as the upper curve in Figure 3-17 (p. 120)) to be piecewise linear. The difference between this curve and the one representing perfect information on the outcome of the risky venture is maximum when the loss ratio r_2/r_1 is equal to the baseline forecast probability p_X.

Many single-stage decision problems can be cast as variants of the newsboy problem where the decision on optimal order size is based on the profit margin of each item sold and the probability that the demand exceeds a given cost-to-profit ratio. A requirement for modeling multi-stage problems in this way is that the commodity under study (e.g., hotel rooms, airline seats, rental autos) be a perishable commodity.

We find that items with high profit margins need good forecasts of the tail of the demand distribution (i.e., the probability of large demands); low profit items need forecasts of the distribution of low demand. This demonstrates that forecasting must be tailored to the decision problem at hand, a point we make repeatedly throughout the book. In some cases point forecasts may suffice, but this is the exception to the general rule that we need to know the shape of relevant distribution functions in particular ranges that are relevant to the decision model and policies under consideration.

Modeling a decision problem under the assumption that you have perfect information on uncertain future events can be of considerable benefit in making good forecasting decisions. Although not often a realistic assumption that will be realized, the expected value of perfect information gives us an upper bound on how much added value one can expect a good forecast to provide. The term *clairvoyant* was coined by Howard (1989).

Sensitivity analyses are essential aids for real decision makers, and we return to this important study in Chapter 5. It is demonstrated in the airline seat problem where we find that the optimal allocation for discount seats depends only on the fare ratios and the demand distribution of full fares, not on the demand for discount seats. If data or actual experience suggest that model predictions are incorrect, then the structure of the model should be questioned and possibly modified. However, surprising and unexpected results are often obtained from the mathematical solutions of correct but parsimonious models.

In the football example in Section 2.3 we emphasize the importance of clarifying what decision problem is being studied and who is the decision maker. The answers to these questions may lead one to model a decision as a random quantity rather than a decision. Thus, for Nancy and Ron, the coach's decision can be modeled as a random variable. In studying hotel room pricing policies we find that if the hotel sets its offering price to the consolidator too high, the consolidator will want to buy fewer rooms than the hotel would like to sell, the reverse being true if the offering price is too low. Equilibrium prices and allocations of discount rooms can be determined by modeling the combined decision problems of the hotel and consolidator to illustrate how new equilibria can be determined in response to such changes.

Yield management problems arise when it is expensive or impossible to store excess production and when commitments need to be made in the presence of uncertain future demand. Most perishable commodities and many services (e.g., hotel rooms, airline seats, auto rentals) have these features. There is probably no single model for decision making under uncertainty that has had a greater impact in improving short-run profits in many service industries with many different services and complex pricing structure than the yield management problems used to obtain short run improvements while maintaining existing production capacities.

Baseline forecasts are often the result of subjective probability assessments; these play a central role in decision modeling and decision making and are discussed in greater detail in Chapter 6. They are also critical in helping the decision maker quantify results that have no obvious numerical measure; utilities are in fact subjective probabilities.

For further reading on the general formulation of decision problems and Bayes decisions the reader should see Smith (1988a), Lee (1989), Lindley (1985), and O'Hagan (1988). There are numerous textbooks that treat the newsboy problem, including Hillier and Lieberman (1994) and Taha (1987).

Articles on yield management have appeared in the scientific and business literature as well as the popular press (New York *Times*, Jan. 2, 1991). For further reading on the modeling of airline seat allocations and fare setting, see Belobaba (1987, 1989), Phillips(1991, 1993a, 1993b) and Phillips, Boyd, and Grossman (1991).

PROBLEMS

2.1 In the coach's decision problem in Section 2.3.1, assume that $Pr\{X = 1 \mid D = K\} = 0.95$ and that $Pr\{X = 1 \mid D = T\} = 0.60$. If this were a professional game where a tie would be resolved in overtime, what would the coach's probability of scoring first in overtime have to be in order for him to prefer K to T? Explain your assumptions.

2.2 For the college football problem with the probabilities given in Problem 2.1, if the coach chooses T, what can you say about the relative value he places on a tie using a scale where a win is valued at 1.0 and a loss at 0?

2.3 Find expressions for the expected value to Nancy of perfect information about the coach's decision when one or the other, but not both, of the probabilities p_k and p_t are less than their indifference probability. Check your answers using (1) $p_k = 0.70$, $p_t = 0.40$, and (2) $p_k = 0.90$, $p_t = 0.25$.

2.4 Draw the influence diagram and decision tree for Nancy's betting problem when she has perfect information on both the coach's decision and the outcome of the play. Does the coach's decision matter if the play outcome random variable X
a. Measures success or failure?
b. Measures points scored?

2.5 Using the following data, repeat the analysis of Section 2.3.3 when Nancy knows the coach's decision before placing the bet and she can place bets with either Ron (decision BR) or George (decision BG), but not both:

	Random outcomes (coach's decision, play result)			
Nancy's decision	$(Y, X) = (k, 1)$	$(Y, X) = (k, 0)$	$(Y, X) = (t, 1)$	$(Y, X) = (t, 0)$
$d = \text{NB}$	0	0	0	0
$d = \text{BR}$	0.25	−1.00	2.00	−1.00
$d = \text{BG}$	0.20	−0.40	1.40	−0.40

2.6   Draw the influence diagrams and decision trees appropriate to Nancy's betting problems when X is the number of points from the play and

a. She makes her bet before knowing the coach's decision.

b. She makes her bet after knowing the coach's decision.

c. How much is it worth to her to know the coach's play selection before betting?

2.7   For the battery problem in Section 2.4 assume that both the demand and number purchased can be any integer from 0 to n.

a. Use the definition of expected value to show that if d are purchased, $0 \le d \le n$:

$$E[R(X,d)] = -dr_C + (p_1 + 2p_2 + \ldots + (d-1)p_{d-1} + d(1 - p_0 - p_1 - \ldots - p_{d-1}))r_S.$$

b. Show that the expression in Part a can be simplified to give Equation (2.6) (p. 58).

2.8   The optimal order quantity d^* for the newsboy problem satisfies $1 - P(d^*) = r_C/r_S$. Show that by substituting this result into Equation (2.11), the maximum expected profit is given by Equation (2.12) (p. 63).

2.9   In the newsboy problem find the optimal solution for d^* and optimal $E[R(\cdot)] > 0$ when $r_C = \$0.15$, $r_S = \$0.25$, $Pr\{X > x\} = e^{-\lambda x}$, and $E[X] = 25$.

2.10   For the demand distribution in Problem 2.9 plot the increase in the relative optimal expected profit, Δ/r_S, resulting from perfect forecasts of newspaper demand as a function of r_C/r_S.

2.11   Draw an influence diagram and decision tree, and formulate the decision problem for a two-period newsboy problem where unsold papers in one period can be carried forward as inventory and used to meet demand in the next period.

2.12*In the allocation of space to first and second class draw an influence diagram and decision tree incorporating the feature that a *known* fraction f of ticketed passengers in second class will not show up to use their tickets. Find the optimal space allocations when the airline has a policy of refunding each unused second-class ticket with an amount $r_R < r_D$.

2.13*Assume that when x is the number of tickets for second class sold, the number of "no-shows" is a random variable Z with a binomial distribution with parameters x and q. Also, assume that no-shows do not have to pay for their tickets. Find the optimal space allocations as a function of q.

2.14   When the ratio of discount fare to full fare is one-half, show that the optimal allocation of discount fares is expressed in terms of the median full-fare demand.

2.15   In the allocation of discount seats assume that both demands are exponentially distributed with $E[X] = 100$ and $E[Y] = 200$ and that the airplane capacity is $C = 300$. Show that the lowest discount fare that should be offered is slightly less than one-quarter the full fare.

2.16*Find graphical solutions for the space allocations to three passenger classes, assuming there are no discounted tickets. Show that four optimality conditions must be satisfied.

2.17*Repeat Problem 2.16 with the assumption that one of the three ticket fares is a discount fare that has to be paid for in advance of plane departure time.

2.18*Consider an airline yield management problem with discount seats (fare: r_D) and regular seats (full fare: $r_F > r_D$) identical in every respect to the problems studied in this chapter but differing only in that there is a probability q that, if an individual requesting a discount fare is denied a ticket, he or she will *immediately* receive an upgrade to a full-fare ticket if it is available.

a. Formulate the problem on a decision tree, and obtain the maximum expected profit decision for the number of tickets to allocate to discount and full fares.

b. How does the solution compare with the case $q = 0$?

2.19*Determine the optimal space allocation to second class in a commercial airplane when each seat in first class requires the space of two seats in second class, C is the total seating capacity of the airplane in units of seats in second class, d is the seat allocation to second class, r_D and r_F are the ticket prices and X and Y are the dependent random demands for second and first class respectively. Assume that the demands are continuous with joint pdf $p_{X,Y}(x,y)$.

a. Reformulate the decision problem on a decision tree, and compare it with one that expresses the total expected costs as a function of ticket prices and the decision d.

b. Find an algebraic and a graphical solution for the optimal space allocation.

c. Find an expression for optimal expected profit.

d. How does the new solution differ from the solution when seats in first and second class take the same amount of space?

e. Repeat the problem when the demand for first class is known before the seat allocation decision is made.

2.20 For the following decision problem where the objective is to maximize the expected return, find

a. The optimal policy and expected return.

b. The value of knowing X, but not Y, before D.

c. The value of knowing Y, but not X, before D.

d. The value of knowing both X and Y before D.

e. Draw the influence diagram that corresponds to this decision tree.

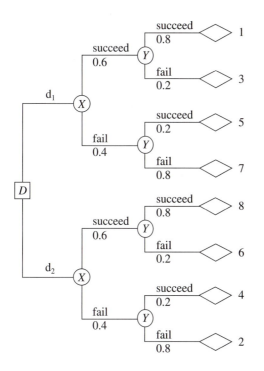

2.21 Suppose that a hotel is considering running its own "consolidator" operation, a separate profit center within the hotel corporation. Let $R(X,Y,d_h)$ be the total return to the hotel when the demands are X and Y and the discount decision $D_h = d_h$. We consider the case where the hotel cannot sell an unoccupied discounted room to fill a demand at the regular rate.

a. Show that

$$E[R(X, Y, d_{\mathrm{h}})] = r_{\mathrm{F}} \int\limits_0^{(C-d_{\mathrm{h}})} [1-P_Y(y)]\,dy + r_{\mathrm{P}} \int\limits_0^{d_{\mathrm{h}}} [1-P_X(x)]\,dx.$$

b. Why does the discount price r_{D} not appear in this expected return equation?

c. Show that the optimal decision d^* satisfies

$r_{\mathrm{F}}[1-P_Y(C-d^*)] = r_{\mathrm{P}}[1-P_X(d^*)]$.

2.22 Using the probability law $\int p_{X|F}(x|f)p_F(f)\,df = p_X(x)$, show that the densities used in the numerical example in Section 2.9 are consistent.

CHAPTER
3

FORECASTS
FOR DECISION
MODELS

3.1   INTRODUCTION

In the preceding chapter we studied several Bayes decision problems where no explicit forecast was offered the decision maker other than his or her belief or knowledge about the uncertainties in question; we called this the *baseline forecast*. We showed how perfect information affects optimal policies and value of the decision problem. Using the newsboy problem as an illustration, we demonstrated how additional information that influenced the baseline forecast could alter not only the optimal policies but also the economic value of these decisions.

The purpose of this chapter is to discuss some of the central issues of forecasting as they pertain to decision making and to address the way in which forecasts bring economic value to decision makers. Even though the role of forecasting is subsidiary to that of decision making, it plays a key role in the overall formulation and analysis of decision models and the actual decisions themselves. Properly incorporating relevant information into forecasts is an art as much as a science and needs to be thought through with great care. Too often, forecasts are made by individuals and organizations far removed from the decision makers who need careful assessments and forecasts of relevant uncertain future events. Although forecasters sometimes make an effort to assess the quality and performance of the forecast, they do not usually place themselves in the shoes of the decision-making client who will end up using the forecast. Thus, we must carefully distinguish between *forecast quality* based on forecast performance isolated from the decision problem and *forecast value* that should be measured in terms of one or more economic benefits to the decision maker, possibly a direct result of good forecasting.

The most important requirement is that forecasting and decision making be a *coherent enterprise with consistent objectives*. Unfortunately, most forecasting systems and techniques with which we are familiar do not pay sufficient attention to this important issue; the design, testing, calibration, and use of forecasting models often proceeds independently of the decision problem. What turns out to be a high-quality forecast from the forecaster's point of view may be of little benefit or value to the decision maker; on the other hand, a low-quality forecast may, under certain conditions, provide great economic value for decisions. The analysis of forecast quality is deferred to Chapter 8; the main purpose of this chapter is to discuss and contrast various classes of forecasts and show how they affect decision policies and expected value of these policies.

Before launching into a more formal approach and illustrating these ideas, consider the elective surgery decision problem described in Section 1.8 and summarized in Table 1.2 (p. 30). Suppose that your doctor tells you that for a person of your age, sex, and condition the chances of success with surgery to alleviate your condition are 70%. This can be thought of as the baseline forecast. Before making your decision you decide to obtain opinions from two surgeons who are specialists in the surgery you must undergo. After a thorough examination the specialists give you advice. The first makes a prediction of an 80% chance of success for you, and the second makes a prediction that the required surgery will be successful in eliminating 80% of your current debilitating condition. The information from each specialist is very different, and your problem is now how to evaluate each one. The first is what is referred to as a *probability forecast* because it estimates the probability of an event of interest, namely, successful surgery. The second is called a *categorical forecast* because it predicts the category of your condition after surgery.

Section 3.2 describes and gives some simple examples of how a forecast can have value in a decision problem. Probability forecasts are introduced and defined in Section 3.3.2 and categorical forecast in Section 3.3.3. Understanding how these differ, how they are related, and how they are used in decision modeling is the primary focus of this chapter. Section 3.4 presents a thorough treatment of how the random variables that represent the forecast of an event and the event itself are related. Section 3.5 extends these results to the case when two or more forecasts of the same event are available. In Section 3.6 we return to the crop protection problem introduced in Section 2.2 (p. 46) but with a probability forecast included. Section 3.7 shows how categorical forecasts can be used in decision models that involve estimating credit risk.

3.2 THE ROLE AND VALUE OF FORECASTS

As mentioned, the role of forecasting is to use any or all relevant information that is available to sharpen our knowledge and reduce our uncertainties on outcomes of quantities that affect our decisions and final results. Information may be in the form of historical observations of the quantities of interest or of some additional variables that influence or are influenced by the quantities we use in our formulation of the decision problem. Knowledge about a random variable F (for forecast) may be very useful in helping us predict the outcome of random variable X, the uncertain quantity that matters

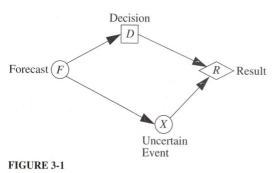

FIGURE 3-1
Influence diagram with forecast of uncertain outcome.

in our decision problem. Figure 3-1 shows a typical influence diagram with a random event X together with a forecast F related to this event. The sequence of events and actions that occur over time is (1) make a forecast, (2) make a decision, (3) observe a random outcome, and (4) calculate a result. The directed arc from F to D indicates that the forecast is known to the decision maker before the decision is made, the one from F to X shows that the probability distribution of X is conditional on the forecast F, and those from D and X to R show that the result depends on both the decision made and the outcome of the random event.

In earlier chapters we assumed that we had a baseline forecast, that is, that we could assess the probability density or mass function $p_X(x|I)$ where I is a limited information set available to the decision maker. Our objective was to find a decision that maximized the expected return (or minimize expected loss) over all choices $d \in \mathcal{D}$ given this information. However, as mentioned, cloudy skies may presage rain, a forthcoming dividend may lead to a stock price increase, the presence of an antigen in the blood may affect the likelihood of prostate cancer, or a photograph of military trucks of a certain type may lead the observer to infer a high probability for the presence of a nuclear weapon or a mobile missile. There are many examples where the observation of a quantity or event different from the one being used in the decision problem has a direct influence on decisions and outcomes.

If we are fortunate enough to possess evidence or observations of some random variable F that is influenced by X or that influences X, we may be able to improve our decision making by obtaining predictions of $\{X = x\}$ occurring, conditional on observing $\{F = f\}$, for $x \in X$ and $f \in \mathcal{F}$. This means that the decision problem can now be written as

$$
\begin{aligned}
r(d^*(f)) &= Max_{d \in \mathcal{D}(f)} E[R(X, d)] \\
&= Max_{d \in \mathcal{D}(f)} \int R(x, d)\, p_{X|F}(x|f)\, dx.
\end{aligned}
\tag{3.1}
$$

It should be understood that in Equation (3.1) the complete information set is $\{f, I\}$; this emphasizes the fact that whatever information was available for the baseline forecast is also available to the new forecast with the additional knowledge that F was observed to be f. In the interests of simplifying notation we shall usually leave out the explicit no-tational dependence on I, but the reader should understand that it is *always* present, even if not explicitly shown. There are two important differences between this mathe-matical problem and the one described earlier. First, the assessment of the conditional distribution must take into account the interrelationship between the random quantities X and F; second, the set of decision alternatives could depend on the forecast outcome f, and so is denoted $\mathcal{D}(f)$; third, the optimal decision d^* is now a *function* of f, which we write as $d^*(f)$. That is to say, for each f there is an optimal decision (a member of the set $\mathcal{D}(f)$).

Consider the newsboy problem in Chapter 2 where we formulated a simple model to decide how many newspapers to buy for the particular day of interest. From a cursory examination of historical demand data one could easily assess the probability distribu-tion $p_X(x)$ of the next-day demand and using this calculate the optimal decision as the solution to

$$1 - P_X(d^*) = \frac{r_C}{r_S}. \tag{3.2}$$

But if a forecast F of demand X, possibly based on observed sales of a competitor jour-nal, were made available, we could use this information to assess $p_{X|F}(x|f)$ and thereby obtain a useful forecast that may lead to a better decision. The event $\{F = f\}$ occurs be-fore and $\{X = x\}$ occurs after the purchase decision is made. With the assumption that F is relevant to X it follows that F is likely to influence the decision. How is it that F can influence a decision when return R depends only on the decision $D = d$ and the uncertain demand $X = x$? The answer lies in understanding what information is known, when it is known, and how it influences choices.

Consider the influence diagrams in Figure 3-2. In the leftmost influence diagram X and F are independent random variables (no arc joining X and F). Neither the result nor the uncertain demand X is influenced by F, which says that F is not informative with respect to X. Even though F is known before the decision is taken, it cannot influence the optimal result, is therefore irrelevant, and can be disregarded. In this case one uses the baseline forecast, $p_X(x)$, in determining the optimal decision.

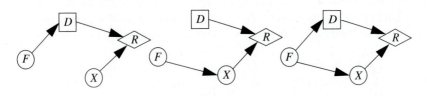

The value of F and $p_X(x)$ are known when D is made.

$p_F(F)$ and $p_{X|F}(x|f)$ are known, F is not observed when D is made.

The value of F, say f, and $p_{X|F}(x|f)$ are known when D is made.

FIGURE 3-2
Influence diagrams and dependence assumptions.

In the middle diagram F influences X (directed arc from F to X), but neither one is known before the decision is made, even though F occurs before and X after D. Even though one might be able to assess and model a complicated interaction between X and F in the form $p_{F,X}(f, x) = p_{X|F}(x \mid f) p_F(f)$, it would not improve decision making if F is not known or observed before the decision is made; the optimal decision depends only on the marginal distribution, $p_X(x)$. Finally, in the right-hand side diagram, we see that F is known before the decision D is made, and because of the (F, X) dependence, directed arcs connect F to both X and D. The optimal decision is found using Equation (3.2) with the conditional probability $p_{X|F}(x \mid f)$ so that the optimal order quantity $d^*(f)$ in the newsboy problem is now the solution of

$$1 - P_{X|F}(d^*(f) \mid f) = \frac{r_C}{r_S}. \tag{3.3}$$

As mentioned, the optimal decision is now a function of f.

A coherent forecast model is one that is consistent with, and an integral part of, the decision-making process and where probability statements satisfy the laws of probability. A decision model based on maximizing expected profit and using objective functions that are piecewise linear in decisions and random quantities should not, by choice, use forecasts based on minimizing quadratic norms that have no direct bearing on the decision problem. A decision problem whose objective gives great weight to the occurrence of extremely large or rare events should not use forecasting models or methods designed for and calibrated against measures of mean squared error or the behavior of central moments. The forecasting literature abounds with examples of forecasts provided for solutions of business decision problems without the least attention paid to the structure of the decision problem or the objectives of the decision maker. Decision makers and model developers must be alert to the unfortunate tendency of forecasters to overemphasize parameter estimation and fitting of standard forecast models that have not been designed as an integral part of the decision problem.

Although we have already noted that the natural order of events is to forecast, make a decision, and then observe outcomes, we may require resources and effort to establish which forecasts to use and how best to obtain them. The type and quality of information obtained from a forecast are both relevant in decision making; an optimal decision may be of little value if it is based on poor-quality forecasts that never materialize. We emphasize that there is much value to be gained from using historical records of the forecast–outcome relationship to improve understanding and forecasts for later decisions. In some problems forecasting and decision making are repetitive, in which case one can access a substantial amount of data that relates to forecast performance. In many cases forecast verification can be scientifically tested, in other cases the forecasts may have to be based on subjective assessments and used only once in solving a decision problem. In order for the decision maker to make best use of forecasts he or she needs to understand the nature and meaning of forecasts and, provided that there are sufficient data and experience, measure and evaluate forecast performance and quality. These can be checked by actual experimentation with forecasts and outcomes in the real world. These important topics will be taken up in some detail in Chapter 8.

3.2.1   Some Examples of Forecasts

Consider a weather forecast that predicts that it will be freezing (f) or non-freezing (nf) the next day. Figure 3-3 contains information that might result from four different forecasters; specifically, the four matrices or tables contain the conditional probability that freezing or non-freezing will occur given a forecast of either one. In the upper left cell of each matrix is the fraction of time that freezing weather occurs when freezing weather is forecast, and in the upper right cell is the fraction that non-freezing weather occurs when freezing weather is forecast; the bottom left cell gives the probability that freezing weather occurs if non-freezing is predicted, and the bottom right cell the probability that non-freezing weather occurs if non-freezing is predicted. In all cases the off-diagonal cells correspond to cases where the forecasts and actual outcomes do not agree. We use quotes around the row names, "f" and "nf," to emphasize that they denote *forecasts* of freezing or non-freezing. The column names f and nf without quotes refer to the *actual condition* itself.

Case (a) corresponds to a perfect forecaster, because the actual weather outcome always agrees with the forecast. Case (b) demonstrates a very helpful but not a perfect forecaster. Note, however, that when non-freezing weather is forecast, we are more likely to realize the predicted outcome than we are when freezing weather is forecast. Case (c) is interesting because it might appear to indicate the worst possible forecaster, one who invariably predicts the wrong outcome! Actually, forecaster (c) would be as useful to a decision maker as forecaster (a) if forecaster (c)'s uncanny ability to be always dead wrong were known. Case (d) demonstrates the worst of the four forecasters for a decision maker because the conditional probability the weather actually freezes is independent of the forecast; the forecast does not give the decision maker any useful information. He or she might as well disregard it as one can do as well by just observing the fraction of time that weather is freezing or non-freezing and always use the baseline forecast in his or her decision model.

3.3   SEVERAL TYPES OF FORECASTS

There are many different types of forecasts, but in what follows we discuss only four: point predictors, probability forecasts, forecasts of odds, and categorical forecasts. We are primarily interested in the latter three; historically, the most popular ones are point predictors usually based on estimates of conditional expectations and least-square measures of performance; unfortunately these may be the least useful for decision makers. Their availability is largely due to the ease of estimation, the popularity of statistical regression in commercial software, and the fact that one can obtain results without think-

$$
\begin{array}{cc}
 & \begin{array}{cc} f & nf \end{array} \\
\begin{array}{c} \text{"f"} \\ \text{"nf"} \end{array} & \begin{bmatrix} 1.0 & 0 \\ 0 & 1.0 \end{bmatrix} \\
 & (a)
\end{array}
\qquad
\begin{array}{cc}
 & \begin{array}{cc} f & nf \end{array} \\
 & \begin{bmatrix} 0.8 & 0.2 \\ 0.1 & 0.9 \end{bmatrix} \\
 & (b)
\end{array}
\qquad
\begin{array}{cc}
 & \begin{array}{cc} f & nf \end{array} \\
 & \begin{bmatrix} 0 & 1.0 \\ 1.0 & 0 \end{bmatrix} \\
 & (c)
\end{array}
\qquad
\begin{array}{cc}
 & \begin{array}{cc} f & nf \end{array} \\
 & \begin{bmatrix} 0.8 & 0.2 \\ 0.8 & 0.2 \end{bmatrix} \\
 & (d)
\end{array}
$$

FIGURE 3-3
Four examples of temperature forecasts.

ing carefully about the structure of the underlying decision problem. Our hope is that decision makers will pay much greater attention to the quality of forecasts, their structure, and use.

As we saw in earlier examples, the random quantities we want to predict may be continuous or discrete. Some examples of continuous random variables are temperature, the amount of rain in a storm, and the intensity of a radar signal return. Outcomes such as the points scored in a football game, the occurrence of rain, or the number of shares sold in a stock transaction are examples of discrete quantities. It is quite common in making predictions to aggregate the levels being measured and thus convert continuous into discrete quantities; for example, we may classify all rainfall between 0 and 0.10 inches as "low," between 0.11 and 0.25 as "medium," and above 0.25 as "heavy." Our forecasts of rainfall could therefore be stated in inches to the nearest hundredth of an inch or in terms of "low," "medium," and "high" categories of rainfall.

As seen in Chapter 2, if the only thing that matters in a football game is whether there is a win, tie, or loss, then point scores can be aggregated into three categories. If the score S_A of team A is greater than that of team B, S_B, we could define a "win" category denoted by +1. If $S_A = S_B$, we have a tie, denoted by 0, and if $S_A < S_B$, we have a "loss" category denoted by −1. It is important in making forecasts to define precisely the outcome space of the forecast itself as well as the random quantity being predicted. These need not be the same.

An area of forecasting familiar to most readers is the weather. The examples we use in this chapter are largely based on weather forecasts to illustrate the important concepts of point, probability, odds, and categorical forecasts. The reader should not conclude that these concepts apply only to the limited area of weather forecasting.

3.3.1   Point Forecasts

Point forecasts are probably familiar to every reader. In a first course in statistics one learns how to estimate (or forecast) the mean of a distribution using an average of past data (the sample average). If one assumes that the future distribution will not change with time, this estimate may serve as a forecast of a future mean. For example, if we assume that the estimated demand for airline passengers this month is likely to grow by 10% next month, then this month's estimate of 150,000 passengers may provide next month's mean forecast of 165,000 passengers. More advanced techniques include exponential smoothing of historical data, moving averages, and other regressionlike methods applied to time-series data, but the important point is that the "forecast" is contained in a single number. In weather forecasting an example of a point forecast is "tomorrow's high temperature will be 85°." Note that a point forecast, such as the expected demand for newspapers, would not be relevant or useful in the newsboy problem; neither would the expected point score be useful in the coach's decision in Chapter 2 or in the bet considered by Nancy and Ron.

In classical statistics one usually states the uncertainty in point estimates by attaching confidence intervals to the estimate. An example of an estimate and forecast using confidence intervals is "we are 95% certain that tomorrow's high temperature will be between 81° and 88°." But it is important to understand that we are also assuming that tomorrow will duplicate or resemble in a statistical sense the weather patterns encountered during the period covered by the estimate; one cannot naively move from estimates based on the past to forecasts of the future without implicitly or explicitly assuming a direct relationship between the model used to obtain estimates and the model that is assumed to hold in the forecast period.

Although confidence intervals and hypotheses testing find important and wide use in many areas, we do not pursue them here. In weather forecasting one is much more likely to hear statements such as "the chance of rain tomorrow is 30%" or "tomorrow's high temperature will be in the 70's." These are examples of probability and categorical forecasts, both of which are widely used in decision problems.

3.3.2   Probability and Odds Forecasts

The weather on a given day or at some particular time of the day can be measured in many ways; for example (1) by temperature, (2) by cloud coverage such as overcast or clear, and (3) by type of precipitation such as drizzle, rain, sleet, or snow. A forecaster, rather than state that rain is predicted tomorrow, will usually state that there is a certain percent chance of rain tomorrow. For example, one might be advised that the chance of rain is 60% and then make a decision based on this information. Such a forecast is called a *probability forecast* and applies to the occurrence or nonoccurrence of some specific event. Thus, in a probability forecast (1) a single event outcome is being predicted, and (2) the forecast includes a probability statement that the event will or will not occur.

An example of a probability forecast for rain is shown in Table 3.1. The first column describes a "forecast state," designated i, which is equivalent to the second column where the numerical probability statement for rain is displayed.[1] Suppose a user of this forecast had neither kept nor had access to historical forecast performance records. If this were the case, the user faced with a decision problem involving rain must use the probability statements in column 2 as the best numbers to use in any calculations. But suppose historical records were available, and assume that the *actual* fractions of time that the (row) forecast results in rain and no rain are those shown in the third and fourth columns, respectively. If the forecast is that rain is unlikely or, equivalently, that there is a 25% chance of rain, the fraction of time that rain actually occurs when this forecast is made is seen to be 0.20, that is, the forecast tends to overestimate the probability of rain occurring. The real data are reflected in this latter fraction.

With probability forecasts there can be many rows, but there are always exactly two columns. These correspond to the occurrence or nonoccurrence of the outcome of interest. In Table 3.1 only the first ($i = 1$) and last ($i = 5$) rows correspond to certain predictions, that is, that rain will or will not occur. Except for $i = 1$ (Certain) and $i = 5$ (No rain), there may not be clarity in the meaning and interpretation of words such as "likely," "moderate," and "unlikely." Of course, even when the forecaster states with certain-

---

[1] The reader should not assume that the phrases used in the first column are standard and will always imply the probability statements in the second column. To the authors' knowledge no such standard phrases exist.

TABLE 3.1
Probability forecast of rain

Forecast state definition		Actual weather outcome		
State i and description	Probability statement	Probability of rain	Probability of no rain	Odds of rain
i	$f(i)$	$Pr\{X = 1 \mid F = f(i)\}$	$Pr\{X = 0 \mid F = f(i)\}$	$O(1)$
1. Certain	1.00	0.90	0.10	9 to 1
2. Likely	0.75	0.65	0.35	13 to 7
3. Moderate	0.50	0.60	0.40	3 to 2
4. Unlikely	0.25	0.20	0.80	1 to 4
5. No rain	0.00	0.15	0.85	3 to 17

ty that it will not rain, there is still a chance of rain, and vice versa. Note that the row sums of conditional probabilities in the third and fourth columns add to 1 as, whatever the forecast, the event either does or does not occur. With probability forecasts one need only specify the probabilities in a single column because the other is easily obtained by subtracting each of the given numbers from 1.

For probability forecasts in general, our event outcome set will often be denoted by $X = \{0,1\}$, where $X = 1$ if the event of interest occurs, and $X = 0$ if it does not. When there are n possible forecast states, we let $f(i)$ be the probability statement associated with state i, then $\mathcal{F} = \{f(1), f(2), \ldots, f(n)\}$[2]. For example, for Table 3.1

$$\mathcal{F} = \{1.00, 0.75, 0.50, 0.25, 0\} \text{ and } Pr\{X = 1 \mid F = 0.5\} = 0.6.$$

We denote by **P** the $(n \times 2)$ matrix of conditional probabilities with

$$p_{X|F}(j \mid f(i)) = Pr\{X = j \mid F = f(i)\}, \quad j \in X, f(i) \in \mathcal{F}. \tag{3.4}$$

The reader must understand clearly in any given situation which probability statements are available to or known by the decision maker. In some cases, only the $f(i)$ probability statements will be known to the decision maker for use in a decision model; in other cases the decision maker may know the conditional probabilities in Equation (3.4) and illustrated in Table 3.1. Also, if the decision maker has been using forecasts from a given forecaster for some time, he or she may know that when the forecaster says that the probability of rain is 75%, the probability that rain will actually occur is 65%.

The concept of a perfect forecast cannot apply to probability forecasts as (with rare exceptions) specific outcomes are not predicted to occur with certainty. But the forecast might be calibrated in the sense that the outcome that is predicted to occur with

---

[2] Some readers may wonder why we do not simply let $\mathcal{F} = \{1, 2, \ldots, n\}$. The reason for the slightly more complex notation will become evident when we discuss the concept of forecast calibration and discrimination in Chapter 8.

a given probability actually occurs that fraction of the time (this is not true in Table 3.1). We discuss this in greater detail when forecast performance and quality are presented.

In many decision-making problems the natural expression of uncertainty is in terms of the odds of an event occurring. There is a natural reason why this is so, and it is useful in business or commercial situations where a gain is realized when an event occurs but there is a loss when the event does not occur.[3]

Consider the decision sapling problem in Figure 1-8 (p. 29) of Chapter 1 where we have assigned result values of r_1, r_2, and r_3. Relabel these instead as g, 0, and $-l$, where g denotes the *gain* obtained when the risky venture is chosen and is successful, and l the level of *loss* when it is not. There is no loss or gain from choosing the riskless alternative. The expected return of the risky venture is $pg - (1 - p)l$ so that the break-even (indifference) point between the two decision choices occurs when the expected return equals zero. Denote the odds of success in the risky venture by $O(1)$. Then

$$O(1) = p/(1 - p) = l/g. \qquad (3.5)$$

In other words, the odds directly express the loss-to-gain ratio. If the gain is $100 when the event occurs and the loss is $300 when the event does not occur, the odds are 3 to 1, and the probability of success in the risky venture must be greater than ¾ for it to have a positive expected return. Thus, the uncertainty has been expressed directly in terms of a monetary reward.

There is another reason why forecasts based on odds may be preferable to probabilities. It turns out that in many financial applications reliable forecasting models can be best expressed in terms of odds or scores; this may also be true in measuring the goodness of the forecast. Fortunately, there is a one-to-one correspondence between the two methods because

$$Pr\{X = 1\} = p = O(1)/[1 + O(1)]. \qquad (3.6)$$

The reader will note that in Table 3.1 the last column corresponds to the odds that rain will occur. Think of these as the dollar amounts one would have to bet on each side of a gamble to make it "fair" when the probabilities of rain and no rain are as indicated in the middle two columns. By using Equation (3.5) the reader should confirm that the odds in Column 5 of Table 3.1 correspond to the probabilities in Columns 3 and 4.

3.3.3   Categorical Forecasts

As the name implies, a categorical forecast is one where a certain category or range of values of an uncertain quantity is predicted to occur. Some examples of categorical forecasts are (1) tomorrow's high temperature will be between 70° and 75°, (2) the demand for air travel on a particular future flight will be at least 300 passengers, (3) the number of sunspots counted next week will be between 4 and 8, and (4) a specific common stock will trade tomorrow between $110 and $120. No probability statement is made, even

[3] For further reading on this important topic, see the excellent treatment by Lindley (1982) or O'Hagan (1988).

though the forecaster and decision maker may be implicitly aware that the predicted out-comes will not occur with certainty.

In most cases the range of values assigned to a particular category is due to an im-portant linkage with the way in which the category is used in a decision problem. In an agricultural decision problem the important uncertain future event may be whether the temperature is or is not freezing. If it is found that whenever the temperature is less than $0°$ C the decision that results is always to protect a crop, then it is useful to replace all temperatures at or below the freezing temperature by a single category called "Freezing" or perhaps by the number "1." Although the particular temperature value may not be im-portant, what matters is whether it does or does not fall below freezing.

Let X be 1 if tomorrow's temperature is at or below freezing and 0 if not. Let F be 1 if tomorrow's temperature is *forecast* to be at or below freezing and 0 if not. A cate-gorical forecast of tomorrow's temperature might make one of two statements: (1) the temperature will be freezing tomorrow, or (2) the temperature will not be freezing to-morrow. If the forecasts are repeated the record will show the fraction of time the fore-cast events materialize. Table 3.2 shows an example of the long-run performance of this categorical forecast. Note that as was the case in Section 3.2.1, quotes are used to dis-tinguish the forecast of the event from the actual occurrence of the event with the same name. The table shows that the forecast of freezing temperatures is correct 80% of the time, but when non-freezing temperatures are predicted, the fraction of the time the fore-cast turns out to be correct increases to 90%. If the actual freezing and non-freezing frac-tions are known to be 0.15 and 0.85 (that is, on average 15% of the days in a year have freezing temperatures), we can easily calculate the fraction of time forecasts of freezing will occur. From the law of total probability in Equation (1.8) (p. 13) we have

$$Pr\{X=1\} = Pr\{X=1|F=1\}Pr\{F=1\} + Pr\{X=1|F=0\}Pr\{F=0\},$$

so

$$0.15 = 0.80Pr\{F=1\} + 0.10(1 - Pr\{F=1\}).$$

This tells us that freezing weather is predicted 7.14% of the time. If, on the other hand, a freezing forecast is made 15% of the time one can show that the (marginal) fractions of time the real weather outcome is freezing or non-freezing must be 20.5% and 79.5%, respectively.

TABLE 3.2
Two-category temperature forecasts

Temperature forecast categories	Actual temperature categories	
	freezing ($X=1$)	non-freezing ($X=0$)
"freezing" ($F=1$)	0.80	0.20
"non-freezing" ($F=0$)	0.10	0.90

In categorical forecasts with a larger number of possible outcomes the forecast states may not necessarily correspond with actual outcomes. For example, if the categorical forecasts were for minimum temperature in degrees Celsius and measured in the intervals (1) −5° or below, (2) −4° to freezing, (3) 1° to 5°, and (4) 6° or above, the categorical forecasts and the actual minimum temperature outcomes might be as shown in Table 3.3. Of course, if the instrument that measures temperature is sensitive enough to record minimum temperatures in tenths of a degree, it would be possible to make and use forecasts with a much larger number of categories. To ensure that the reader interprets the data correctly we illustrate again by referring to Table 3.3: Of the past occasions when the forecast called for a minimum temperature in the range 1° to 5°, 5% of the time it was between −4° and 0°, 90% of the time it was between 1° and 5°, and 5% of the time it was 6° or above.

In what follows we assume that forecasts and event outcomes are represented by the same categories. Let there be n categories indexed by i for rows and j for columns. In Table 3.3, n is 4 and $i = 3$ represents "Minimum temperature between 1° and 5°." For categorical forecasts the sets X and $\mathcal{F}$ each contain all of the categories, so that $\mathcal{F} = \{1,2,3, \ldots ,n\}$ and $X = \{1,2,3, \ldots ,n\}$. The entries in Table 3.3 give the conditional probabilities of X given F; for example, $Pr\{X = 2|F = 1\} = 0.20$, and $Pr\{X = 3|F = 4\} = 0.04$. As in probability forecasts the table of probabilities is the matrix $\mathbf{P}$. The element at the intersection of row i and column j of $\mathbf{P}$ is

$$p_{X|F}(j \mid i) = Pr\{X = j \mid F = i\}, \qquad i \in \mathcal{F}, \qquad j \in X. \tag{3.7}$$

The reader must take care to keep clear the differences between expressions (3.4) and (3.7) and not confuse the events $\{F = f(i)\}$ in a probability forecast with $\{F = i\}$ in a categorical forecast. Whereas Equation (3.4) refers to elements of an $n \times 2$ matrix, the matrix with elements defined by Equation (3.7) is square. Because the categories cover the whole range of possible outcomes of X, in all cases each row of $\mathbf{P}$ must add to 1.

In categorical forecasts perfect forecasts would result in a matrix with ones on the diagonal (the forecast is always correct) and zeros off the diagonal. In this ideal case $\mathbf{P}$ would be an n-dimensional identity matrix. An example of a perfect forecast when we have only two categories is shown in Figure 3-3(a).

TABLE 3.3
Four-category minimum temperature forecasts

Forecasted minimum temperature		Actual minimum temperature category j			
Category i	Interval	1	2	3	4
1	−5° or below	0.75	0.20	0.05	0
2	−4° – 0°	0.05	0.80	0.10	0.05
3	1° – 5°	0	0.05	0.90	0.05
4	6° or above	0	0.01	0.04	0.95

3.4   DECISION PROBABILITIES, LIKELIHOODS, AND BAYES' RULE

Up to this point we have discussed several types of forecasts and the notation we use for categorical and probability forecasts. To make good use of these ideas, we need to have a clear understanding of the relationship between the two random quantities F and X. We write the joint probability mass function (pmf) of X and F as

$$p_{F,X}(f, x) = Pr\{F = f, X = x\}, \qquad f \in \mathcal{F} \text{ and } x \in \mathcal{X}. \tag{3.8}$$

Let the forecast random variable F have a pmf $\mathbf{p}_F$, where the ith element of this vector is (1) for probability forecasts,

$$p_F(f(i)) = Pr\{F = f(i)\}, \qquad f(i) \in \mathcal{F},$$

and (2) for categorical forecasts

$$p_F(i) = Pr\{F = i\}, \qquad i \in \mathcal{F}.$$

This n vector tells us the relative frequency of each forecast and is found by summing the joint pmf in Equation (3.8) over all values of x. The distribution depends not only on the forecaster or forecasting system being used *but also on the environment in which the forecasts are being made*. One would expect the marginal distribution of freezing weather forecasts for a town in northern Canada to be very different from the distribution of forecasts for one in the southern United States even if the same forecasting model were used.

Let the actual outcome random variable X have a pmf $\mathbf{p}_X$, where the jth element is

$$p_X(j) = Pr\{X = j\}, \qquad j \in \mathcal{X}.$$

This vector tells us the relative frequency of each event outcome and is found by summing the joint pmf in Equation (3.8) over all values of f. The notation is the same for both categorical and probability forecasts and is what we have referred to as the baseline forecast. It is the probability that we assign to X taking on the value j based only on the information set I. However, it is important to keep in mind that (1) for probability forecasts $\mathcal{X} = \{0,1\}$, and (2) for categorical forecasts $\mathcal{X}$ enumerates all the categories and can therefore be represented by $\{1, 2, \ldots, n\}$.

The vector $\mathbf{p}_X$ tells us how frequently each category or outcome occurs. It should be emphasized that the distribution $\mathbf{p}_X$ is the only description we have that is *independent of the forecasting system or technique used*. In weather forecasting it describes the frequency of outcomes that characterize the environment, the geographical region, the events, or the situation to which the forecasts are being applied. All other joint and conditional probabilities depend on judgment and/or forecasting methodology, but $\mathbf{p}_X$ does not. There are situations where $\mathbf{p}_X$ is influenced by decisions; in the football example in Chapter 2, where X was 1 if the team scored and 0 if it did not, the distribution depended on the play decision made by the coach. The important point is that $\mathbf{p}_X$ is not something we calculate from a forecasting model; it corresponds to reality, and for this reason, verification and calibration of forecasting techniques and systems must ultimately depend on how well forecasts perform compared with actual outcomes of the random quantity X.

In weather forecasting $\mathbf{p}_X$ is known as the climatological forecast; in other applications it is often called the baseline forecast because all conditional forecasts attempt to improve on our estimation of the distribution of X. Returning to our temperature example, if we use and compare two different forecasting models in the same town in northern Canada over the same period, the $\mathbf{p}_X$ would stay the same, but the two marginal distributions of forecasts $\mathbf{p}_F$ could differ.

3.4.1 Decision Probabilities and Likelihoods

Two very important conditional distributions play a major role in decision models with forecasts. First is the probability of a particular outcome x given a particular forecast f, $p_{X|F}(x|f)$; second is the probability of a particular forecast f given a particular outcome x, $p_{F|X}(f|x)$. These are denoted by the matrices $\mathbf{P}$ and $\mathbf{F}$, respectively, where each row represents an outcome of the forecast, and each column an outcome of the actual event. Note that the rows of $\mathbf{P}$ sum to 1 as do the columns of $\mathbf{F}$. From Equation (1.20) (p. 21) recall that the multiplicative law of probability tells us how to calculate these from joint probabilities. For probability forecasts the elements of $\mathbf{P}$ are given by

$$p_{X|F}(j|f(i)) = \frac{p_{F,X}(f(i),j)}{p_{F,X}(f(i),1) + p_{F,X}(f(i),0)}, \; f(i) \in \mathcal{F} \text{ and } j \in \mathcal{X} \quad (3.9)$$

and the elements of $\mathbf{F}$ by

$$p_{F|X}(f(i)|j) = \frac{p_{F,X}(f(i),j)}{\displaystyle\sum_{f(k) \in \mathcal{F}} p_{F,X}(f(k),j)}, \qquad f(i) \in \mathcal{F} \text{ and } j \in \mathcal{X}. \quad (3.10)$$

For categorical forecasts simply replace $f(i)$ by i in these expressions.

   Examples of conditional probabilities for X given F ($\mathbf{P}$ matrices) have been shown in Tables 3.1, 3.2, and 3.3. It is important to understand the differences between $\mathbf{P}$ and $\mathbf{F}$. The conditional probability of X given F is referred to as a *decision probability*, because when these are known by the decision maker, they are often the probabilities to use on a decision tree to find optimal policies. The conditional probability of F given X is referred to as the *likelihood* of the forecast f, given the outcome x. These are typically derived when testing the forecast procedure or method against known outcomes.

   The reader has already seen the use of unconditional probabilities $p(x)$ (baseline forecasts) on decision trees when no expert forecast F is available. We see later in this chapter that the probability statements from a probability forecast may be all that are known by the decision maker and that, as a result, they are used in formulating the decision problem and finding optimal policies. Our use of the term *decision probability* for the conditional distribution of X given F is intended to emphasize the difference between it and the conditional distribution of F given X. It has been our experience that students frequently confuse the two.

   The sets $\mathcal{X}$ and $\mathcal{F}$ (i.e., the values that X and F can assume) will depend on the particular problem and whether probability or categorical forecasts are being used. If we know the joint distribution we can obtain whatever marginal and/or conditional distributions are needed to measure forecasting performance. Bayes' Rule is used to compute the decision probabilities from the likelihoods and vice versa.

3.4.2 Bayes' Rule with Probability Forecasts

Using Equation (1.11) (p. 14) with $\{X = j\}$ substituted for B and $\{F = f(i)\}$ substituted for C, we see that the forecast likelihoods and decision probabilities are related, and this relationship is given by either

$$p_{F|X}(f(i)|j) = \frac{p_{X|F}(j|f(i)) p_F(f(i))}{p_X(j)}, \qquad f(i) \in \mathcal{F} \text{ and } j \in X \qquad (3.11)$$

or

$$p_{X|F}(j|f(i)) = \frac{p_{F|X}(f(i)|j) p_X(j)}{p_F(f(i))}, \qquad f(i) \in \mathcal{F} \text{ and } j \in X. \qquad (3.12)$$

The same expressions can be obtained using Equation (1.21) (p. 21) with appropriate substitution of notation.

These equations may look somewhat complex, but their application is often quite straightforward and simple. We again illustrate with a numerical example using the data displayed in Table 3.4. We use the same probability statements for forecasting rain as we used in Table 3.1, but in Table 3.4 we record the actual data, that is, the counts of days, when both a given state was forecast and the given weather outcome occurred. For example, for the 1,000 days when measurements were taken, on 70 there was no rain when the forecast was that rain was likely (75% chance, so $i = 2$ and $f(2) = 0.75$). Forecasts as to certainty of rain or no rain were made on only 200 days, and of these the forecast was correct on 175. These records show that the probability forecasts are not well calibrated.[4]

The joint probability that a particular probability (statement) forecast will be made and a particular outcome of rain or no rain will occur can be estimated by dividing the entries in the body of Table 3.4 by 1,000.[5] From the row sums the marginal pmf of F is

$$\mathbf{p}_F = (0.1, 0.2, 0.4, 0.2, 0.1). \qquad (3.13)$$

These are the fractions of the time that each probability statement was made in the region or environment where these forecasts were made. Similarly, from the column sums, the marginal pmf of X is

---

[4] Forecast calibration is discussed in Chapter 8.

[5] We do not go into the important subject of statistical estimation in this book as many excellent textbooks are available. Our focus is on understanding forecasting as it applies to modeling and solving decision problems, and we use simple frequencies to estimate probabilities.

$$\mathbf{p}_X = (0.515, 0.485). \tag{3.14}$$

These are the fractions of time that rain did and did not occur. Recall that these fractions in Equation (3.14) are the only numbers that *do not change* when either different probability statements or different forecasting models are used. Using Equation (3.9) the decision probabilities $p_{X|F}(x \mid f)$ are found to be (compare with Table 3.1)

$$\mathbf{P} = \begin{bmatrix} 0.90 & 0.10 \\ 0.65 & 0.35 \\ 0.60 & 0.40 \\ 0.20 & 0.80 \\ 0.15 & 0.85 \end{bmatrix}. \tag{3.15}$$

Using Equation (3.10) the forecast likelihoods $p_{F|X}(f \mid x)$ are

$$\mathbf{F} = \begin{bmatrix} 0.175 & 0.021 \\ 0.252 & 0.144 \\ 0.466 & 0.330 \\ 0.078 & 0.330 \\ 0.029 & 0.175 \end{bmatrix}. \tag{3.16}$$

Let us interpret what our estimates mean and how they are related to the probability forecasts themselves. First, in the area where the data were taken it rains on 51.5% of the days. Second, the forecast makes a prediction of at least a 50% chance of rain 70% of the time. Third, on days when the forecast calls for a 50% chance of rain it will rain 60% of the time; on days when the forecast calls for a 75% chance of rain it will rain 65% of the time. Finally, of all the days when it actually rained, the forecast predicted rain for certain 17.5% of the time, no rain 2.9% of the time, a 50% chance of rain 46.6% of the time and so on.

What the data clearly show is that the actual fraction of rainy days differs from the probability forecasts; in some cases the forecasts overestimate the chance of rain, and in other cases they underestimate them. If a decision maker using the probability forecasts does not have available the data records or measures of performance of the

TABLE 3.4
Data for rain forecast example

Forecast state definition			Weather outcome		
State i and description	Probability statement $f(i)$	Rain ($X = 1$)	No Rain ($X = 0$)	Total days	
1. Certain	1.00	90	10	100	
2. Likely	0.75	130	70	200	
3. Moderate chance	0.50	240	160	400	
4. Unlikely	0.25	40	160	200	
5. No chance	0.00	15	85	100	
Total days		515	485	1,000	

probability forecasts against actual weather, then he or she can only use the probability forecasts, that is, the $f(i)$ statements in a decision problem. This is why, as decision makers, we must become interested in and concerned with the quality of forecasts, their calibration, validation, and performance. These topics are taken up in greater detail in Chapter 8.

3.4.3   Bayes' Rule with Categorical Forecasts

Equations (3.11) and (3.12) hold for categorical forecasts as well as probability forecasts except that $f(i)$ is replaced by i. For categorical forecasts $\mathcal{F}$ and X are both given by $\{1, 2, \ldots, n\}$. Again using either Equation (1.11) or (1.21) (i.e., Bayes' rule), we obtain

$$p_{X|F}(j|i) = \frac{p_{F|X}(i|j)\,p_X(j)}{p_F(i)}, \qquad i \in \mathcal{F} \text{ and } j \in X. \qquad (3.17)$$

Again, we illustrate with a numerical example. Using the same categories for minimum daily temperature as in Table 3.3, suppose we have historic data on 1,000 days. The frequency of occurrence of the actual category of minimum temperature and the forecasts are shown in Table 3.5 together with row and column totals. For example, on 360 of the 1,000 days, the minimum temperature agreed with the forecast that it would be between $1°$ and $5°$, whereas on only 8 days was it forecast to be at least $6°$ and was actually between $1°$ and $5°$. The reader should check that the actual minimum category disagreed with the forecast category on 135 days.

The joint probabilities are estimated by dividing the number of days in each cell by 1,000. The marginal pmf of the forecast is found by dividing the row sums by 1,000 to find

$$\mathbf{p}_F = (0.1, 0.3, 0.4, 0.2). \qquad (3.18)$$

The marginal pmf of the actual minimum temperature is found by dividing the column sums by 1,000 to find

TABLE 3.5
Data for the minimum temperature example

Forecasted minimum temperature		Actual minimum temperature category j				Total days
Category i	Interval	1	2	3	4	
1	$-5°$ or below	75	20	5	0	100
2	$-4°-0°$	15	240	30	15	300
3	$1°-5°$	0	20	360	20	400
4	$6°$ or above	0	2	8	190	200
Total days		90	282	403	225	1,000

$$\mathbf{p}_X = (0.090, 0.282, 0.403, 0.225). \tag{3.19}$$

The decision probabilities are found using Equation (3.9). The reader should check that they agree with those in Table 3.3, that

$$\mathbf{P} = \begin{bmatrix} 0.75 & 0.20 & 0.05 & 0.00 \\ 0.05 & 0.80 & 0.10 & 0.05 \\ 0.00 & 0.05 & 0.90 & 0.05 \\ 0.00 & 0.01 & 0.04 & 0.95 \end{bmatrix}. \tag{3.20}$$

The forecast likelihoods are found using Equation (3.10),

$$\mathbf{F} = \begin{bmatrix} 0.833 & 0.071 & 0.012 & 0.000 \\ 0.167 & 0.851 & 0.075 & 0.067 \\ 0.000 & 0.071 & 0.893 & 0.089 \\ 0.000 & 0.007 & 0.020 & 0.844 \end{bmatrix}. \tag{3.21}$$

Finally, the reader should check that $\mathbf{P}$ can be obtained from $\mathbf{F}$ or vice versa using Equation (3.17) (see Problem 3.1).

Before leaving this example, let us interpret what our calculations mean. First, the actual weather in the area where the data were taken had a minimum temperature between $-4°$ and $0°$ 28.2% of the time and between $1°$ and $5°$ 40.3% of the time. We stress again that the probability vector $\mathbf{p}_X$ is independent of the forecasting system. It is derived using only the historical minimum temperature records. Second, our forecast method has predicted a minimum temperature between $-4°$ and $0°$ 30% of the time and between $1°$ and $5°$ 40% of the time. It overestimates the first of these ranges and slightly underestimates the second. Third, if the forecast calls for a minimum temperature at or below $-5°$ there is a 75% chance that this will occur, a 20% chance that it will be between $-4°$ and $0°$, and a 5% chance it will be between $1°$ and $5°$. Finally, of all the days that the minimum temperature was actually above $5°$, the forecast predicted that it would be 84.4% of the time, that it would be in between $1°$ and $5°$ 8.9% of the time, and so on. Clearly, the forecast is not perfect as $\mathbf{P}$ and $\mathbf{F}$ are not identity matrices, but the high probabilities on and close to the diagonal do indicate a useful forecast.

Many decision problems are formulated using categorical forecasts with only two categories so that $\mathcal{F} = \{0,1\}$ and $X = \{0,1\}$. Because both $\mathbf{F}$ and $\mathbf{P}$ are 2×2 matrices, it is desirable to simplify the notation for the decision probabilities and the forecast likelihoods in these cases to

$$p_1 = Pr\{X = 1 \mid F = 1\} \text{ and } p_0 = Pr\{X = 1 \mid F = 0\}, \tag{3.22}$$

$$f_1 = Pr\{F = 1 \mid X = 1\} \text{ and } f_0 = Pr\{F = 1 \mid X = 0\}. \tag{3.23}$$

Using these, we obtain

$$\mathbf{P} = \begin{bmatrix} p_1 & 1 - p_1 \\ p_0 & 1 - p_0 \end{bmatrix}$$

and

$$\mathbf{F} = \begin{bmatrix} f_1 & f_0 \\ 1-f_1 & 1-f_0 \end{bmatrix}.$$

3.4.4 Summary of Results in Matrix Notation

If $\mathbf{F}$ and $\mathbf{p}_X$ are known, the joint probability of forecasts and outcomes and $\mathbf{p}_F$ can both be easily calculated. Using Bayes' rule we can then calculate the unique values in $\mathbf{P}$. On the other hand, suppose that the decision probabilities have been estimated and are represented by the $n \times 2$ matrix $\mathbf{P}$. The relationship between the marginal distribution of forecasts, $\mathbf{p}_F = (p(f(1)), p(f(2)), \dots, p(f(n)))$, and outcomes $\mathbf{p}_X$ is the one described earlier, namely,

$$\mathbf{p}_F \mathbf{P} = \mathbf{p}_X.$$

Because $\mathbf{p}_X$ is always a two-element vector and, in general, $\mathbf{p}_F$ has more than two elements, there is more than one vector $\mathbf{p}_F$ that yields the desired $\mathbf{p}_X$ for a given $\mathbf{P}$. For example, with the Rain/No Rain data given in Table 3.4 we have $\mathbf{p}_F = (0.1, 0.2, 0.4, 0.2, 0.1)$. The reader can check that the vector $\mathbf{p}_F = (0.45, 0, 0, 0.55, 0)$ also yields the same climatological vector $\mathbf{p}_X = (0.515, 0.485)$. In general, if $\mathbf{P}$ and $\mathbf{p}_X$ are known, $\mathbf{p}_F$ is a solution of

$$\mathbf{p}_F \mathbf{P} = \mathbf{p}_X$$

$$\mathbf{p}_F \mathbf{1} = 1,$$

$$\mathbf{p}_F \geq 0,$$

where we have used $\mathbf{1}$ to represent a vector with every element equal to 1.

The first equality is the equation that relates marginal and joint distributions, the second is the requirement that marginal probabilities sum to 1. The inequality ensures that the marginal probabilities of the probability statements are nonnegative.

In general, $\mathbf{P}$ has rank 2 and has many rows so that there are many vectors $\mathbf{p}_F$ that yields a desired $\mathbf{p}_X$ for a given $\mathbf{P}$.[6] For a fixed forecasting environment identified with $\mathbf{p}_X$, each solution for $\mathbf{p}_F$ can be identified with a different forecaster or forecasting system. For the same forecaster or forecasting system in a different forecasting environment, a unique solution for $\mathbf{p}_F$ is obtained from each $\mathbf{p}_X$.

3.4.5 Sensitivity of Decision Probabilities

We use five 2-category categorical forecasts to demonstrate how decision probabilities in a given problem depend on the likelihoods and the baseline probabilities. A forecaster

---

[6] $\mathbf{P}$ has rank 2 except in the unusual case where both columns of $\mathbf{F}$ are equal, in which case the likelihoods do not discriminate and $\mathbf{P}$ has rank 1. Forecast discrimination is a measure of forecast performance described in greater detail in Chapter 8.

is asked to state whether or not an event of interest will or will not occur. Recall that $F = 1$ if the forecaster predicts that the event will occur and $F = 0$ otherwise, with $p_F = Pr\{F = 1\}$. From historical records of the forecaster and data on actual outcomes of the event, we can estimate the forecast likelihoods (using the notation in Equation (3.23))

$$f_1 = Pr\{F = 1 \mid X = 1\} \text{ and } f_0 = Pr\{F = 1 \mid X = 0\}.$$

Suppose we have available five different forecasts designated I through V; the values f_1 and f_0 for each are shown in Table 3.6. Forecast I is highly discriminating because 95% of the time it correctly identifies the occurrence of the event when it happens and only 0.9% of the time does it incorrectly forecast the occurrence of the event when it does not happen. As the forecast number increases, the ability of the forecaster to discriminate decreases. Using the same notation as shown in Equation (3.22), we define the decision probabilities p_1 and p_0 to be

$$p_1 = Pr\{X = 1 \mid F = 1\} \text{ and } p_0 = Pr\{X = 1 \mid F = 0\}.$$

These can be calculated from the forecast likelihoods using Bayes' rule to give the relationships

$$p_1 = \frac{f_1 p_X}{f_0 (1 - p_X) + f_1 p_X}, \qquad p_0 = \frac{(1 - f_1) p_X}{1 - f_0 (1 - p_X) - f_1 p_X}. \tag{3.24}$$

Note that the values for p_1 and p_0 depend on both the forecast likelihoods *and* the probability p_X of occurrence of the event being forecast. A forecaster who correctly predicts adverse weather 90% of the time when it actually is adverse and mistakenly forecasts adverse weather 9% of the time when it is not would have a value of 0.87 for p_1 when adverse weather occurs 40% of the time. This value would fall to 0.71 when adverse weather occurs only 20% of the time. At the same time the value of p_0 would drop from 0.07 to 0.03. Figure 3-4 shows plots of Equations (3.24) for the five forecasts shown in Table 3.6. The top five (concave) curves show p_1 and the bottom five (convex) curves show p_0. Notice that if our prior probability p_X is small, a poorly discriminating forecaster would have both a p_0 and p_1 close to zero; if it were large they would both be close to 1. As the reader might expect, the difference between p_1 and p_0 is largest for all five forecasts when $p_X = 0.5$, the case of maximum uncertainty in the event being predicted. Notice that the best of the forecasts (number

TABLE 3.6
Forecast likelihoods for five forecasts

Forecast	f_1	f_0
I	0.95	0.009
II	0.90	0.090
III	0.85	0.255
IV	0.80	0.400
V	0.75	0.525

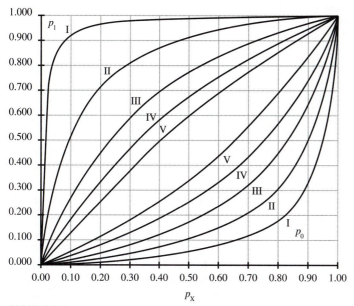

FIGURE 3-4
Decision probabilities versus p_x for five forecasts.

I) does poorly as p_X approaches 0 or 1. It takes an extremely good forecast to be useful in predicting a rare event.

3.4.6   Node and Arc Reversal with Bayes' Rule

To give the reader more insight into the use and meaning of the mathematical notation and formulation, we illustrate Bayes' rule on a two-node chance influence diagram and the associated event trees. A chance influence diagram is a special case of the influence diagram used heretofore, in that it contains only chance nodes without decision or result nodes. An event tree is the corresponding decision tree without decision nodes. In it branches represent outcomes of the chance nodes in the chance influence diagram.

Consider the forecast F and random event X in the chance influence diagram of Figure 3-5 where the marginal and conditional distributions are shown alongside the chance nodes. In (a) the directed arc from X to F indicates that the marginal distribution $\mathbf{p}_X$ and the forecast likelihoods $\mathbf{F}$ are known; in (b) the arc is reversed, indicating that the marginal distribution $\mathbf{p}_F$ and the decision probabilities $\mathbf{P}$ are known. Associated event trees are shown in Figure 3-6 for a case where both X and F have only two possible outcomes. The one that corresponds to (a) starts with a single node X from which branches emanate, one for each of the elements of the set X. Each of these terminates at a node F so that there are as many F nodes as there are elements of X. From each F node there is a branch for each element of the set $\mathcal{F}$. In the event tree that corresponds to (b) the role of X and F are reversed.

(a) $p_X(x)$ and $p_{F|X}(f\,|\,x)$ known   (b) $p_F(f)$ and $p_{X|F}(x\,|\,f)$ known

FIGURE 3-5
Chance influence diagrams with forecast and event.

In Chapter 5 we study a number of two-stage decision models in which the first decision is whether or not to obtain more information about an uncertain outcome before making an operational decision. One example requires a decision on whether or not to test an existing (possibly defective) aircraft part. If it is tested and the test indicates that the part is defective, a second decision must be made to either repair it or replace it with a new one. One can think of the test as providing a categorical forecast. Naturally, the optimal decisions depend on knowledge the test gives of the current part being defective or not as well as on the historical data observed and gathered over time on similar parts; both are used to infer the decision probabilities required in model solution. The natural way to calibrate the test equipment is to see how well it performs against a part of known condition. If X is the condition of the part and F is the forecast of part condition based on the outcome of a test, then we observe $p_{F|X}(f|x)$ and Figure 3-5(a) applies. It correctly indicates that the condition of the part influences the test outcome or forecast (note the direction of the arrow). However, when time comes to make a decision on a tested part (whose condition is unknown), the decision maker needs to know the probabilities that a test will correctly or incorrectly reveal the true condition of the part, which is unobservable at the moment of the test as shown in Figure 3-5(b). To find these, we construct the event trees corresponding to the chance influence diagrams in Figure 3-5. These are shown in Figure 3-6.

Let us assume that one-eighth of the parts have been defective historically. The forecast likelihoods are also given; the probability that the test correctly reveals the part condition is given to be 3/4. This means that for parts known to be defective, 3/4 of the time the test will indicate a defective part; the same holds for nondefective parts appearing to be nondefective. Let $F = 1$ if the test shows "defective" and $F = 0$ if it shows "nondefective." Let $X = 1$ if the part is defective and $X = 0$ if it is nondefective. In our notation $p_X = 1/8$, $f_1 = p_{F|X}(1\,|\,1) = 3/4$, and $f_0 = p_{F|X}(1\,|\,0) = 1/4$. Coherent conditional probabilities can be obtained in both event trees through the use of Bayes' rule.

In Figure 3-6(a) the four paths correspond to the four possible pairs of outcomes for X followed by F. The probabilities on the branches of this tree are all given in the statement of the problem. The joint probabilities at the terminal nodes are calculated from these using Equation (1.20) (p. 21). In Figure 3-6(b) the four paths correspond to the four possible pairs of outcomes for F followed by X. Now we know that the path probabilities *must be the same regardless of the ordering of the chance nodes in the event tree*; therefore, the joint probabilities at the terminal nodes must be the same as those in Figure 3-6(a). We can calculate the probabilities on the branches corresponding to $F = 1$ and $F = 0$ using Equation (1.20) (p. 21). These are

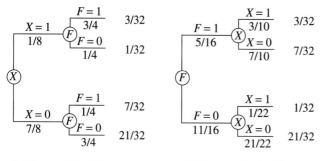

(a) Forecast Likelihoods   (b) Decision Probabilities

FIGURE 3-6
Node reversal in the aircraft part event trees.

$$p_F(1) = (1/8)(3/4) + (7/8)(1/4) = 5/16,$$

$$p_F(0) = (1/8)(1/4) + (7/8)(3/4) = 11/16.$$

The only remaining task is to calculate the desired four decision probabilities $p_{X|F}(x|f)$. By again using Equation (1.20) these are easily found. Of course, these numbers are identical with the conditional probabilities obtained from Bayes' rule.

More complex event trees are shown in Figure 3-7 using the rain probability forecasts illustrated in Equations (3.13) through (3.16). The probabilities in (a) are found in Equations (3.14) and (3.16), and those in (b) are found in (3.13) and (3.15). As before, path probabilities do not depend on the ordering of the chance nodes. For example, using event tree (b) in Figure 3-7, the joint probability that it both rains and the forecast calls for a 50% chance of rain is $(0.4) \times (0.6) = 0.240$; using the event tree in (a) we see it is also $(0.515) \times (0.466) = 0.240$. Coherent conditional probabilities ensure that the path probabilities remain unchanged because the path probability is the product of the marginal probability and the conditional probability of each branch leaving the second node.

The concepts illustrated in these chance influence diagrams and event trees for two events extend to influence diagrams with many nodes and large numbers of outcomes. The notion of arc and node reversal in directed graphs with large numbers of chance nodes, though more complicated to analyze and implement, is based on the notion that the chance conditioning tree used to describe and formulate the original model may require a reordering of events to make forecasts that are used in decision making. The conditions that must be satisfied for arc reversal in influence diagrams with more than three nodes is complex and is studied in some detail in Chapter 9.

3.4.7 Using Bayes' Rule with Odds

We now consider a probability forecast that arises in the field of consumer credit. The forecast problem is to predict the creditworthiness of an applicant using information on a credit application form. An application usually contains data on many factors, such as age, occupation, years in current job, whether a residence is owned or rented, types and sizes of bank accounts, major credit cards held, and so on. These factors are treated as

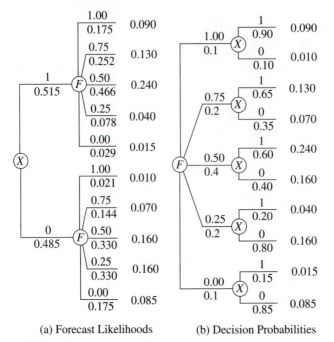

(a) Forecast Likelihoods   (b) Decision Probabilities

FIGURE 3-7
Node reversal in the rain forecast event trees.

random variables and their outcomes used to evaluate the applicant. We consider a simplified example with only two pieces of data, age and homeownership, to illustrate the concepts of conditional independence and the relationship of prior and posterior odds.

Consider three random variables, X, Y, and Z, that indicate the applicant's creditworthiness, age, and homeownership status, respectively. We assume X has two possible outcomes, G and B denoting good and bad credit risk outcomes, respectively, so that $X = \{G, B\}$.[7] The directed arcs in Figure 3-8 indicate an assessment of the probabilities $p_X(G)$ and $p_X(B)$ along with the conditional probabilities $p_{Y|X}(y \mid G)$, $p_{Y|X}(y \mid B)$, $p_{Z|X}(z \mid G)$, and $p_{Z|X}(z \mid B)$. Our objective is to reverse both arcs to determine

$$p_{X|Y,Z}(G \mid y, z) \text{ and } p_{X|Y,Z}(B \mid y, z) = 1 - p_{X|Y,Z}(G \mid y, z).$$

The effect of the random quantities Y and Z on the probability of being good or bad is often measured by a scoring system. The problem is to forecast X after observing Y and Z; in developing a procedure, it is natural to look at past records segregated into those that proved to be creditworthy and those that did not; such records in combination

---

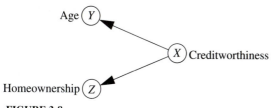

FIGURE 3-8
Chance influence diagram for the credit example.

with a scoring system lead directly to estimates of the probabilities $p_X(G)$, $p_X(B)$, $p_{Y|X}(y \mid G)$, $p_{Y|X}(y \mid B)$, $p_{Z|X}(z \mid G)$, and $p_{Z|X}(z \mid B)$.

The reader might suspect that age and homeownership are not independent events; for example, the older you are the more likely you are to own your own home. The absence of any type of arc directly connecting Y and Z in Figure 3-8 indicates that the following assumption is being made: Given a person's creditworthiness (i.e., good or bad), age and homeownership are considered independent events. This concept is referred to as *conditional independence*; sometimes we write that the random variables $\{Y \mid X\}$ and $\{Z \mid X\}$ are independent. In terms of a probability statement, we assume that

$$Pr\{Y = y, Z = z \mid X = x\} = Pr\{Y = y \mid X = x\}Pr\{Z = z \mid X = x\}, \tag{3.25}$$

for all $x \in X$, $y \in Y$, *and* $z \in Z$.

The odds of an event occurring are the ratio of the probability of it occurring to it not occurring; thus, the odds of a good is the ratio

$$O_X(G) = p_X(G)/(1 - p_X(G)) = p_X(G)/p_X(B),$$

which is often referred to as the population odds. If 80% of the population being measured are good credit risks for the particular application, then the odds that a person drawn at random from this group is a good credit risk is 4. How would we use records of goods and bads, broken down by age and income, to calculate the conditional odds that we have a good, given observed values of age $Y = y$ and income $Z = z$? From Bayes' rule and the conditional independence assumption leading to Equation (3.25), we know that the conditional probability of good given age and income is

$$p_{X|Y,Z}(G \mid y, z) = p_{Y|X}(y \mid G)p_{Z|X}(z \mid G)p_X(G)/p_{Y,Z}(y, z). \tag{3.26}$$

Similarly, for bad we have

$$p_{X|Y,Z}(B \mid y, z) = p_{Y|X}(y \mid B)p_{Z|X}(z \mid B)p_X(B)/p_{Y,Z}(y, z). \tag{3.27}$$

If we divide Equation (3.26) by (3.27) and cancel the common $p_{Y,Z}(y, z)$ terms, we obtain

$$O_{X|Y,Z}(G \mid y, z) = O_X(G)\left(\frac{p_{Y|X}(y \mid G)}{p_{Y|X}(y \mid B)}\right)\left(\frac{p_{Z|X}(z \mid G)}{p_{Z|X}(z \mid B)}\right). \tag{3.28}$$

It is important to emphasize that in Equation (3.28) we have assumed that Y and Z are conditionally independent given X. The term on the left is referred to as the conditional odds. The first term on the right is the population odds. This is followed by two factors that are often called *information odds*. Strictly speaking, they are not odds but likelihood ratios, or the ratio of two conditional probabilities whose conditions are the binary credit risk outcomes. The product of these two ratios is also referred to as the Bayes' factor.

Equation (3.28) can be rewritten as

$$O_{X|Y,Z}(G|y,z) = \frac{O_{X|Y}(G|y)\, O_{X|Z}(G|z)}{O_X(G)}.$$ (3.29)

Taking logarithms[8] of both sides gives

$$log\, O_{X|Y,Z}(G|y,z) = log\, O_{X|Y}(G|y) + log\, O_{X|Z}(G|z) - log\, O_X(G).$$ (3.30)

What is important about Equation (3.30) is that many forecasting formulas for the occurrence of an event, such as bankruptcy, payoff of a loan, presence of a target, are reliably expressed in terms of scores that equate to log odds. As seen, scores on individual factors can be added to give a total score that is equivalent to posterior odds.

If Y and Z are dependent, so that Z is conditional on both X and Y, then the ratio in Equation (3.28) involving Z must be replaced by the ratio

$$p_{Z|X,Y}(z|\, G,y)/p_{Z|X,Y}(z|\, B,y),$$

which includes the dependence of Z on Y. The posterior odds conditioned on both Y and Z,

$$\begin{aligned} O_{X|Y,Z}(G|y,z) &= O_X(G)\frac{p_{Y|X}(y|G)\, p_{Z|X,Y}(z|G,y)}{p_{Y|X}(y|B)\, p_{Z|X,Y}(z|B,y)} \\ &= \frac{p_{Z|X,Y}(z|G,y)}{p_{Z|X,Y}(z|B,y)}O_{X|Y}(G|y), \end{aligned}$$ (3.31)

can be expressed in terms of the posterior odds conditioned only on Y multiplied by the conditional likelihood ratio of Z given X, and Y and either G or B. These ideas are used in several problems in this and later chapters.

3.5 MULTIPLE LIKELIHOODS AND DEPENDENT FORECASTS

Up to this point we have been concerned with the solution of decision problems where we have a single forecast for the uncertain quantity of interest. Real problems often require multiple forecasts and multiple decisions, possibly obtained from different fore-

---

[8] All logarithms referred to in this book are natural logarithms.

casters or different forecasting systems. In the remainder of this chapter we consider problems that may require simultaneous or sequential forecasts. In some cases the forecasts depend on the alternatives selected by earlier decisions. This relationship between decisions and forecasts and the relationships between forecast likelihoods and decision probabilities become much more complicated as the number of decisions and uncertain quantities increases. We begin by examining two forecasts and one event of interest and then consider a medical diagnosis problem where four separate (but dependent) tests or forecasts are obtained. As one might expect, conditional dependencies become a most important aspect of the forecasting problem.

3.5.1   Two Forecasts for a Single Event

We now consider problems where two forecasts or sensor inputs are available. In the descriptions that follow, think of the word *target* as a generic name that might represent the status of an individual in a credit report or an individual who might be a target for a promotion campaign; alternatively, it might represent the presence or absence of a medical condition in which the "forecasts" might be symptoms associated with the condition. Let

$X = 1$ if an event occurs or the object or individual under consideration is a target, 0 otherwise.

Each forecast or sensor output is a random quantity that we denote by

$F = 1$ if forecast 1 predicts that the event will occur or the first sensor predicts an object is a target, 0 otherwise,

$G = 1$ if forecast 2 predicts that the event will occur or the second sensor predicts an object is a target, 0 otherwise.

The assumed dependencies of the likelihoods are illustrated in the chance influence diagram of Figure 3-9. Notice that the diagram is drawn so that X is known before F and G are observed, that is we have at hand the forecast likelihoods. If X is the uncertainty on which the decision problem is based, we would like to reverse the influence of

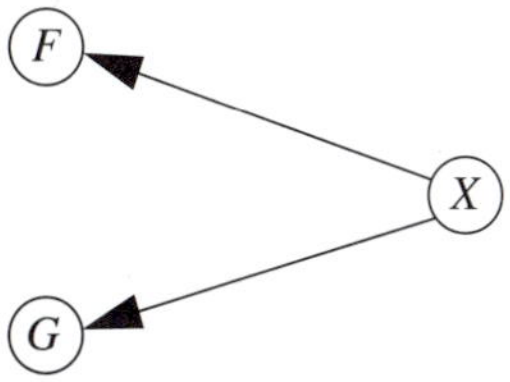

FIGURE 3-9
Influence of known target on forecasts.

X on F and G. At first glance it may appear that we are assuming F and G to be independent random variables as no arc is drawn between these nodes. Obviously this would be a very poor assumption because one would expect and hope that F and G are very highly correlated. If both forecasts or sensors observe a known target (e.g., a person with cancer, a good credit risk, a particular military target, etc.) so that $X = 1$, we would hope that there is a high probability of both F and G being 1; and when both are observing a non-target ($X = 0$), we would hope that there is a high probability of both of them being zero. What the diagram shows is that both F and G depend statistically on X, but that we are assuming F and G to be *conditionally independent, given* X. Mathematically, our assumption can be stated as

$$Pr\{F = i, G = j \,|X = k\} = Pr\{F = i \mid X = k\}Pr\{G = j|X = k\}, \tag{3.32}$$

$$i, j, k = 0, 1.$$

Using our earlier notation, this equation becomes

$$p_{F,G|X}(f, g|x) = p_{F|X}(f \,|x)p_{G|X}(g|x).$$

This assumption would be reasonable if the errors made by each sensor in identifying the target were caused by some random phenomena that are quite different for the two sensors. If, for example, both sensors were used to detect cloud cover, they would surely be sensitive to the same particular atmospheric conditions and so both likely to misidentify under similar conditions; this assumption may be a poor one. Similarly, if two loan application review procedures used common data such as age, income, or car ownership, one might not expect statistical independence of the factors that will eventually be used to predict the performance of the individual.

Because the decision maker makes a decision after both forecasts are known, the arcs in Figure 3-9 need to be reversed, yielding four possible forecast combinations $\{F = 1, \; G = 1\}$, $\{F = 1, \; G = 0\}$, $\{F = 0, \; G = 1\}$ and $\{F = 0, \; G = 0\}$. Let $p_{ij} = Pr\{X = 1|F = i, G = j\}$, $i = 0, 1$ and $j = 0, 1$. These are the decision probabilities that are required in solving a decision problem using both forecasts; they are the two-forecast counterparts to Equation (3.22) (p. 100). To calculate them from the forecast likelihoods

$$Pr\{F = 1| X = 1\} = f_1, \qquad Pr\{G = 1| X = 1\} = g_1,$$

$$Pr\{F = 1| X = 0\} = f_0, \qquad Pr\{G = 1| X = 0\} = g_0, \tag{3.33}$$

we use Bayes' rule together with Equation (3.32) to obtain

$$p_{11} = \frac{f_1 g_1 p_X}{f_1 g_1 p_X + f_0 g_0 (1 - p_X)},$$

$$p_{01} = \frac{(1 - f_1) g_1 p_X}{(1 - f_1) g_1 p_X + (1 - f_0) g_0 (1 - p_X)},$$

$$\tag{3.34}$$

$$p_{10} = \frac{f_1(1-g_1)p_X}{f_1(1-g_1)p_X + f_0(1-g_0)(1-p_X)},$$

$$p_{00} = \frac{(1-f_1)(1-g_1)p_X}{(1-f_1)(1-g_1)p_X + (1-f_0)(1-g_0)(1-p_X)}.$$

Because both sensors discriminate targets from nontargets, we have $f_1 > f_0$ and $g_1 > g_0$. If the G sensor discriminates targets better than F, then $g_1 > f_1$ and $g_0 < f_0$. These inequalities imply that $g_1/1 - g_1 > f_1/1 - f_1$ and that $g_0/1 - g_0 < f_0/1 - f_0$. These in turn imply that the ratio of the odds of $G = 1$ conditional on $X = 1$, relative to the odds of $G = 1$ when $X = 0$, is larger than the same ratio for the F forecast, or

$$\frac{\dfrac{g_1}{(1-g_1)}}{\dfrac{g_0}{(1-g_0)}} \geq \frac{\dfrac{f_1}{(1-f_1)}}{\dfrac{f_0}{(1-f_0)}}. \tag{3.35}$$

Another way to state this result is that the G sensor or forecast is superior in performance to the F sensor or forecast if the odds of the prediction of the object when it is know to be present, relative to the odds when it is known to be absent, are greater for the G sensor than for the F sensor. The inequality (3.35) implies the following ordering of the decision probabilities:

$$0 \leq p_{00} \leq p_{10} \leq p_{01} \leq p_{11} \leq 1.$$

3.5.2 Likelihoods for Four Colon Cancer Tests

Important decision problems often arise in medical diagnosis; one such example is to find a cost-effective way of detecting colon cancer. We defer formulation of the complicated decision problem until Chapter 4, but at this point we formulate the problem of calculating the appropriate decision probabilities through arc reversal. The reader will find that results obtained in this section are required for the decision problems found in Section 4.6.

There are assumed to be four tests for colon cancer, which in increasing order of cost are occult blood, digital examination, sigmoidoscopy and barium enema examination. We define the four random variables OB, DE, SIG and BaE to represent the outcomes of these four tests, and the random variable C to represent the presence or absence of colon cancer. The data that are available for a problem of this type are often in the form of probabilities; for example, that a given test will show positive or negative given that cancer is present or not present. We use the convention that a letter enclosed in quotes is an outcome of the forecast of the event, and without quotes it is the outcome of the event itself. Thus the outcome sets of the four random variables that represent the tests each contain two elements, "c" and "nc", both in parentheses, to denote that the test result shows cancer and no cancer respectively. The outcome set for C contains the elements c and nc to denote the presence or absence of an actual cancer.

We observe that:

1. Colon cancer (*C*) is present in 1 of every 1,000 in the particular target population.
2. A digital examination (*DE*) finds all of the 10% of cancers in the very low end of the colon.
3. Sigmoidoscopy (*SIG*) finds all of the 85% of cancers found in the first 25 to 30 cm.
4. An occult blood test (*OB*) will find 80% of any cancers present but will indicate that 10% of those with no colon cancer have the disease.
5. A barium enema (*BaE*) finds all cancers with no false indications.

Our problem here is to calculate the required conditional probabilities from these historical results of the tests on patients whose condition was known (colon cancer present or not present). One can see that the output of a chosen sequence of the diagnostic tests may not be independent of one another, even though the tests are performed independently of one another. For example, if it is decided to follow an *OB* test with a *SIG* test, the conditional probability is very high that the *SIG* test will find a cancer *given* that the *OB* test has reported a detection, even though the *OB* test has a 10% chance of indicating the presence of cancer when it is absent (called a false positive) and a 20% chance of missing a cancer that is present (a false negative). Conditional dependencies make analysis of cost-effective procedures difficult because of the effect of false-positive and false-negative test results.

As stated, for each chance node representing a test let the random variable take on the value "c" if the test indicates cancer, and "nc" if it does not so that each outcome set is {"c", "nc"}. For the *C* node it is {c, nc}. Figure 3-10 shows the structure of a chance influence diagram consistent with the observations. Four directed arcs start at *C* and end at one of the tests; it is assumed that we have estimated the fraction of people in the population with colon cancer and the probability that each test will show a positive result, given that the person tested does or does not have colon cancer. The fifth arc connects *DE* to *SIG*. This is required because *DE* and *SIG* are *not* conditionally independent, given *C*. Given that *DE* = "c", then with probability 1 *SIG* = "c". The important problem to be addressed is how to convert the conditional probability distributions implied by Figure 3-10 into the distributions that are required to solve the decision problems of Section 4.3.4 (p. 156) that consider optimal testing sequences. The chance influence diagrams for those problems are shown in Figures 3-11 and 3-12.

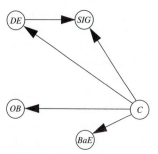

FIGURE 3-10
Colon cancer test chance influence diagram.

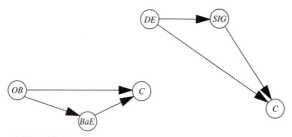

FIGURE 3-11
DE-SIG and *OB-BaE* colon cancer test sequences.

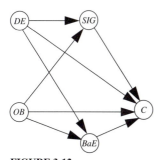

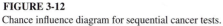

FIGURE 3-12
Chance influence diagram for sequential cancer tests.

When each of the four tests is used separately on a patient who has colon cancer, the likelihood that it will find the cancer is shown in Figure 3-13; each forecast likelihood matrix **F** carries the appropriate subscript for *OB*, *DE*, *SIG*, and *BaE* tests. The rows of each matrix correspond to the test (or forecast) outcome, "c" for cancer or "nc" for no cancer. The columns correspond to the presence or absence of a true cancer, c or nc. The least expensive *OB* test is the poorest discriminator whereas the other three discriminate perfectly if a cancer is known to be present in the region covered by the test.

The *DE* test examines only a small part of the lower colon and is not useful for prediction of cancers throughout the colon since not all cancers are located in the lower section. With any perfectly discriminatory test (*DE* and *SIG* are assumed to be such tests),[9] the limited range of the tests in the lower part of the colon means that only a fraction of all colon cancers will be detected. Medical experts believe that 10% of the colon cancers are in the lower region reachable by a digital examination and 85% are reachable by the sigmoidoscopy exam. Thus, if a digital exam shows positive, so will a sigmoidoscopy exam, and

---

[9] The reader should understand that the likelihood data are highly simplified to illustrate how relevant forecasts are obtained from likelihoods. In real life a tissue sample would have to be obtained in order to provide completely specific tests (i.e., tests with likelihoods having 0 or 1 in their cell entries).

	c	nc
"c"	0.8	0.1
"nc"	0.2	0.9

$$\mathbf{F}_{OB}$$

	c	nc
	0.1	0
	0.9	1.0

$$\mathbf{F}_{DE}$$

	c	nc
	0.85	0
	0.15	1.0

$$\mathbf{F}_{SIG}$$

	c	nc
	1.0	0
	0	1.0

$$\mathbf{F}_{BaE}$$

FIGURE 3-13
Likelihoods for four cancer diagnostic tests.

$$Pr\{SIG = \text{"c"}|\ C = c \text{ and } DE = \text{"c"}\} = 1. \tag{3.36}$$

But if the digital exam shows negative, there is high probability that the cancer will be found by the sigmoidoscopy exam,

$$Pr\{SIG = \text{"c"}|\ C = c \text{ and } DE = \text{"nc"}\} = 0.75/0.9 = 0.833. \tag{3.37}$$

We now have all the distributions consistent with the influence diagram in Figure 3-10. To further illustrate the need to include the directed arc from DE to SIG, let us find the joint probability of cancer and all four tests showing a positive result. Without the arc from DE to SIG we would calculate this by

$$Pr\{\text{Cancer and all tests positive}\}$$

$$= Pr\{DE = \text{"c"}|\ C = c\}Pr\{SIG = \text{"c"}|\ C = c\}Pr\{OB = \text{"c"}|\ C = c\}$$

$$Pr\{BaE = \text{"c"}|\ C = c\}Pr\{C = c\} = 0.000068.$$

But with the arc from DE to SIG included, the second term, $Pr\{SIG = \text{"c"}|\ C = c\} = 0.85$, must be replaced by $Pr\{SIG = \text{"c"}|\ DE = \text{"c"}, C = c\} = 1$ to give the correct joint probability value 0.000080.

If we assume that the chance of a colon cancer in a patient is 1 in 1,000 (i.e., the odds are 1 to 999 in favor of finding a colon cancer in a given population), the decision probabilities for each test made independently of all others are given by the **P** matrices shown in Figure 3-14. For example, the conditional probability that a cancer will be present if $OB = \text{"nc"}$ is 0.00022. These conditional probabilities are found by applying Bayes' rule to the **F** matrices in Figure 3-13. Because the BaE test is the only perfect discriminator, when this test reports a cancer, it is certain that a true colon cancer is present; likewise, if no cancer is reported, we are guaranteed that no colon cancer is present.

We emphasize that the decision probabilities in Figure 3-14 only apply if a single test is used in isolation of all others. Therefore they will not be the decision probabilities we need to analyze a sequential testing program if one of the tests has been preceded by the finding of another test (as is the case in both Figures 3-11 and 3-12). Neither the **F** nor the **P** probabilities given can be used directly when one makes two or more sequential tests. Equations (3.36) and (3.37) show this to be the case. We can use the likelihoods given in Figure 3-13 to obtain the numbers we need for making decisions, but it requires careful attention to known dependencies and the structure and location of the examination procedures to ensure that (often unstated) conditional dependencies are properly included. These are easily missed and must be given special attention and careful thought by medical experts and decision analysts. Note that in Figure 3-15, the col-

$$
\begin{array}{cc}
 & \text{c} \qquad \text{nc} \\
\begin{array}{c} \text{``c''} \\ \text{``nc''} \end{array}
\begin{bmatrix} 0.00794 & 0.99206 \\ 0.00022 & 0.99978 \end{bmatrix} \\
\mathbf{P}_{OB}
\end{array}
\qquad\qquad
\begin{array}{c}
\text{c} \qquad\quad \text{nc} \\
\begin{bmatrix} 1 & 0 \\ 0.00090 & 0.99910 \end{bmatrix} \\
\mathbf{P}_{DE}
\end{array}
$$

$$
\begin{array}{c}
\begin{bmatrix} 1 & 0 \\ 0.00013 & 0.99987 \end{bmatrix} \\
\mathbf{P}_{SIG}
\end{array}
\qquad\qquad
\begin{array}{c}
\begin{bmatrix} 1.0 & 0 \\ 0 & 1.0 \end{bmatrix} \\
\mathbf{P}_{BaE}
\end{array}
$$

FIGURE 3-14
Decision probabilities for four independent cancer tests.

$$
\begin{array}{c}
\quad\ \text{``c''} \qquad \text{``nc''} \\
\begin{array}{c} \text{``c''} \\ \text{``nc''} \end{array}
\begin{bmatrix} 1 & 0 \\ 0.00075 & 0.99925 \end{bmatrix} \\
\mathbf{P}_{SIG\backslash DE}
\end{array}
\qquad\qquad
\begin{array}{c}
\text{``c''} \qquad \text{``nc''} \\
\begin{bmatrix} 0.00794 & 0.99206 \\ 0.00022 & 0.99978 \end{bmatrix} \\
\mathbf{P}_{BaE\backslash OB}
\end{array}
$$

FIGURE 3-15
Conditional probability of a follow-up test.

umns as well as the rows have been relabeled to include quotes as the columns correspond to reports of the second test, not the presence or absence of cancer.

3.5.3 Forecasts for Sequential Tests

In Chapter 4 we shall need conditional decision probabilities that can be used when we find optimal sequential testing decisions. For example, sequential testing procedures are used in the decision tree of Figure 4-33 (p. 178); the decision probabilities depend on the particular sequence of tests and must be calculated from the forecast likelihoods in this section.

In any program where one is interested in cost-effective policies, the main idea is to use one of the inexpensive tests (*DE* and *OB*) followed by one of the more expensive and more discriminatory tests (*SIG* and *BaE*). Not every patient will need every test, but we might be interested in having each patient undergo one of the less expensive tests and, based on the results of this first test, then decide whether further testing is warranted and, if so, which test(s) to perform.

The chance influence diagrams in Figure 3-11 corresponds to the conditional probabilities that are required to evaluate the testing procedures shown in the first column of Table 4.1 (p. 177); they are used in the decision tree of Figure 4-31 (p. 176). Since we know ahead of time that the only sequences being considered are (i) *DE* alone, (ii) *OB* followed by a *BaE*, and (iii) *DE* followed by a *SIG*, there is no linkage between *OB* and *SIG* tests nor between *DE* and *BaE* tests. It is therefore possible to obtain the $\mathbf{P}_{C\backslash DE,SIG}$ and $\mathbf{P}_{C\backslash OB,BaE}$ matrices by two separate and independent calculations. Of course we must add directed arcs from *DE* to *SIG* and from *OB* to *BaE* to recognize the dependencies that exist when those paired tests are used.

Consider the chance influence diagram in Figure 3-11 containing the three nodes *OB*, *BaE*, and *C*. Note that a *BaE* test yields a positive result if and only if a cancer is present. Thus Pr{*OB* = "c"| *BaE* = "c"} is equal to Pr{*OB* = "c"| *C* = c}. Reverse the arc from *OB* to *BaE* using Bayes' rule; the conditional probability matrix $\mathbf{P}_{OB|BaE}$ is identical to the transpose of the likelihood matrix $\mathbf{F}_{OB}$ since no information is gained by knowing that both *BaE* has a positive report and a cancer is known to be present. Under our assumptions, it is not possible for a cancer to be present or absent and to have the *BaE* test yield a contrary report.

The chance influence diagram in Figure 3-12 corresponds to the conditional probabilities that are required to evaluate the sequential tests in the decision tree of Figure 4-33 (p. 178). In general, there will be dependencies between all four tests because the conditional probabilities of detection of the two more expensive tests depend on the outcomes on the two less expensive ones. Thus, we add arcs from *DE* to *BaE* and *SIG* and from *OB* to *BaE* and *SIG*, but not between *DE* and *OB* or between *BaE* and *SIG*. Although the details of arc reversal in large chance influence diagrams are deferred to Chapter 9, we have illustrated how dependencies in the colon cancer problem can be explicitly recognized and used to evaluate the decision probabilities that may be required in a decision tree that recognizes sequential testing policies.

3.6 OPTIMAL CROP PROTECTION

In this section we study how one uses probability forecasts, decision probabilities, and likelihoods using the crop protection problem first described in Section 2.2 of Chapter 2 but now assuming that a probability forecast of freezing temperatures is available to the decision maker.

Assume that a forecasting expert can provide the decision maker with one of the four freezing predictions from Column 1 in Table 3.7 together with his or her association of an appropriate probability statement in column 2; for this example $\mathcal{F} = \{0.8, 0.6, 0.4, 0.2\}$. Note that we assume in this problem that no "certain" forecasts are made, such as "Freezing will occur" or that "Freezing will not occur."

The decision tree for this problem is shown in Figure 3-16; note that it consists of four decision saplings connected to a forecast node *F*. The top decision sapling corresponds to a forecast outcome $f(1) = 0.8$. Keep in mind that this is *not* the probability

TABLE 3.7
Probability forecasts for freezing temperatures

State *i* and description of chance of freezing temperatures	Probability statement *f(i)*
1. High	0.8
2. Good	0.6
3. Moderate	0.4
4. Low	0.2

or fraction of time that this probability statement is made by the forecaster. There is no way to estimate the distribution of F from the forecast probability statements in Table 3.7 unless we know much more about the environment in which the forecasts are made. How is the probability forecast used by the decision maker? We assume initially that no additional information on weather or observed outcomes is available to the decision maker other than the probability statements $\{f(i)\}$. The probability statements $f(i)$ and $1 - f(i)$ are assigned to the branches of the ith decision sapling in the decision tree as shown in Figure 3-16. Without further data, we would assume that the probability statements assigned to the branches agree with the observed fre-

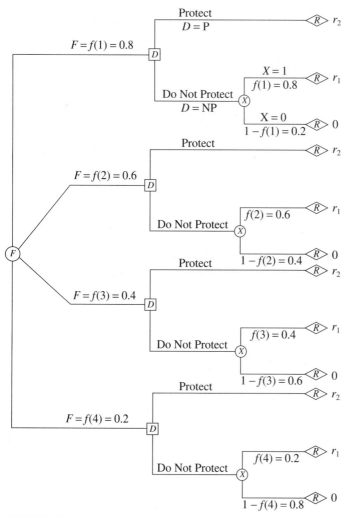

FIGURE 3-16
Decision tree using probability forecasts.

quencies of freezing and non-freezing weather if they were available. The reader should not confuse the probability statements $f(i)$ either with elements of the likelihood matrix $\mathbf{F}$ or with the elements of the vector $\mathbf{p}_F$, which is the marginal distribution of the $f(i)$'s.

We consider four subproblems, each identical in structure to the baseline forecast problem studied in Section 2.2 (p. 46); rather than use the single value of p_X for the probability of freezing weather, we assign four different $f(i)$ probabilities from Table 3.7. With each different probability statement $f(i)$ the optimal decision is

$$d* = \text{P (Protect)} \qquad \text{when } r_2/r_1 \leq f(i) < 1,$$
$$d* = \text{NP (Do Not Protect)} \qquad \text{when } 0 < f(i) \leq r_2/r_1.$$

Whether or not it is optimal to protect depends on the size of the cost–loss ratio and the value of the forecast probability statement. Once the optimal policy has been obtained for each possible probability statement, we only need to know the distribution of F in order to calculate the unconditional minimum expected loss.

Order the n probability statements so that

$$0 \leq f(n) < f(n-1) < \ldots < f(1) \leq 1,$$

and let the cost–loss ratio (r_2/r_1) lie between the probability statements $f(k+1)$ and $f(k)$, that is, $f(k+1) < r_2/r_1 \leq f(k)$. The optimal expected loss l_P* is given by

$$l_P* = [p_F(f(1)) + p_F(f(2)) + \ldots + p_F(f(k))]r_2$$
$$+ [p_{X|F}(1|f(k+1))p_F(f(k+1)) + \ldots + p_{X|F}(1|f(n))p_F(f(n))]r_1, \qquad (3.38)$$

where $p_F(f(i))$ is the marginal probability that the probability statement $f(i)$ occurs in the forecasting environment under consideration. If $r_2/r_1 > f(1)$, then $l_P* = p_X r_1$, and if $r_2/r_1 < f(n)$, $l_P* = r_2$.

Table 3.8 shows the optimal policy for all cost ratios between 0 and 1 with each forecast probability statement $f(i)$; recall that P means Protect and NP means Do Not Protect. For example, if protection costs one-third the potential loss of the crop, it is optimal to make the "Do Not Protect" decision only when the forecast is "freezing weather unlikely."

TABLE 3.8
Optimal policies using probability forecasts

Range of cost ratio r_2/r_1	Forecast probability statement, $f(i)$			
	$f(4) = 0.2$	$f(3) = 0.4$	$f(2) = 0.6$	$f(1) = 0.8$
(0.8, 1.0]	NP	NP	NP	NP
(0.6, 0.8]	NP	NP	NP	P
(0.4, 0.6]	NP	NP	P	P
(0.2, 0.4]	NP	P	P	P
[0, 0.2]	P	P	P	P

TABLE 3.9
Historical data for the probability forecast

Probability forecast		Temperature outcome			
Probability statement	$f(i)$	Below freezing $(X=1)$	Above freezing $(X=0)$	Total days	Fraction of days below freezing
1. High	0.80	68	12	80	0.85
2. Good	0.60	74	56	130	0.57
3. Moderate	0.40	163	197	360	0.45
4. Low	0.20	67	363	430	0.16
Total days		372	628	1,000	

Unfortunately, Table 3.7 by itself does not contain the information required to estimate the $\mathbf{p}_F$ distribution. But suppose that the decision maker can gain access to historical records on how well the forecaster has performed in the past; in particular, let us assume that the decision maker obtains the data in Table 3.9. This table gives data for 1,000 days showing the frequency counts for all eight combinations of the four probability statements with the two observed states of the weather. From these we can calculate the joint, conditional, and marginal probabilities.

Note from the total for column 3 that the fraction of time the weather freezes is 0.372; this agrees with the p_X value used in Section 2.2. But note that the fractions in the last column do not agree with the $f(i)$'s in column 2. For example, a 40% chance of freezing was forecast on 360 days, but freezing occurred on 45% of these days. To be "well calibrated" these percentages should agree (i.e., columns 2 and 6 should be identical). If the forecasts are not well calibrated, the observed frequencies of actual weather will disagree to a greater or lesser extent with the probability statements.

In general we shall not be so fortunate as to find complete agreement between the forecasts and the fraction of time the forecasts are realized. In Table 3.10 we normalize the counts in Table 3.9. The second column gives the fraction of time each probability statement was made, $\mathbf{p}_F$, and columns 3 and 4 give the fraction of time the weather was below or above freezing for each probability statement. The last two columns of Table 3.10 are the elements of the probability matrix $\mathbf{P}$.

TABLE 3.10
The marginal vector $\mathbf{p}_F$ and decision probabilities P

Probability statement, $f(i)$	Marginal distribution $p_F(f(i))$	Fraction of days	
		Below freezing	Above freezing
0.80	0.08	0.85	0.15
0.60	0.13	0.57	0.43
0.40	0.36	0.45	0.55
0.20	0.43	0.16	0.84

Using the estimates obtained from Table 3.10 we can then estimate the optimal expected loss as

$$l_P^* = r_2 \qquad\qquad \text{when } 0 < r_2/r_1 \leq 0.2,$$
$$= 0.07r_1 + 0.57r_2 \quad \text{when } 0.2 < r_2/r_1 \leq 0.4,$$
$$= 0.23r_1 + 0.21r_2 \quad \text{when } 0.4 < r_2/r_1 \leq 0.6,$$
$$= 0.30r_1 + 0.08r_2 \quad \text{when } 0.6 < r_2/r_1 \leq 0.8,$$
$$= 0.37r_1 \qquad\qquad \text{when } 0.8 < r_2/r_1 \leq 1.0.$$

The result is shown plotted (with $r_1 = 1$) by a heavy line in Figure 3-17. Notice that l_P^* is a piecewise linear function of the cost/loss ratio with discontinuities at the values in the forecast probability statements. It lies below the expected loss function when the cost–loss ratio is between 0.2 and 0.8 and a baseline forecast is used (see Section 2.2); in this range there is an economic advantage to using the probability forecast rather than the baseline forecast. The diagonal line corresponds to the case where we have perfect information on freezing and non-freezing outcomes. Even with improved forecasts, the decision maker cannot attain the increase in economic value provided by perfect information forecasts.

The reader may wonder why the curve obtained using a probability forecast has discontinuities. In Chapter 8 we show how these are caused by poor forecast calibration. Had the numbers in columns 2 and 6 of Table 3.9 been identical, these discontinuities would not have occurred.

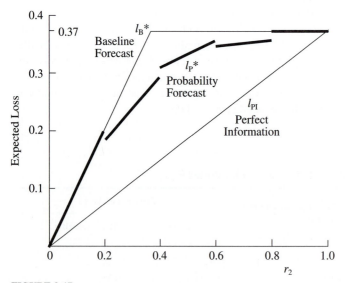

FIGURE 3-17
Optimal expected cost with probability forecast.

3.7   CREDIT SCORING DECISIONS

We now use probability forecasts in a credit scoring problem where individual applicants apply for credit at a bank or other lending institution. The objective is to accept or reject an applicant for credit based on the likelihood of his or her paying or not paying credit bills in a timely fashion. As suggested, there are a number of important decision problems in credit and finance where an individual is evaluated by a system that ranks his or her creditworthiness with a numerical score. Although the score is usually designed so that it predicts the odds of the applicant being a good risk, it can also be designed so that it provides the lender of funds with a probability forecast that the lender will make a profit.

The basic idea in developing a score is to evaluate factors such as age, address, job title, income, and other individual or institutional data, assign weights to the values of each of these factors, and then compute a scalar score that provides the decision maker with the probability that each individual is good or bad; thus, the score ranks the risk of the individual within a population of interest. When a financial institution determines its accept-reject policies so as to maximize the expected profit or yield on a portfolio of scored applicants, it is easy to show that under certain conditions the optimal policy is to use a single cutoff score; only those applicants whose score is at or above the cutoff are given credit.

Optimal policies and the expected profits also depend on the extent to which coherent predictions of good/bad performance of each applicant discriminate well and are well calibrated, two measures of forecast quality that are discussed in Chapter 8, where we show how optimal policies and portfolio profits are affected by miscalibration and poor discrimination and how the expected return and risk of the portfolio can be decomposed into calibration and discrimination components. In the following sections we assume that the predictions obtained from a scoring model are well calibrated.

3.7.1   Notation for Scores and Forecasts

The eventual performance of an individual is denoted by the random quantity X with possible outcomes $X = \{G, B\}$, where G and B denote good and bad, respectively. The precise meaning of good or bad varies considerably depending on the financial application and context of the decision problem. In some examples it may represent fraud or no fraud, bankruptcy or no bankruptcy. In others it could refer to individuals whose bills are paid or not paid within 90 days or whose credit terms have been violated. In what follows, scores of individual applicants are denoted by the outcomes of the random quantity S that can take on values s in $S = \{s_1, s_2, \ldots, s_n\}$.

A decision D is made with possible alternatives $\mathcal{D} = \{A, R\}$ for Accept or Reject the credit application. The result $R(X, D)$ is a random variable that depends on whether the application is accepted or rejected and on whether the credit applicant is good or bad. The payoffs in the result set are given by $\mathcal{R} = \{0, g, -l\}$, where g is a gain, l the size of the loss (a positive number), and

$$R(G, A) = g, R(B, A) = -l,$$

$$R(G, R) = R(B, R) = 0.$$

As before, expectations are denoted by E with or without a subscript to denote unconditional expectation or conditional expectation over the subscripted random variables. An example is the optimal expected result (profit in this example)

$$r^* = E_S[E_{X|S}[R(X, d^*)|\, S]]$$

$$= \sum_s E_{X|S}[R(X, d^*(s))|\, S = s]p_S(s), \tag{3.39}$$

which shows that r^* is the expectation over the distribution of scores of the conditional expected profit for a given score. The marginal distribution of scores is a vector $\mathbf{p}_S$ with elements $p_S(s)$; the marginal distribution of performance outcomes is a two-element vector $\mathbf{p}_X = (p_X(G), p_X(B))$ denoting the probability of being good and bad, respectively.

The conditional distributions of score given performance are elements of the matrix,

$$\mathbf{F} = [p_{S|X}(s|x)],$$

and those for performance given score are elements of

$$\mathbf{P} = [p_{X|S}(x|s)].$$

Both $\mathbf{F}$ and $\mathbf{P}$ are $n \times 2$ matrices; these are the forecast likelihood and decision probability matrices we discussed earlier in this chapter, where the columns of the former and the rows of the latter add to 1. The only difference is that the rows of each matrix are stated in terms of an individual's score rather than a probability statement. The elements of the $\mathbf{F}$ matrix express the conditional probabilities that particular scores will obtain given that the individual is good or bad. These likelihoods are often referred to as information odds even though they do not represent either prior or posterior odds. As before, $\mathbf{P}$ and $\mathbf{F}$ matrices are related by Bayes' rule.

As already mentioned, only $\mathbf{p}_X$ represents the real world and is independent of the particular scoring or forecasting system in use. All other marginal and conditional distributions described are affected to a greater or lesser degree by the forecast system in use and the environment in which it operates. For example, if some or all of the elements of $\mathbf{F}$ change with the use of a new forecasting technique or system, then some or all the marginal and conditional probabilities, except $\mathbf{p}_X$, will change. Only the latter reflects the operational environment and is invariant to the scoring system being used.

Because scores are equivalent to probability forecasts, we shall use the score, s, and the probability forecast, $f(s)$, interchangeably. From the definition of a score as log odds of an event (see Equation (3.30) (p. 108)), it follows that an applicant with score s has a probability

$$f(s) = \frac{1}{1 + e^{-s}} \tag{3.40}$$

of being good. Either s or $f(s)$ is transmitted by the forecaster to the decision maker. As was pointed out in Section 3.4.7 and expressed mathematically in Equation (3.30) (p. 108), scores on individual factors (such as age, marital status, etc.) of a credit applicant can often be expressed as the logarithm of the posterior odds of being good. The scores or weights of individual factor can be added to give a log-odds score, denoted by s, and equal to the logarithm of the posterior odds of being good. Rather than work with log-odds scores, it is customary to find an equivalent, s', that is a score used by the client that imposes a scale of odds doubling useful to the decision maker. It always involves a linear transformation of log-odds score. For example, a score of 200 may correspond to 20 to 1 odds or a log-odds score of 3.00, or a score of 100 may correspond to the population odds of the development sample under investigation.

Once the scoring model has been designed and estimates obtained, it is customary to identify each applicant of a given population with his or her log-odds score and then group applicants in score intervals as shown in Table 3.11. Column 1 groups the transformed scores. Columns 2 and 3 give the number of goods and bads, respectively, for a development sample that fell into each group, with the totals shown in column 4. Column 5 gives the log-odd score of a hypothetical applicant whose score lies in the middle of the interval of Column 1. We assume throughout that Equation (3.31) (p. 108) is a valid model. With score expressed as the logarithm of the conditional odds of being good, given the score,

$$ s = log\, O\,(G|s) = log\frac{p_{X|S}(G|s)}{p_{X|S}(B|s)} = log\frac{p_{X|S}(G|s)}{1 - p_{X|S}(G|s)}. \tag{3.41}$$

For example, consider the applicants with scores s' (column 1) between 180 and 189. The odds of an applicant being good are 851/63, and the logarithm of this ratio is 2.60, the entry found in column 5. These columns are equivalent, as s' is a linear transformation of s. Probability and expected profit calculations made in this chapter are based on the log-odds score s given in the fifth column. The conditional likelihoods that make up the **F** matrix are shown in the last two columns. They are calculated by dividing the counts in columns 2 and 3 by the respective column totals.

TABLE 3.11
Scoring data for credit applicants

Applicant score (s')	Number of goods	Number of bads	Total applicants	Log-odds score (s)	Forecast likelihoods $p_{SlX}(s|G)$	$p_{SlX}(s|B)$
Below 170	1386	281	1,667	1.60	0.147	0.493
170–179	760	70	830	2.38	0.080	0.123
180–189	851	63	914	2.60	0.090	0.111
190–199	940	48	988	2.97	0.099	0.084
200–209	905	39	944	3.14	0.096	0.069
210–219	920	25	945	3.61	0.097	0.044
220–229	931	17	948	4.00	0.099	0.030
230–239	857	12	869	4.27	0.091	0.021
240–249	770	7	777	4.70	0.081	0.012
250–259	586	5	591	4.76	0.062	0.009
260–289	433	1	434	6.07	0.046	0.002
290 and above	111	1	112	—	0.012	0.002
Total	9,450	569	10,019	—	1.000	1.000

We assume, as we did in the crop protection problem described in the preceding section, that the forecaster provides the decision maker with the scores or the corresponding probability forecasts from a development sample. In a development sample the way in which raw scores are grouped and in which these groups are transformed into log-odds scores (in other words, the design of the scoring system) are chosen so that the forecast is well calibrated, which means that $p_{X|S}(G|s) = f(s)$. With repeated use of the forecast, of all the times an applicant is predicted to be good with some probability p, the applicant will turn out to be good a fraction p of these times. Thus, in an identical forecasting environment, the decision maker can rely on and use these probability statements for the decision probabilities in a decision problem. In our example a log-odds score 2.97 (that corresponds to a raw score in the range 190–199) is equivalent to $f(s) = p_{X|S}(G|s) = 0.95$.

When these forecasts are applied to a new environment or portfolio that differs too greatly from that used in the development sample, the decision probabilities may not remain well calibrated, and optimal policies based on the original estimates of being good may not remain optimal when applied to a new portfolio. We discuss this situation in Chapter 8.

3.7.2 Expected Profit and Risk of an Individual

We can couch the decision problem in terms of an individual being either a good or bad risk. Recall that good results in a gain g whereas bad yields a loss $-l$. In real applications we would probably include random gains and losses that are score-dependent; for purposes of this chapter we assume that they are deterministic and constant. The decision tree for this problem is shown in Figure 3-18, where only four of the n decision saplings are labeled. The forecast results in a log-odds score s_i shown on the branches leaving the forecast node F. For a given s_i the conditional probability that an individual is good is $f(s_i)$ and the probability of bad is $1 - f(s_i)$, where $f(s)$ is the deterministic logistic function of Equation (3.40). These are used as decision probabilities in Figure 3-18.

There is an indifference probability $\bar{f} = f(\bar{s})$, that is equivalent to a cutoff score $\bar{s}$ (Equation (3.40)) such that the expected profit from an individual applicant (the risky investment) equals zero. This occurs when

$$\bar{f} = f(\bar{s}) = \frac{l}{l+g}. \tag{3.42}$$

With scores larger than the cutoff score, the expected profit for accepting the applicant is positive. Thus, if the decision maker wants to construct such a portfolio, the optimal policy, d^*, would reject applicants with scores less than or equal to $\bar{s}$. For example, the cutoff score when $l = 1$ and $g = 0.05$ yields $\bar{f} = 0.9524$, an odds of $O(G) = 20$ and a log-odds score of $\bar{s} = 2.996$, or a scaled score of 200 in Table 3.11. Using (3.42), the optimal expected profit can be expressed as a function of the score:

$$r(s)^* = E_{X|S}[R(X; d^*)|\ S = s] = 0 \qquad \text{if } s \leq \bar{s},$$
$$= f(s)g - (1 - f(s))l = (l + g)(f(s) - \bar{f}) \qquad \text{if } s > \bar{s}. \tag{3.43}$$

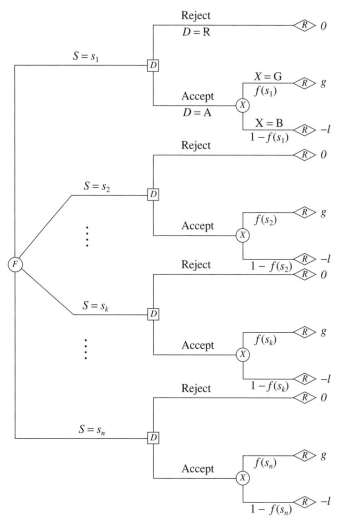

FIGURE 3-18
Tree using probability forecasts from credit scores.

We see that above the cutoff score, expected profit is linear in the sum of gain and loss multiplied by how much the probability of good differs from the indifference probability. It is common practice for institutions to determine subjectively a single cutoff score $\bar{s}$, refusing applicants with scores below cutoff and accepting those with scores greater than the cutoff. Here we think of cutoff scores and indifference probabilities in the context of zero expected profit so that acceptance of an individual with a score greater than $\bar{s}$ is equivalent to positive expected profit.

The risk of the individual to the institution under the cutoff score policy can be measured in several ways. One such measure is the conditional variance or standard deviation of the random return of that individual. Under the policy discussed, the conditional variance of profit is score-dependent and given by

$$V(s)^* = Var_{X|S}[R(X; d^*)|\ S = s] = 0 \qquad \text{if } s \le \bar{s},$$

$$= (g + l)^2 f(s)(1 - f(s)) \qquad \text{if } s > \bar{s}. \tag{3.44}$$

In portfolios of interest to credit-granting institutions the probability of each individual being good is so close to 1 that Equation (3.44) is dominated by the $1 - f(s)$ term; therefore, both the risk and the expected return decrease rapidly with increasing score.

A simple calculation shows that, when using this measure, a single individual is a very risky investment. Suppose that $l = 1$ and $g = 0.05$ as before, and the net return for the risk-free rate per dollar invested is 5%. The indifference probability is 0.9524, which means that the expected increase in profit over the risk-free rate for an individual with a log-odds score s of 3.22 corresponds to the probability $f(3.22) = 0.9616$ and optimal expected profit $r(3.22)^* = 1.05(0.9616 - 0.9524) = 0.0097$; by accepting this individual, the portfolio manager can increase the expected return by about 1% over the risk-free rate. On the other hand, the standard deviation of this profit is a very large number. Using Equation (3.44) we find that the standard deviation is

$$S.D.[R(X; d^*)|\ 3.22] = 1.05\ \sqrt{0.9616}\ \sqrt{0.0384} = 0.2018. \tag{3.45}$$

This level of risk for such a small increase in expected return would surely be intolerable for a portfolio manager. However, the expectation increases linearly and the standard deviation increases as the square root of portfolio size so that in a portfolio of 10,000 accounts the same calculations provide us with the numbers $r_P(3.22)^* = 97$ and $S.D.[R(X; d^*)|\ 3.22] = 21$. Thus, we see that a moderate-size portfolio can provide reward–risk ratios much greater than 1.

3.7.3 Expected Profit for the Portfolio

Thus far we have discussed only the rejection or acceptance of a single scored individual. The unconditional expected profit for the portfolio of booked accounts (i.e., those accepted) depends on (1) the marginal distribution, $p_S(s) = Pr\{S = s\}$, of scored individuals in the population of interest, (2) the probability of good given a score, $f(s) = p_{X|S}(G|s)$, (3) the gain or loss for good and bad accounts, and (4) the accept/reject policy $d(s)$.

Assume that the decision maker accepts any applicant with a score that yields $f(s) > \bar{f}$; this, together with our assumptions of score-independent gains and losses, ensures positive expected profit for each accepted individual and thus for the entire portfolio. If the forecasts $f(s)$ are applied to the population represented by the development sample, we must weight the expected profit of each individual in Equation (3.43) by the marginal distribution of scores to obtain unconditional optimal expected profit

$$r^* = \sum_s E_{X|S}[R(X;d^*)\ |s]\ p_S(s)$$

$$= (l + g)\left[\sum_{s > \bar{s}} f(s)\ p_S(s) - \bar{f}\sum_{s > \bar{s}}(s)\ p_S(s)\right]. \tag{3.46}$$

Again $p_S(s)$ is the score distribution and $\bar{s}$ corresponds to the cutoff score obtained from Equation (3.42).

We now return to the data in Table 3.11. These data are for a development sample representative of a population of goods and bads used to estimate scores and the likelihoods. Table 3.12 displays the distribution of scores in Column 1, the corresponding log-odds scores in column 2, and the fraction of the population having each score in column 3. The conditional probabilities of good and bad for each score (entries of the **P** matrix) are shown in columns 4 and 5. We emphasize that since $f(s) = p_{X|S}(G|s)$ these can be used interchangeably in this development sample credit portfolio. The sixth column contains the odds that an individual with a score falling in the interval specified by the row will eventually result in a good. If the loss to gain ratio l/g for this population is 20:1, the indifference probability is equal to 0.9524, which corresponds to an optimal cutoff score of 200 in Table 3.12. Thus, in this example applicants with scores below 200 would be rejected and those with scores greater than 200 would be accepted.

If the optimal cutoff score is used to accept incoming applicants, expected portfolio profits are obtained from Equation (3.46) multiplied by the total number of applicants in the population seeking credit. These calculations should be compared against two other cases, one representing the expected profit if every applicant is accepted, the other being the expected profit when, as a clairvoyant, we have perfect information on each applicant (i.e., whether the individual will be good or bad; see Problem 3.9). If every applicant is accepted, expected portfolio return is given by

$$r = gp_X(G) - l\, p_X(B). \tag{3.47}$$

If we have perfect information and only accept those we know to be goods, the expected return r_{PI} is

$$r_{PI} = gp_X(G).$$

TABLE 3.12
Score distribution and good/bad performance

Applicant score (s')	Log-odds score (s)	Score distribution $p_S(s)$	Decision probabilities		Odds $f(s)/(1-f(s))$
			$p_{X|S}(G|s)$ $= f(s)$	$p_{X|S}(B|s)$ $= 1-f(s)$	
Below 170	1.60	0.166	0.831	0.169	4.92
170–179	2.38	0.083	0.916	0.084	10.90
180–189	2.60	0.091	0.931	0.069	13.49
190–199	2.97	0.099	0.951	0.049	19.41
200–209	3.14	0.094	0.959	0.041	23.39
210–219	3.61	0.094	0.974	0.026	37.46
220–229	4.00	0.095	0.982	0.018	54.56
230–239	4.27	0.087	0.986	0.014	70.43
240–249	4.70	0.078	0.991	0.009	110.11
250–259	4.76	0.059	0.992	0.008	124.00
260–289	6.07	0.043	0.998	0.002	499.00
Over 290	—	0.011	0.991	0.009	110.11

The expected return in the portfolio in Equation (3.46) can also be expressed in terms of the forecast likelihoods as

$$r^* = gp_X(G) \sum_{s \leq \bar{s}} p_{S|X}(s|G) - lp_X(B) \sum_{s \leq \bar{s}} p_{S|X}(s|B), \qquad (3.48)$$

which can then be rewritten as the expected return obtained in accepting every applicant plus the positive contribution due to the ability of the forecasts to discriminate between goods and bads. By adding and subtracting terms in Equation (3.47), we can express the optimal expected portfolio profit as

$$r^* = (gp_X(G) - lp_X(B))$$
$$+ \left(lp_X(B) \sum_{s \leq \bar{s}} p_{S|X}(s|B) - gp_X(G) \sum_{s \leq \bar{s}} p_{S|X}(s|G) \right). \qquad (3.49)$$

The terms in the second line of Equation (3.49) tell us how much profit is added by discriminating forecasts because each sum is the cumulative likelihood function for scores larger than $\bar{s}$ conditioned on being good or bad. Because the portfolio manager can manage the portfolio by accepting the applicant or investing funds at a risk-free rate, the success or failure of the risky investment rests not only on the ability of scoring models to provide reliable predictors of individual performance but also on their ability to discriminate well between goods and bads.

3.8   SUMMARY AND INSIGHTS

A forecasting model can often use additional expertise or information that improve decision making. It is very important for the decision maker, the model builder, and the forecaster to understand how forecast models are integrated with decision making models; it is therefore important for the forecaster to understand the context and environment in which the forecasts will be used and how they can improve on a baseline forecast. While many forecasting *techniques* such as regression, moving average, exponential smoothing, and other time series techniques, are important and useful in some contexts, they are seldom designed with a decision-making problem in mind. Our focus is on understanding the different contributions made by probability, odds and categorical forecasts, with emphasis on how these differences affect the use and implementation of a forecast in making decisions. The most important requirement is that forecasting and decision making be a *coherent enterprise with consistent objectives*.

A *probability forecast* is a forecast of the *probability that an uncertain event will occur*. This is in contrast to a *categorical forecast*, where the item that is forecast is the *particular category that the event of interest will fall into*. Using the familiar example of weather, a probability forecast of rain might say that there is a 40% chance of rain tomorrow, whereas a categorical forecast might say that tomorrow the rainfall will be between zero and one-tenth of an inch; it is usually unstated but assumed that this will occur with probability 1.

The reader must clearly distinguish between the probability that a forecasted outcome is correct given the actual outcome of the uncertain event, and the probability that the event of interest will occur given the forecasted outcome. The first of these is called a *forecast likelihood* with an associated matrix of conditional probabilities **F**. The second is called a *decision probability* with an associated matrix of conditional probabilities **P**. One can think of the **F** matrix as summarizing the historical performance of the forecast after actual outcomes have been observed, whereas the **P** matrix gives the expected performance of the forecast in predicting the future. It is the probabilities in the **P** matrix that are used in a decision problem. As a simple example, suppose that a diagnostic screening test for a medical condition correctly identifies the condition 99% of the time when it is present but that 5% of the time it will indicate that the medical condition is present when it is not. The 99% and 5% when divided by 100 are two elements of the **F** matrix. This may seem like an excellent test, but when used in a population where it is believed that only 1 in 1,000 has the condition, if a test on a random person shows positive, the chances they have the condition are about 2% (0.0194 is an element of the **P** matrix). This says that on average about 98 out of every 100 people that test positive do not have the condition, so for this population it not a very useful screen. If the same test (forecasting procedure) were used in a population where 1 in 100 are thought to have the condition, the chances that a person who tests positive has the disease increases to 16.7%. Clearly the usefulness of the test or forecast depends on the particular application.

The decision probabilities **P** and the forecast likelihoods **F** are related by the laws of probability using Bayes' rule. When finding **P** from **F** or **F** from **P**, the same procedures apply for both probability and categorical forecasts; only the detailed mathematical notation is different. The way that Bayes' rule manifests itself in an influence diagram is in the reversal of a directed arc between two chance nodes. In complex problems where more than two random variables are present, influence diagrams with only chance nodes can be of considerable benefit in finding a feasible and efficient sequence of applications of Bayes' rule to determine a particular conditional probability (see Chapter 9).

Just how a forecast is used in a particular decision problem depends on many factors, including the type of forecast (probability or categorical) and the connection, if any, between forecaster and decision maker. We demonstrated the application of probability forecasts to two decision problems. The first was in agriculture where weather plays a crucial role in the outcome. In this case the forecaster is likely to be an agency that provides forecasts to a very wide variety of users for an even wider variety of applications. In such a case it is not realistic to assume that the agency can tailor its results to a particular decision problem. Rather, the decision maker and modeler must determine how to best use the available forecast. The second was an example of predicting risk in credit applications, a more specialized field with a smaller number of potential users. It is usual in such cases for the forecaster to work much more closely with a client to integrate a forecast procedure with a particular application. This possibility raises the general question of how such integration should be accomplished, and that in turn suggests that we know when one forecasting procedure performs better than another. This interesting and important topic is covered in Chapter 8.

There is an extensive forecasting literature with thousands of journal articles and over 100 books on subjects ranging from statistical techniques, curve fitting, time-series methods, regression, to the use of large data bases and numerical procedures for estimating distributions. There has also been considerable interest in the combination of forecasts to achieve improved predictors although the decision making context of such models is unclear. A very small body of the literature deals with forecasting and decision making in a coherent formulation and with the development of forecasting models that are an integral part of the overall decision making problem.

The use of weather forecasts in decision making has been extensively studied by Murphy and Winkler (1987, 1989, 1991, 1992).

The medical literature is rich in examples of decision making and in the use of forecasting and prediction methods that are derived from medical data. See, for example, Weinstein and Fineberg (1980). The reader is also encouraged to refer to the *Journal of Medical Decision Making* and additional references given in Chapter 8.

For an excellent exposition and nonmathematical treatment of scoring methods in credit management see Lewis (1991). For a survey of papers dealing with various topics in the analysis and use of credit scoring systems see Thomas, Crook, and Edelman (1992) and Crook and Edelman (1992).

PROBLEMS

3.1 Starting with the forecast likelihoods **F** for the minimum temperature problem (Equation (3.21) (p. 100)) use Equation (3.17) (p. 99) to find the decision probabilities **P**, and check your results against Equation (3.20).

3.2 From Equation (3.12) (p. 97) show that

$$p_{F|X}(f(i)|j) = \frac{p_{X|F}(j|f(i)) p_F(f(i))}{p_X(j)},$$

and use this to calculate **F** from **P** for the rain example in Table 3.4 (p. 98). Check your answer against Equation (3.16) (p. 98).

3.3 In Table 3.3 (p. 94) suppose that the category "minimum temperature in the interval $(-4°, 0°)$" is of particular interest, and let $Y = 1$ if this statement holds, $Y = 0$ if it does not. Also let $G = 1$ if the statement is forecast to hold, $G = 0$ if not. Find $Pr\{Y = 1|G = 1\}$ and $Pr\{Y = 1|G = 0\}$.

3.4 Verify the conditional probability formulas in Equation (3.34) (p. 110).

3.5 Suppose that we select a pair of tests for detecting colon cancer: (1) *SIG* test following a *DE* test if *DE* = "nc", and (2) *BaE* test following an *OB* test if *OB* = "c". You are given the test outcomes and asked to calculate the conditional probabilities of the presence of colon cancer given the possible outcomes. These are contained in the following matrices $\mathbf{P}_{C|SIG,DE}$ and $\mathbf{P}_{C|BaE,OB}$.

a. Show that the probabilities of cancer under this testing sequence are given by the decision probability matrices shown below. Note that we have three rather than four rows because we do not make a follow-up *SIG* test if the *DE* test is positive or a *BaE* test if the *OB* test is negative.

Let the costs of the tests be *OB* – $1, *DE* – $4, *SIG* – $50 and *BaE* – $160.

b. Show that test sequence (1) finds 85% of cancers at a cost per patient of $54.00, whereas test sequence (2) finds 80% of cancers at a cost per patient of $17.11.

$$\begin{array}{cc} & \text{c} \qquad\quad \text{nc} \\ \begin{array}{c} \text{``c,\_''} \\ \text{``nc,c''} \\ \text{``nc,nc''} \end{array} & \begin{bmatrix} 1.0 & 0 \\ 1.0 & 0 \\ 0.00015 & 0.99985 \end{bmatrix} \end{array}$$

$$\begin{array}{cc} & \text{c} \qquad\quad \text{nc} \\ \begin{array}{c} \text{``c,c''} \\ \text{``c,nc''} \\ \text{``nc,\_''} \end{array} & \begin{bmatrix} 1.0 & 0 \\ 0 & 1.0 \\ 0.00022 & 0.99978 \end{bmatrix} \end{array}$$

(1) $\mathbf{P}_{C|DE,SIG}$ \qquad\qquad (2) $\mathbf{P}_{C|OB,BaE}$

Conditioning on results from two diagnostic tests.

3.6 For the colon cancer problem, construct the event tree that corresponds to the chance influence diagram in Figure 3-10 (p. 112), calculate all the conditional and joint probabilities, and verify the structure of the influence diagram structure.

3.7 From the original colon cancer data (p. 112), calculate all decision probabilities used on branches of the second decision subtrees in Figure 4-33 (p. 178). *Hint*: Use the answer to Problem 3.6.

3.8 a. Using the data in Table 3.9 (p. 119), verify the plot of optimal expected cost l_P^* shown in Figure 3-17 (p. 120).

b. Plot the conditional cdf $Pr\{F \le f(i)|X = 1\}$ versus the conditional cdf $Pr\{F \le f(i)|X = 0\}$ for the data in Table 3.4 (p. 98), and for the data in Table 3.9 (p. 119). On the basis of these two plots can you decide which forecast discriminates best.

3.9 Calculate the expected profit in the credit portfolio (data in Table 3.12 (p. 127)) with $l = 1$ and $g = 0.05$

a. using the optimal cutoff score,

b. when all applicants are accepted without regard to score,

c. with perfect information and only those known to be good are accepted.

What is the increase in profit due to the use of forecast scores? (See Equation (3.49) (p. 128))? How much more profit could be obtained with perfect information?

3.10*The following table lists the number of days in each month on which a pollution exceedance is recorded at two sites 1 and 2. Each site is downwind of a large refinery. On careful examination of the data it is found that exceedances at Site 2 occur *only* on the very same day an exceedance has occurred at Site 1.

Counts of pollution exceedances

Month	J	F	M	A	M	J	J	A	S	O	N	D	Total days
Site 1	0	7	0	1	3	9	18	7	2	15	5	3	70
Site 2	0	1	0	1	1	3	8	4	1	1	0	0	20

These data suggest a model in which there is a daily exceedance at Site 1 with unknown probability θ and, given an exceedance at Site 1, an exceedance at Site 2 occurs with probability π. In what follows assume that each month has 30 days, that a priori π and θ are unconditionally independent of each other, and that each has the uniform distribution. Let X and Y denote the observed cumulative number of exceedances at Sites 1 and 2 after N days and ΔX and ΔY be the unobserved future exceedance counts in the following K days.

a. Draw the chance influence diagram that illustrates how the parameters π and θ, the historical cumulative counts and future counts at both sites influence each other before X and Y have been observed.

b. Redraw the diagram in (a) after X and Y have been observed.
c. Draw the chance influence diagram showing how X and Y influence ΔX and ΔY after integrating out the parameters π and θ.
d. Derive the *predictive distribution* $Pr\{\Delta Y=y|\Delta X = k, X = m, Y = n\}$.
e. Find the conditional expectation and variance of Site 2 August counts given only the historical counts at both sites through July. How does this prediction agree with the actual count in August?

CHAPTER
4

MODEL
BUILDING

4.1 INTRODUCTION

In this chapter our goal is to show how, starting with a blank sheet of paper, one can use a set of tools and follow a sequence of steps that lead to a model of a decision problem and insights into the likely solutions. Because model building is to a large extent an art, the reader will not find a rigid set of steps in the form of a recipe but, rather, a structured procedure that will assist in maintaining a logical thought process while including the important goals and constraints of the problem.

A book on decision making would be deficient without attempting to help the reader in the difficult but crucial area of model building. We believe that becoming proficient in the art of model building is at least as important as learning and practising the techniques of mathematical analysis and algorithms to which the majority of the decision theory literature is devoted. One must strike a balance between the efforts placed on developing and improving solution methods and those placed in model building.

It is our belief that in modeling a real decision problem one should employ the simplest model that captures the important aspects of the problem. This efficiency in modeling is a central feature of the scientific method, and is perhaps best summarized in the principle of parsimony as stated by Tukey [1961]:

> It may pay not to try to describe in the analysis the complexities that are really present in the situation.

It is often the case that a relatively simple model with only a few parameters can be found that closely represents the real world problem. The need for this is particularly important in decision modeling. As we have stated before, the obvious but often forgotten fact is that decisions are made by decision makers. A decision model is not intended to replace the decision maker. Its usefulness is limited to aiding this person understand the structure and implications of alternate decision choices. If the model is so complex, or if it contains so many parameters and built-in dependencies, that the decision maker finds it difficult to understand, it will probably have no impact in the decision making process. We have no interest in building models for their own sake.

It is important to study and apply solution techniques that yield insight about the structure of the problem as well as serve to calculate the best course of action. The two graphical tools we use in building models are the influence diagram and the decision tree. The reader has already had some introduction to these in Chapters and . The basic elements of influence diagrams were shown in Figure 1-6 (p. 27), and simple influence diagrams were illustrated in Figures 1-7, 2-2, 2-4, and 2-6. Simple decision trees are shown in Figures 1-8, 2-3, 2-5, and 2-7. In this chapter these basic notions are extended so that more complex problems can be modeled.

Section 4.2 describes a general procedure for constructing and modifying influence diagrams to model the *structure* of a decision problem. By structure we mean not only the timing of events and the identification of uncertainties but also the probabilistic relationships among decisions and uncertain quantities or events. Decision models are illustrated extensively in Section 4.3 where we illustrate examples in the areas of hurricane seeding, bank credit accounts, naval mobile basing systems, and colon cancer diagnosis. Our intent is to introduce the reader to the iterative model building process which is crucial to producing useful and accurate abstractions of a real situation. It is rare that one's first attempt at model building is successful. Indeed, one of the reasons for using influence diagrams is to ensure accurate modeling of dependencies through a graphical representation and to have a framework in which to confirm or reject assumptions.

Section 4.4 describes how to construct decision trees from influence diagrams and how to obtain the "best" decisions, those that maximize or minimize some objective given the data and assumptions for the problem. We thought of naming this chapter "Model Building and Model Solution" but felt that this would reinforce the prevalent view that models produce "the optimal solution." If this were the case, why have a decision maker? Such a person could simply endorse the analyst's solution. The model builder and analyst should take the far more humble view that they can find the optimal decisions, *given the data and assumptions they have used* (which may or may not be accurate), and so help the decision maker narrow down the set of alternatives to be considered. Our definition of model solution includes in-depth sensitivity and policy analysis to important model parameters, which is the subject of Chapter 5. To avoid misleading the reader into thinking that finding the best solution for a given decision tree is in fact the solution to the problem, we left "Model Solution" out of the chapter's title, although we include what is known as the *rollback* algorithm in Section 4.4. Sections 4.5 and 4.6 illustrate the design and use of decision trees that are consistent with the influence diagrams used in Section 4.3; these examples include the bank credit and colon cancer test with multiple decisions as well as a popular game show quiz in Section 4.7 that is formulated for the first time as a decision problem.[1]

We often find it necessary to introduce a problem in an early section and revisit the same problem as we progress through the chapter. Section 4.8 is included to partially offset these breaks by presenting the formulation and solution of a problem from start to finish in one section. The example chosen arises in government agency operations where a financial plan must be developed and agreed on at the beginning of the fiscal year with considerable uncertainty as to the actual level of financial resources that will materialize as the plan is carried out through the year.

4.2 CONSTRUCTING INFLUENCE DIAGRAMS

Influence diagrams help the decision maker and analyst visualize graphically how decisions, uncertain events, and results are interrelated. They ensure an appropriate model *structure*, serving as precursors to the more familiar techniques of describing decision problems with decision trees, event trees, or fault trees. Influence diagrams offer a compact form and global look at the probabilistic structure of decision problems, the timing of available information and the interdependence of decisions that can be taken and uncertain outcomes that may arise. Although they do not give detailed information on the outcomes of chance events or the specific decision choices, they make three important contributions.

The first, and probably most important contribution, is that influence diagrams provide a framework in which experts and decision makers can discuss problem structure and dependencies without invoking formal mathematical or statistical notation and analysis. Thus, they serve to link the real world of decision making with that of the scientifically trained analyst. The decision maker focuses on the problem and the possible implications of his or her decisions. The analyst is usually most comfortable in the world of mathematical modeling, giving advice on techniques and tricks that can be used to solve well-posed mathematical problems, or techniques of inference, prediction, and optimization. Such a person may not spend much time wrestling with real decision problems or have much experience in designing and describing different model structures that can be used to help solve them. To quote Howard (see Chapter 1 of Oliver and Smith (1990)):

> *Over the years, experience with the influence diagram has shown that it is an effective means of communicating with both decision-makers and computers. The influence diagram has proved to be a new 'tool of thought' that can facilitate the formulation, assessment, and evaluation of decision problems.*

---

[1] Many of the decision trees shown in this chapter are produced by the TreeAge© software DATA©.

The second important contribution is in the reduction of large volumes of data to that portion essential and relevant to decision making. In today's world the availability of large data sets on every imaginable topic has become a reality. The ease, efficiency, and low cost of information processing present new and difficult challenges to both the decision analyst and decision maker. It is more important than ever before to discard what is irrelevant and keep only what is relevant for decision making. Even with their high speeds and high-capacity storage capabilities, computers cannot reveal how decision models should be structured and what portions of the available data are relevant for making decisions. Insights on such issues can be revealed through the use of influence diagrams.

Finally, we must acknowledge the important contribution of algorithms and numerical techniques that can be used to reduce the complexity and size of influence diagrams. An influence diagram is able to provide sensitivity analyses and to reveal the importance of alternative model structures, forecasts, and policies to the decision maker. In the remainder of this chapter we illustrate the design and construction of several decision problems in terms of influence diagrams and offer some simple procedures for their formulation and analysis.

To design and draw an influence diagram one needs first to agree on a logic and a notation to represent the uncertain events, decisions, and outcomes of the problem. We start with the notation introduced in Chapter 1. There are three types of nodes:

1. Circular nodes that represent uncertain events,
2. Square nodes that represent decisions,
3. Diamond nodes that represent results.

We represent the influence or relevance between two events by connecting their nodes with a directed arc. The arc direction corresponds to the perceived influence. The reader is referred to Figure 1-6 (p. 27) with its six parts (a) through (f). Part (a) shows two uncertain events X and Y, that the probability distribution of Y can be influenced by the random event X, and that the value of X is known when the distribution of Y is assessed. Alternatively we say that the random variable Y is dependent on the random variable X, or that X is relevant to Y, or that X influences Y. This does *not* mean that X *causes* Y, rather that the knowledge of X alters the conditional probability of Y. Part (b) shows that an uncertain event X occurs prior to a decision D; the directed arc shows that the value of X is known to the decision maker when D is made, and that the value of X probably influences the decision. If the order of X and D are reversed in time so that X occurs after the decision, the directed arc in part (c) shows that the probability distribution of X is influenced by the decision. In part (d) we represent the influence of the first of two distinct decisions, D_1, on a later decision D_2. Finally in parts (e) and (f) the directed arcs indicate that the result of the decision process is affected by the outcome of the uncertain event in (e) and by the decision alternative taken in (f).

4.2.1 The Procedure

The following is a simple sequence of steps for drawing an influence diagram that describes the structure of a decision problem:

1. Start with a preliminary list of the decisions and a list of random events or random quantities of interest whose outcomes you believe to be important in problem formulation. Identify the attributes and objective(s) that will be used to measure the result (often cost or value are used) of the decisions and outcomes.

2. Name each random quantity and decision; pay special attention to the clear meaning, definition, and measurement units of each such quantity. Give each one a unique number, and, where possible, order them in time with respect to each other.

3. Represent each random quantity with a round node and each decision with a square node, and draw them in time order of occurrence from left to right.

4. Identify any influences or dependencies between random quantities and decisions that you believe are important. Define as clearly as possible what each outcome or action means and what it depends on. Insert directed arcs between nodes that influence one another with the direction corresponding to what you believe to be the natural influence, using the meanings of directed arcs as outlined.

5. Quite often, a random event or decision A is conditionally independent of (is not influenced by) another decision or event B, given a third event or decision C; in such a case one need not draw an arc from B to A, even though C and B and C and A are connected by directed arcs. Always confirm your belief that the conditional independencies represented by the influence diagram are correctly stated.

6. Check to see that there are no directed cycles in your diagram (a connected set of arcs in a directed path that leads out of one node and back to itself).

7. Be certain that any decision that occurs before a later decision has a directed arc from the former to the latter. Also, any chance node known to a given decision node must be known to a later decision node. This is a requirement of the principle of coherence.

If during the design phase you are unsure about the internal connections or dependencies between nodes in your influence diagram, draw alternative diagrams and try to identify where and how the graphs differ and what it is about these different assumptions that make you prefer one diagram over another. *If you are acting as an analyst rather than the decision maker, do this together with the decision maker to ensure meaningful formulation*s. Make a list of the different conditional independencies and the timing of available information for decision making and their implications in the diagrams. Do the conditional probability assessments change? If so, what do they mean and how do they affect decisions? If there is no agreement about the definition of certain random quantities, it may be important to reformulate the problem in terms of newly defined quantities on which there is agreement. Compare these with the original.

Keep the formulation as simple as possible by reducing to a minimum the number of nodes and directed arcs. Whenever possible and appropriate, assume conditional independence rather than conditional dependence. Greater detail and more complex dependencies can be added once the underlying structure of the problem is well understood.

The spirit of the model builder should be that of an experimentalist, trying different designs, arguing for and against different assumptions, and constantly testing the implications of different assumptions about the structure on conditional probabilities. One should not be married to a particular influence diagram structure any more than one should believe that all forecasting techniques are represented by regression models, that all optimization problems are linear programs, or that there is only one correct way to formulate and explain a decision problem.

The reader is cautioned that the conventions used in influence diagrams by different authors and software developers are not standard. In particular, we point out that our depiction of the timing of events from left to right is not always standard in influence diagrams, although it is in decision trees. We use the same timing convention for both so that the reader can better relate these two important but different aspects of the modeling procedure.

We end this section with a simple example. Consider an engine maintenance problem in which we are trying to decide whether or not to keep as is, to maintain, or to replace the engine with a new one. The proposed procedure is:

1. Perform a diagnostic test on the engine. With the result of this test known,
2. The decision maker must make one of the above three choices.
3. After action has been taken, the performance of the engine is observed in operation, and
4. The result of the decision is calculated.

We need four nodes to represent these four items. The result of the test (node 1) and the performance of the engine (node 3) are both represented as uncertain events with round nodes, the decision (node 2) with a square node, and the result with a diamond node (node 4). Figure 4-1 shows these ordered in time from left to right.

A directed arc is drawn from 1 to 2 to indicate that the result of the test is known to the decision maker before the decision is made. The arc from 1 to 3 indicates that the result of the test is known when the probability of engine performance is assessed and will probably affect this assessment. The arc from 2 to 3 shows that the probability of engine performance depends on the decision made. The two arcs leading into 4 show that the result depends on both the performance of the engine and the decision taken. Note that Figure 4-1 includes five of the six influence diagram elements shown in Figure 1-6 (p. 27).

There are other possible influence diagrams for this simple problem. Suppose that the result is measured in units of monetary loss and that an engine failure leads to very high losses compared to the cost of performing maintenance. In this case the arc from 2 to 4 could be removed as the result depends only on the true performance of the engine. Note that adding an arc from 1 to 4 would indicate that the outcome of the test affects the result. But given the decision made and the actual performance of the engine, the result is conditionally independent of the test result. Even if the cost of the test were significant compared to other costs in the problem, because it is always incurred, it is a constant and need not be included in the result.

Another feature sometimes makes problem formulation more difficult than one might expect. The information about probabilistic structure, that is, dependencies and

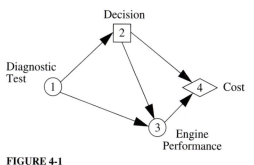

FIGURE 4-1
Influence diagram for engine maintenance problem.

conditional probabilities, may not be given or known in an order that seems natural or useful in obtaining a solution. For example, in Figure 4-1 we may only have data on engine performance that, for each action taken, gives us the fraction of times (conditional probability) that a certain test result was obtained. This suggests that the arrow from node 1 to node 3 should be reversed. If the analyst structures the problem in this way, note that we have a cycle in which we can follow the arrow directions in going from 1 to 2 to 3. This is not allowed (see step 6 on p. 137) because the diagram would say that we knew the engine performance (node 3) before the decision maker chose an alternative (node 2); obviously this contradicts the rationale for making the decision in the first place. Arc reversal (discussed in Section 3.4.6) may lead to cycles, and we therefore devote a short section to this potential difficulty.

4.2.2   Arc Reversal and Cycles

In this section a number of properties of influence diagrams are discussed that lead to proper formulations in addition to ensuring correct depiction of the timing and knowledge of events. Some of these have been mentioned or illustrated already, but they are gathered here for emphasis and clarity.

Consider a decision problem where two or more decisions are to be made sequentially. Figure 4-2 illustrates a problem where first a forecast is obtained on some chance event of interest and where, depending on the outcome, an initial decision is to be made as to whether to take action or obtain a second forecast. An example of this might be a visit to your doctor as a result of some disturbing and/or debilitating chest pains. After examination your doctor may diagnose clogged arteries (F) and recommend a number of procedures such as bypass surgery or angioplasty, predicting a certain probability of success in relieving the condition for each one. You must now decide (D_1) whether or not to proceed with one of the suggestions or seek a second opinion (G) from another doctor. If you decide on the latter you must then make a decision as to what action to take. The chance node X represents your true (but unknown at the time either decision is made) arterial condition.

First, the reader will note that the direction of the arcs connecting X to G and F appear to be in the wrong direction. Because X occurs later in time than either F or G,

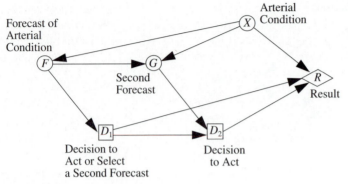

FIGURE 4-2
An improper two-stage influence diagram.

why have we drawn the directions of the arcs as shown? The reason is that often the probabilities available or assessments that can be made are in the form of a forecast having a certain outcome, given the true condition of the event being predicted; in our case Figure 4-2 shows that we can assess the conditional probabilities $Pr\{F = f \mid X = x\}$ and $Pr\{G = g \mid X = x, F = f\}$. This last probability indicates our judgment as to the qualifications of the doctors in making a correct diagnosis. It says that we expect the diagnosis of the second doctor to be dependent on that of the first. In fact, if diagnostic procedures were perfect and doctors infallible, then $Pr\{G = g \mid F = g\}$ would be 1 and $Pr\{G = g \mid F \neq g\}$ would be 0.

Before Figure 4-2 can be used as a basis for building a decision tree, we must reverse the arcs from X to F and G. As pointed out in Chapter 3, the order in which this is attempted is important. Note that if we attempt to reverse the arc from X to G first, that is, if we try to calculate $Pr\{X = x \mid G = g\}$, a directed cycle would be created. Such cycles are not allowed as they represent infeasible factorizations of joint probabilities. To avoid creating a cycle, we first reverse the arc from X to F (find $Pr\{X = x \mid F = f\}$), and then the one from X to G. Influence diagrams can be a great help in determining a feasible sequence of conditional probability calculations when more than two random variables are involved. Although it is usually easy to reverse directed arcs by direct examination in simple two- or three-node problems, it becomes much more complicated in larger influence diagrams. A more formal mathematical treatment of this difficult problem is deferred to Chapter 9.

4.2.3 No-Forgetting Arcs

The influence diagram in Figure 4-2 cannot describe a meaningful decision problem because there is no directed arc from the forecast node F to the second decision node D_2. This means that information that was known at the time of the first decision is "not remembered" when the second decision is taken. This is an unrealistic rejection of information that may be relevant at the time of the second decision; one of several requirements for a proper influence diagram is that information known at an earlier de-

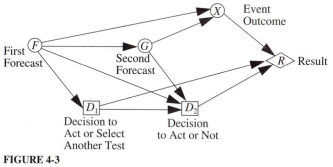

FIGURE 4-3
A proper two-stage influence diagram.

cision is always remembered when a later decision is taken. To be consistent with the principle of coherence recall that appropriate "no-forgetting" arcs must be included in influence diagrams:

> If a chance or decision node is connected by a directed arc into a decision node in an influence diagram, then it must be connected to all nodes corresponding to subsequent decisions.

In Figure 4-3 we introduce the arc from F to D_2. We now have a model that is a valid representation of the decision problem because information known when the first decision is made is also known to the second decision.

4.2.4   Perfect Information

As part of building and testing a model an analyst is often interested in the effect of knowing the values of one or more uncertain quantities that follow a decision. Thus, uncertainty in the quantity is replaced by perfect information. Can one analyze the problem assuming such quantities are known before the decision is made? When this is possible, one can formulate the problem on an influence diagram closely related to the original.

We have already illustrated this idea with Nancy and Ron's betting problem in Section 2.3 (p. 48). When the coach's decision is known the Y node lies to the left of the decision node D as shown in Figure 2-4 (p. 51). When Nancy does not know the coach's decision (Y is uncertain) it lies to the right of the decision node as shown in Figure 2-6 (p. 53). In neither of these figures is there an arc connecting nodes D and Y which means that the probability distribution of Y is *independent* of the decision.

Returning to our engine performance example in Figure 4-1, suppose that the decision (node 2) does not influence performance (node 3). The correct influence diagram is now shown in Figure 4-4. Suppose we wish to investigate the expected cost of having perfect information about engine performance. Move node 3 in front (to the left) of node 2 and add arc (3,2) so that engine performance is known before the decision is made. This yields the influence diagram in Figure 4-5(a) where the action and outcome sets remain unchanged. A moment's reflection should convince the reader that one can delete

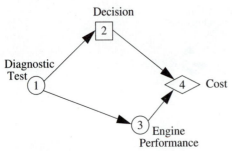

FIGURE 4-4
Engine performance independent of decision.

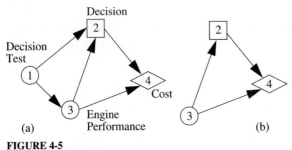

FIGURE 4-5
Perfect information on engine performance.

node 1 (Diagnostic Test) and the two directed arcs out of it; if one has perfect information on engine performance, one can no longer make use of the test that predicts engine
performance! Thus, an "equivalent" influence diagram is one where only the subgraph
of Figure 4-5(b) with nodes 2, 3, and 4 remains but node 1 is deleted.

In the original problem depicted in Figure 4-1 there is an arc from node 2 to node
3; this indicates that the probability distribution of engine performance may depend on
the decision. Consider the effect of knowing the engine performance (node 3) before the
decision to overhaul or replace (node 2) is made. A new difficulty arises. If we were to
move node 3 before node 2, as we have done in Figure 4-5(a,b), we would create a cycle
with nodes 2 and 3; as we know from Section 4.2.2, cycles are not allowed. What is the
meaning of this apparent contradiction, and can we get around the difficulty?

Suppose that there are three outcomes for engine performance at node 3 corresponding to completely satisfactory, minor failure, and catastrophic failure. Because we
are assuming that the decisions influence the outcomes, we draw three distinct influence
diagrams, one for each known outcome. Again, we only need concern ourselves with
nodes 2, 3, and 4 as a diagnostic test or forecast of engine performance plays no role
when there is perfect information about that performance.

With decision-dependent performance probabilities, each outcome is represented
by its own influence diagram in Figure 4-6; the random node for engine performance is
shaded to emphasize that its outcome is fixed and known to the decision maker. Because
it is no longer possible to assign prior probabilities to engine performance outcomes, as

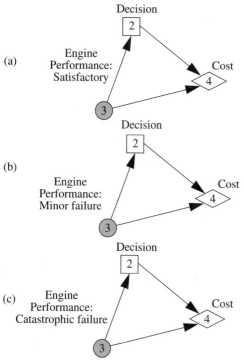

FIGURE 4-6
Perfect information given engine performance.

was the case in Figure 4-5 when the outcome was assumed to be independent of the decision taken, the calculation of expected value of information is not meaningful. With each outcome the decision maker selects the least costly decision among deterministic alternatives. This construction is clarified further when we consider how the associated decision trees are solved.

4.3 EXAMPLES OF MODEL FORMULATIONS

This section includes a number of examples in which we construct influence diagrams. We demonstrate the design decision models in the areas of hurricane seeding, bank credit accounts, naval mobile basing systems, and colon cancer diagnosis. Unlike the simple example in the previous section, we do not propose and draw one diagram for a problem, but illustrate alternative models and model simplifications. Our intent is to introduce the reader to the iterative model-building process that is crucial to the design and development of useful abstract descriptions of a real situation. The iterative aspect is necessary for producing acceptable models as it is rare that one's first model-building attempt is successful. One should not strive for "the" model. Indeed, one of the reasons for using influence diagrams is to ensure accurate modeling of dependencies through a graphical representation that allows different conditional beliefs and assessments. Whenever possible, influence diagrams should be developed in close cooperation with the decision maker.

4.3.1   A Decision to Seed Clouds in Hurricanes

Consider the problem of seeding a hurricane in the hopes of reducing its intensity. This problem has been studied in a well-known paper by Howard, Matheson, and North (1972). It is an example of a decision to be made at government policymaking levels that could have significant economic and social impact; although their original formulation did not make use of influence diagrams, we introduce them here to illustrate their use and great value in the model-building process.

Briefly stated, the problem is as follows: an offshore hurricane is being tracked and its position, wind speed, and direction are measured. Twelve hours later the hurricane is expected to pass over land at a densely populated area; the hurricane winds and resulting tide surge may result in loss of life and serious property damage. Experiments have shown that cloud seeding may lower wind speeds (also, possibly alter the direction) and thereby reduce the serious consequences that would occur if the hurricane were allowed to continue unabated. Unfortunately, cloud seeding may also increase wind speeds and make matters worse. One important point to keep clear as we build a model is the specific problem being considered. There is the strategic problem of setting government policy on whether or not to allow seeding hurricanes in general, and the tactical decision as to whether or not to seed in a particular case. We are discussing the first of these problems.

At the outset it is important to think through how the result of the decision is to be measured. In most real problems results are measured using a number of criteria. Three criteria that could be used for this problem are (1) property damage, (2) lives lost, and (3) legal liability losses of the government. The second of these depends significantly on other policies such as evacuation. Quantification and estimation of the third is difficult. Even with seeding, hurricane winds will probably increase in the twelve hours preceding landfall; it is expected that the increase will be less than it would be with no seeding, but arguing or proving such a point in legal proceedings would no doubt prove difficult. For simplicity in our example we measure results by property damage.

After initial discussions with meteorological experts it is decided that three important variables must be included in the formulation and analysis of the decision problem. A decision variable, say D, offers a choice of whether to seed or not seed. Two random quantities that need to be included are the wind speed observed offshore prior to seeding, say U, (such observations may be used for forecasting), and the wind speed at some future time when the hurricane hits land, say V. The possible outcomes of U and V might be the set of nonnegative numbers or a small number of speed intervals such as low, medium, or high. The choice of how to measure these outcomes does not have to be made at this stage; an influence diagram can be drawn without such details. As was stated, we also assume that the consequence of seeding or not seeding is measured by the cost attributed to damage done by a hurricane, say R. This in turn is related to the actual wind speed at landfall.

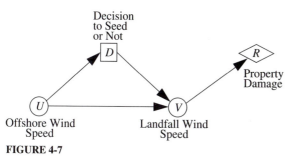

FIGURE 4-7
Initial hurricane seeding influence diagram.

The time sequence of events is: Measure the *offshore wind speed U*, make decision *D*, observe *the landfall windspeed V*, and measure *the property damage R*. Figure 4-7 shows four nodes representing these events in this sequence from left to right. The directed arc from *U* to *D* indicates that the offshore wind speed *U* is known when the seeding decision *D* is made. The arcs from *U* to *V* and *D* to *V* indicate that the probability distribution of landfall wind speed will depend on the offshore windspeed and the decision to seed or not seed. Note that the only arc drawn into the result node is from V. This indicates that, given the landfall speed, the result (property damage loss) is conditionally independent of the decision taken and the offshore wind speed. If the cost of seeding is significant enough to be included in the losses, an arc should be drawn from *D* to *R*. Typically this is not the case.

The influence diagram in Figure 4-7 is associated with a formal statement of the probabilistic structure of the important variables in the decision problem. It tells us that the joint probability of *U* and *V* is given by[2]

$$p(u,v \mid d) = p(u)\, p(v \mid u, d). \tag{4.1}$$

Because property damage is only a function of final windspeed, $R = R(V)$, the conditional expectation of the property damage is a function of the offshore windspeed. Note that the seeding decision is influenced by the observed values $U=u$ of offshore windspeed, so that

$$r(d(u)) = E_{V|U,D}[R(V) \mid u, d]. \tag{4.2}$$

Even though the reader is probably familiar with the use of conditional probability and conditional expectation, the dependencies illustrated in Figure 4-7 should help the reader understand the meaning of the dependencies in Equations (4.1) and (4.2). What we want to emphasize in this chapter is the interpretation and implication of the different models that might be constructed for a given problem.

It appears that the effects of seeding can best be measured by considering the *fractional* changes in wind speeds. Let *Z* denote the fractional change in natural wind speed

---

[2] As the meaning is obvious, we use $p(u,v \mid d)$ in place of $p_{U,V|D}(u,v \mid d, I)$.

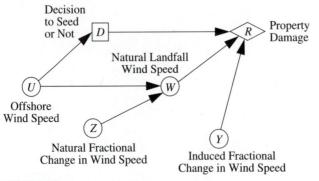

FIGURE 4-8
Second hurricane seeding influence diagram.

and Y the fractional change due to seeding. Figure 4-8 shows an influence diagram with these new random variables included. Note that Z can be placed before or after D but that there is no arc from Z to D as the former is unknown to D. Similarly, Y can be placed to the left or right of W, the uncertain landfall windspeed that would occur without seeding. There are no directed arcs from U to either Y or Z because (by assumption) Y and Z are independent of U. If the decision is not to seed, then by including a directed arc from D to R we can force R to include or exclude the contribution of Y. In this case there is no need for an arc from D to Y. There is no node to denote the landfall wind speed V. There are now a total of four chance quantities, one decision node and six directed arcs. This second influence diagram is substantially more complex than the first. We argue later in the chapter that it pays to keep an influence diagram as simple as possible without sacrificing important problem structure. The addition of a single node in an influence diagram usually leads to considerable growth in the size and complexity of the corresponding decision tree.

We can write an equation for final wind speed that expresses V as the product of three factors

$$V = W(1 + Y) = U(1 + Z)(1 + Y), \qquad (4.3)$$

which implies that even though the mathematical relationship for property damage is identical to what it was in the first formulation, the probabilistic structure has changed partly because the conditional dependencies have changed in this formulation and also because the assessment of these probabilities will now depend on different beliefs and different data sets.[3] Note also that given the decision d property damage is now $R(W, Y \mid d)$. For example, the measured offshore windspeed might be 100 mph. In a particular hurricane the natural change in wind speed by landfall might average a 30% increase($Z = 0.3$), and the change due to seeding might be a 12% reduction

---

[3] For example, U and Z may be conditionally independent if Z is the fractional change in wind speed; we might not be justified in making this assumption if Z were the absolute change in wind speed.

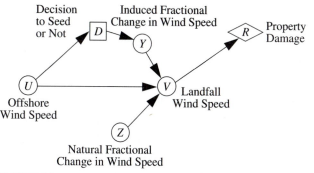

FIGURE 4-9
Third hurricane seeding influence diagram.

$(1 + Y = 0.88)$. If this were the case, the landfall wind speed with seeding would be 114 and without seeding 130.

Figure 4-9 shows a third influence diagram for this problem where V has been explicitly included and W removed. These diagrams are presented in order of increasing complexity and number of variables under consideration. Even though we might assume the same $R(v)$ function in Figure 4-9 as we did in Figure 4-7, we now have

$$p(u, v, y, z|d) = p(u)\, p(y \mid d)\, p(z)\, p(v \mid u, y, z), \qquad (4.4)$$

where the conditional expectation of property damage is

$$r(d(u)) = E_{V|U,D}[R(V) \mid u, d]$$

$$= \Sigma_{v,y,z} R(v)\, p(y \mid d)\, p(z)\, p(v \mid u, y, z)$$

$$= \Sigma_{v,y} E[R(v)|u, y, z]\, p(y \mid d(u)). \qquad (4.5)$$

Again, for the sake of simplicity, we have left off the conditioning subscripts on E. Only the conditional distribution of induced fractional change in wind speed is influenced by the seeding decision; the others are not. The reader should be able to write out the expectation for property damage conditioned on the observation of offshore wind speed.

One must always ask whether the additional complexity brings greater realism to problem formulation and how the results obtained from the different modeling assumptions yield different insights into the structure of the problem. In this case the additional variables seem necessary in order to model accurately the probabilistic dependence of V on U. The cost of building them into the influence diagram is a considerable increase in the size of the associated decision tree. As we see later in the chapter the number of paths through a decision tree grows extremely fast as the number of nodes in an influence diagram increases, and depends on the numbers of decisions or outcomes each node can take on. For this particular problem, there appears to be nothing gained in understanding the *decision* process by adding additional variables to the influence diagram. The conditional distribution of V, given U and D, can be found directly using Equation (4.3) and the much smaller and simpler influence diagram in Figure 4-7. We have included these different influence diagrams not to argue that one is correct and the others not; rather, our intent is to illustrate the model-building process, emphasize the principle of parsimony, and stimulate the reader to think through the assumptions of his or her model carefully with particular attention to the definition and structural relationship of decisions and random quantities.

4.3.2   Keeping Good Credit Accounts at a Bank

A bank that manages a large number of credit accounts has paid for a marketing study that reports bank charges and services are not competitive with other banks in the region. The bank has concluded that unless it offers better service to its existing customers, many of them will leave and take their accounts elsewhere. After reviewing the survey the bank's staff prepares a new offer to its existing customers that hopefully contains incentives that discourage its "better" customers from moving their accounts to a competitor.

Management must decide whether or not to make a new offer and how it should be made. Their first thought is that a single offer should be made to all current customers, no matter what the record of individual account loan repayments has been. The important elements of their decision problem appear to be (in order of occurrence) (1) the Offer/No Offer decision D, (2) the response of each customer to stay or leave, (3) whether or not each customer that stays has good (G) or bad (B) performance, and (4) the payoff resulting from the offer and different performance outcomes.

The initial attempt at modeling the problem suggests the influence diagram in Figure 4-10. In general, deciding on possible values for D, $\mathcal{D}=$ {Make Offer, Do Not Offer} and outcomes of the random result X, $\mathcal{X}=$ {stay, leave} may be easy to define,[4] but how to measure the good or bad performance of a customer is less obvious. The actual performance of each customer is assumed to be a random variable Y with $\mathcal{Y}=$ {G, B}.[5] Even though the precise definition of good or bad might be difficult to determine in a

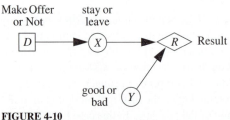

FIGURE 4-10
First influence diagram for bank credit.

---

[4] Note that the decision or action of the individual applicant to stay or leave is shown as a circular chance node in Figure 4-10 even though it surely requires an overt decision on his or her part. Why?

[5] We again point out that the use of upper case letters G and B differs from our usual lower case convention for outcomes of a random quantity; we retain upper case letters in this application to conform with standard usage in the credit and finance literature.

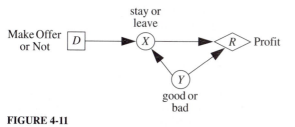

FIGURE 4-11
Second influence diagram for bank credit.

given situation, for purposes of this discussion think of a good customer as one who pays off his or her credit balances within 60 days; a bad is one who does not. The payoff to the bank will depend not only on whether a customer leaves or stays but also on eventual customer performance; for this reason a directed arc from Y to R is included.

The bank's analyst points out that, as drawn, there is no directed arc connecting X and Y or one connecting D to Y. This implies two assumptions that may have to be justified: that customer performance (good/bad) does not depend on the act of staying or leaving nor on the offer made by the bank. It should be noted that this model, with its strong independence assumptions and its reliance on the same offer to all customers, may have the unfortunate effect of encouraging good customers to leave and bad ones to stay. What we really want to do is discriminate on the basis of performance and tailor an offer to encourage *differential* behavior among goods and bads. In addition to the fact that we may want to make use of forecasts of individual risk, several structural changes may have to be made to the influence diagram to be consistent with observed behavior and performance.

Suppose we now add a directed arc from X to Y in Figure 4-10 to indicate that X and Y are dependent; it would also indicate that the conditional probability that a customer is a good or bad, given whether he or she stays or not, $p_{Y \mid X}(y \mid x)$ is known. Before the offer has been made (remember we are building a model to help decide this), some industry records suggest that we can estimate the distribution of X given Y, that is $p_{X \mid Y}(x \mid y)$; Figure 4-11 correctly indicates this belief. However, we are now faced with a dilemma: if we use our convention of placing the nodes[6] in a left-to-right order corresponding to their real occurrence over time, we see that a (backwards-in-time) directed arc leads from Y to X; on the other hand, we know that in real life the status of an individual being good or bad is not known until many months or even a year after the new offer has been accepted whereas the stay or leave event usually occurs shortly after the offer is made.

As we progress from influence diagrams to decision trees the reader will see the importance of reversing arcs directed from right to left. We know from the preceding chapter that if we can assess $p_Y(y)$ and $p_{X \mid Y}(x \mid y)$, then we can also calculate the condition-

---

[6] In Chapter 9 we define and discuss the important role of an extensive form influence diagram (EFID) whose arcs are pointed in the direction of time-ordered events as they actually occur.

al probability distribution for Y given X when the arc from Y to X is reversed. This requires the use of Bayes' rule that we described in Chapter 3. As shown in Figure 4-11 the conditional probability structure is

$$p(x, y \mid d) = p(x \mid y, d)\, p(y). \tag{4.6}$$

The expected return, given a decision, is only a function of whether the person stays and whether that individual ends up being a good or bad. It is irrelevant whether X is thought to influence Y or vice versa because both occur *after* the decision is made. The expected profit is

$$r(d) = E_{X, Y|D}[R(X, Y) \mid d] = \sum_{x,y} R(x,y)\, p(x \mid y, d)\, p(y). \tag{4.7}$$

On reviewing Figure 4-11 the analyst points out that, if the cost of making an offer is substantial, there should also be a directed arc from D to R. Management decides this is not the case, but asks why there is no arc joining D to Y since different offers might influence goods and bads differently. On further reflection they both agree that, although performance is *indirectly* influenced by an offer, it is better to think of the offer explicitly affecting individuals in terms of whether they stay or leave but not explicitly influencing performance. For example, if good customers tend to be more aware of competitive offers at other banks, then the probability of staying or leaving will affect $p_{X|Y}(x \mid y)$ and this, because of Bayes' rule, will affect $p_{Y|X}(y|x)$. But as long as a single offer is made to all individuals, then the offer cannot discriminate different individuals in terms of their good/bad performance.

As the discussion and design of the model proceeds it is pointed out that, although the new offer will result in different behavior by different individuals, what is needed is a forecast that discriminates between being good or bad.[7] Also, we note that although the cost of an offer is still small relative to potential profits or losses, it is nevertheless more expensive than the original (single) offer that used a baseline forecast of the group to predict that all individuals have the same probability of being a good or a bad. The question then arises whether or not the offer should be made only to an individual whose forecast indicates he or she is likely to be a good. Because the true performance of each individual cannot be predicted with certainty, some goods who are mistakenly classified as bads will not receive an offer, while some bads mistakenly classified as goods will.

As the result of further management discussions, it is decided that we should forecast individual account performance and make the most appropriate offer. Let F denote a forecast of good or bad performance for a given customer in the influence diagram shown in Figure 4-12. An F node has been added and arcs drawn from F to D and F to Y because F is a forecast of performance, not a forecast of whether the individual will stay or leave. The first new arc indicates that the forecast is known to the decision maker and that it may influence the decision. The second indicates that the probability of actual customer performance is conditional on the forecast of performance. Customer performance may also depend on whether the customer leaves or stays. Whereas in Figures 4-10 and 4-11 the

---

[7] Such as the scoring methods described in the preceding chapter.

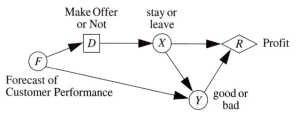

Make Offer   stay or
or Not   leave

FIGURE 4-12
Third influence diagram for bank credit.

appropriate probability to use for Y is $p_Y(y|x)$, in Figure 4-12 it is $p_{Y|F,X}(y|f, x)$. Expected return depends on the decision $d = d(f)$, which implicitly depends on the performance forecast. Thus, we must now condition on $F = f$ to obtain

$$r(d(f)) = E_{X,Y|D,F}[R(X, Y) \mid d(f), f] = \Sigma_{x,y}R(x, y) \, p(x, y \mid d(f), f), \qquad (4.8)$$

where it is understood that the joint probability can be factored as

$$p(x, y \mid d(f), f) = p(y \mid f, x) \, p(x \mid d(f)). \qquad (4.9)$$

This equation is identical to Equation (4.7) except that the conditioning is explicit. Study Equation (4.9) carefully, and convince yourself that the notation accurately depicts the influence diagram in Figure 4-12. Once again we emphasize the obvious: the conditional expectation depends on f through the decision $d(f)$, which is forecast-dependent; the expectation also depends on the distribution of Y, which is conditioned on f. If we were interested in knowing the unconditional expectation of the $d(f)$ policy, we would multiply Equation (4.8) by $p_F(f)$ and sum over $f \in \mathcal{F}$.

No arc leads from F to X because F is a forecast of performance, not a forecast of whether the individual stays or leaves. An increasingly important decision problem for banks is the inclusion of a forecast for the individual's behavioral response to each new offer as well as the individual credit performance once an offer is accepted. If one wanted to include a forecast that the individual would stay or leave after receiving an offer, one should include another forecast node, say G^8, whose directed arcs would lead from G to D and from G to X. There are many circumstances where such arcs are appropriate, but it should be clear that this requires an additional forecast of a different event. (See Problem 4.8.)

If staying or leaving is thought to be independent of customer performance, the arc from Y to X can be removed, and the appropriate influence diagram is shown in Figure 4-13. Note, however, that although the conditional expected return is still

$$E_{X,Y|D,F}[R(X, Y) \mid d(f), f] = \Sigma_{x,y}R(x,y)p(x, y \mid d(f), f), \qquad (4.10)$$

the joint probability of X and Y can now be factored as

---

[8] not to be confused with the outcome G, denoting good performance.

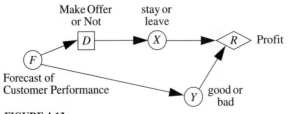

FIGURE 4-13
Performance independent of stay or leave.

$$p(x, y \mid d(f), f) = p(x \mid d(f)) \, p(y \mid f). \tag{4.11}$$

Both the hurricane seeding and bank credit problems illustrate the use of the principle of parsimony. In each case we design an influence diagram that starts with simple relationships; as we begin to understand the implications of these changes we add nodes and arcs to bring in more complex assumptions about timing of events and conditional independence. The focus is on clarity of definitions and simple relationships. If a simple model does not capture a sufficiently realistic picture of the situation, complexity can be introduced in an iterative learning process. In our experience, initial attempts at model building start with much confusion and complexity; the designer, analyst, or decision maker soon becomes lost in a sea of detail and unresolved dependencies. Numerical output then substitutes for insight and the art and science of model building makes little or no contribution.

4.3.3   A Navy Mobile Basing Decision Problem

As different nations' alliances and prosperity change, their willingness to allow foreign military bases on their territory changes. The recent losses of the Subic Bay Naval Base and Clark Air Force Base in the Philippines by U.S. forces are examples. To meet the U.S. Navy's continual requirement to maintain a forward presence in support of regional alliances, the United Nations, or national interests, it has been proposed that mobile ocean basing systems (MOBS) be built. These are large floating units that can be broken into segments, towed, and re-assembled at a given location. Once in place they could support ship repair, maintenance, and re-supply and serve as air stations from which operations could be carried out. Proposals vary from very large units with runways that could handle most conventional aircraft and facilities to service and support large naval vessels, to much smaller units that would be limited to aircraft capable of short takeoff and landing (STOL) and relatively limited ship size. Having proved the feasibility of constructing such units, the decision as to whether they would be cost-effective assets and should be built has not been made. An example of how to model this decision is discussed in this section.

Suppose that a MOBS of a given configuration were available. Would we deploy it to a given area to help in area stability or counter a threat should it arise? To answer this, first we must try to write down those factors that are likely to affect the requirement for a MOBS as well as those factors that its presence might affect. The requirement is

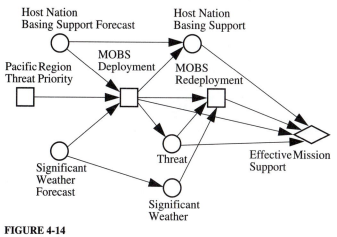

FIGURE 4-14
Influence diagram for MOBS.

derived from an assessment of its potential to support perceived regional threats. There are four U.S. regional commands that have responsibility for assessing threats to peace and stability throughout the world. If a specific threat erupts, estimates (forecasts) are available of the extent to which a host nation in the area will support U.S. forces and its allies in combating the threat. Both the threat and the possibility of obtaining host nation support should be important factors in making the decision to deploy a MOBS. If a MOBS were deployed to a given area, it may well have an effect on whether or not host nation support is forthcoming, as such support would be less necessary. It might also act as a deterrent affecting whether or not a given threat occurs. It also seems important to consider the weather. Since a MOBS would float in the ocean much like an oil drilling platform, severe weather such as typhoons or hurricanes could seriously affect its usefulness.

An initial attempt to draw an influence diagram is shown in Figure 4-14. The initial plan was to create a single influence diagram for four regions of the world. Before doing this, it was decided to concentrate on ensuring that the structure of one region was indeed a reasonable model.

According to Figure 4-14 an input to the MOBS deployment decision is the decision of the regional commander as to threats and their priorities. Although it is true that the commander makes these priority decisions, in terms of the MOBS deployment decision it is an (uncertain) input. As in the football example in Chapter 2, although the coach makes the play decision, to solve Nancy's betting problem she treats his action as a random variable. Thus, the first correction or improvement in Figure 4-14 is to convert the square node on the left-hand side to a chance node.[9] Notice also that, in addition to deciding whether or not to deploy a MOBS, a second decision is shown to possibly re-

---

[9] There are as many of these as there are regions. We have only shown the one for the Pacific region.

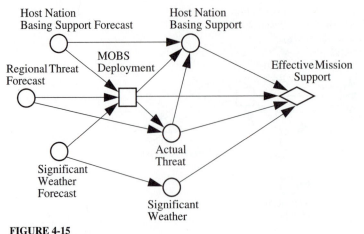

FIGURE 4-15
Second influence diagram for MOBS.

deploy it to a different location. At this point it is advisable to rethink which of the many possible decision problems associated with building a major new system we are trying to model. Although redeployment may be very important in tactical decision making on how best to use one or more MOBS if and when they exist, our problem is to help with the initial decision for a major design and procurement program. It is extremely important to keep in mind that such a decision will be made at very high levels (not only in a defense problem such as this but also for any significant capital expenditure by government or industry). The people involved in such decisions are extremely busy and can never become acquainted with detailed aspects of possible operations of a large system such as MOBS. What they need to understand are the major economic and structural implications of their decision. Briefly stated, the analyst should build models that are as simple as possible without sacrificing critical aspects of the problem.

Figure 4-15 shows a revised influence diagram. Based on the comments just made, we concentrate on understanding the problem for a given region and have modeled the regional threat forecast as a random event. The MOBS redeployment decision has been eliminated. Notice that in Figure 4-14 there was no arc connecting the regional threat priority node and the threat node. Being more careful in our definitions we have included a directed arc from the regional threat forecast node to the actual threat node to indicate the conditional dependence. Also there was no arc connecting the actual threat to host nation basing support. It is highly likely in many situations that the outbreak of a particular threat could change our assessment of the chances of obtaining basing support in a host nation (the reverse of this could also be true as is mentioned later). To model this aspect, we add a directed arc between the two nodes.

The diagram in Figure 4-15 appears to capture the structure of the problem, but a little thought by the readers should convince them that it is still more appropriate as a tactical rather than a strategic decision model. The problem we are attempting to solve is whether or not to build a MOBS and, if so, what type. For this decision problem the threat forecast can be thought of as an initial set of information; we denoted this by I in

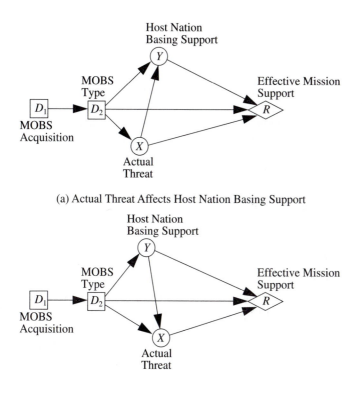

(a) Actual Threat Affects Host Nation Basing Support

(b) Host Nation Basing Support Affects Actual Threat

FIGURE 4-16
Strategic MOBS decision influence diagrams.

Chapter 1. Our current model is structured more like Nancy's betting problem where the coach makes the kick/try decision with every touchdown. For our current problem it is more appropriate to think of whether or not we should build a MOBS for a given area to be used in a whole variety of possible threats. The weather should not be a factor in this decision if, as would certainly be the case, the MOBS has been designed to withstand most foreseeable weather states. The weather would be more appropriately included in a decision to deploy in a given situation, for example, in the western Pacific in typhoon season. The model in Figure 4-15 is more appropriate for the case where we would first wait and see what threat was about to arise (there are usually, but not always, warning signs of imminent threats), and then decide whether or not to move an existing MOBS into the area.

Based on these ideas we simplify our influence diagram to one of those in Figure 4-16. Note that we have changed the name of the MOBS deployment decision and split it into two well-defined and less confusing decisions. With these simpler diagrams the decision maker can concentrate on the strategic issues. The directed arc from MOBS type to the actual threat models the important concept of deterrence. The arc to host nation basing support indicates our uncertainty about the effects that placing a

MOBS in an area will have on a given country providing support. In (a) we model the effect that an actual threat will have on gaining host nation support, whereas in (b) the reverse is shown, what effect gaining host nation support for basing might have on a threat breaking out.

As was the cases for both the hurricane seeding and bank credit account examples, each node in the influence diagrams in Figure 4-16 has been given a symbol. It is very important to define clearly each uncertain event and each decision. In the next section the reader will see that sets of possible decisions and uncertain outcomes need to be carefully specified before a decision tree can be constructed. Influence diagrams should not be thought of merely as flow charts where there is often vagueness as to what a directed arc connecting two nodes means. We continue to stress the need for clarity of definition and meaning in constructing the influence diagram.

4.3.4   Colon Cancer Diagnosis

In Section 3.5.2 we first introduced an important example of finding a cost-effective way of detecting colon cancer. The effective treatment of most cancers relies on early detection, usually before the appearance of obvious symptoms. Because the frequency of occurrence often increases with age, doctors often recommend a person in a given age group have one or more tests performed on a regular basis to ensure early detection. Four different tests to detect colon cancer are in common use. Each test has a different probability of detecting a true cancer, and each test has a different cost, with the most expensive being approximately 160 times as expensive as the cheapest. Each test examines different portions of the colon; only two are able to test the entire colon for cancer. The incidence of colon cancer is small (only 1 out of a 1,000 patients may have this disease); as one might expect, the less expensive tests are less likely to detect a true cancer than are the more expensive ones. The decision problem is to determine which tests to give and in what sequence.[10]

The four tests in order of increasing cost are occult blood (*OB*), digital examination (*DE*), sigmoidoscopy (*SIG*) and barium enema (*BaE*) examination. The costs are *OB* - \$1, *DE* - \$4, *SIG* - \$50 and *BaE* - \$160. A well-thought-out strategy for making tests has important economic consequences and benefits if one thinks in terms of testing hundreds of thousands or possibly millions of patients.

The objectives of the decision maker are not completely clear. From the patient's point of view, it would be most desirable to detect every cancer or establish that there is no cancer. On the other hand, the medical and insurance professions are interested in cost-effective programs that reduce or minimize some measure of costs of testing or costs to the health providers. We shall see how the choice of objective can change an influence diagram.

---

[10] We have modified the cost data from Weinstein and Fineberg (1980) so that the least expensive test (*OB*) has unit cost and all other tests have the same cost relative to *OB* that they did in the original formulation.

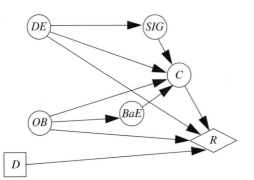

FIGURE 4-17
Initial colon cancer influence diagram.

In its simplest form the problem can be formulated as: Perform one or more tests, possibly in sequence, and observe the result. We first consider the following three possible test sequences:[11]

1. *DE* only,
2. *OB* followed by *BaE* if *OB* is positive,
3. *DE* followed by *SIG* if *DE* is negative.

Assume that we have not yet decided on a measure of effectiveness. Three possible objectives are (1) maximize the expected number of cancers found, (2) minimize the expected cost, and (3) some combination of the previous two such as minimize the expected cost per cancer detected.

Our first attempt at an influence diagram for this problem is shown in Figure 4-17. The order of the nodes agrees with our starting assumptions: *D* designates the decision (which of the three sequences to use). The chance nodes *DE*, *OB*, *SIG*, and *BaE* denote the tests with uncertain outcomes. *C* denotes the uncertain presence or absence of cancer; *R* is the result, as yet unspecified but that might represent cost, number of cancers detected, or some combination of the two. The arcs connecting the chance nodes to each other show the conditional dependencies among the tests. For example, given that a digital examination is negative, the probability distribution assigned to *SIG* is affected by this result. A similar statement can be made for *OB* and *BaE*. The reader should think through why each of the arcs shown has been included. It is true that the outcome of *DE* would also affect the distribution of *BaE*, but because it has been decided ahead of time not to include the possibility of a *DE* test followed by a *BaE* test, there is no need to include an arc from *DE* to *BaE*. There are four arcs entering the result node. Those from *DE* and *OB* are added to indicate that the cost depends on the outcomes of these tests.

[11] Weinstein and Fineberg (1980) specify the makeup of the sequences a priori and suggest five combinations, three of which are identical to those in our list.

The arc from *C* indicates that the number of cancers detected affects the result and depends on the outcome of every test.

The influence diagram in Figure 4-17 is correct if the objective is to minimize the expected cost per cancer detected. Suppose that we decided that our measure is the expected number of cancers detected. In this case the direct arcs from *DE* and *OB* to *R* can be removed, because their only purpose is to show that the *cost* depends on the outcomes of these tests, and cost is no longer part of the result. The resulting influence diagram is shown in Figure 4-18.

Notice that the joint probability of the four tests and the presence or absence of cancer (using first letter of test) is given by

$$p(c, d, o, s, b) = p(c \mid d, o, s, b) \, p(s \mid d) \, p(b \mid o) \, p(o) \, p(d). \tag{4.12}$$

If the objective is to minimize expected costs, we could start with Figure 4-17 and remove the arc from *C* to *R* because the result will no longer depend on the number of cancers found but only on the tests selected. The resulting influence diagram is shown in Figure 4-19.

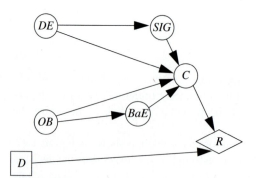

FIGURE 4-18
Influence diagram for number of cancers detected.

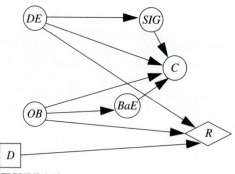

FIGURE 4-19
First influence diagram for cost.

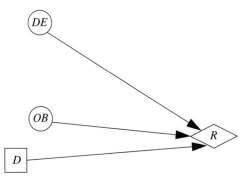

FIGURE 4-20
A simplified influence diagram for cost of testing.

Note the special structure at node *C*. All four of the arcs that connect *C* to other nodes are directed *into C*, but none are directed out of *C*. A node with this property is referred to as a *barren node*, and any time one is found in an influence diagram it can be removed, together with all arcs leading into it. We say that such a node has no *direct successors*; by contrast all arcs leading into a node come from *direct predecessors* of that node. Sometimes in model formulation, or as a result of arc reversal that arises during a Bayesian analysis, a chance node in an influence diagram will be found to have no direct successors. When such a barren node is found, it and all the arcs leading to it from its direct predecessors can be removed as their presence has no effect on the result. We also find that removing a barren node can lead to other nodes becoming barren, in which case the process can be repeated until no more barren nodes are present.[12]

If the barren *C* node is removed in Figure 4-19, *SIG* and *BaE* become barren nodes and can also be removed. After the removal of these nodes and arcs, the resulting influence diagram shown in Figure 4-20 has a much simpler structure. As we see later in the chapter, it leads to a much smaller and simpler decision tree than the one derived from Figure 4-18. Perhaps the reader, having decided ahead of time that the objective is to minimize expected costs, could have produced Figure 4-20 directly. The only factors that affect cost are the particular testing sequence chosen and the outcomes of *DE* or *OB*. It should now be clear to the reader that the structure of an influence diagram can depend on the particular attribute(s) and objectives used to measure the result.

In these examples we have considered several different preselected test sequences to determine the "best" for some measure of effectiveness. We now consider a different model formulation in which sequential decisions are an essential ingredient of the decision problem. For example, suppose that a decision is made to perform an *OB* test. Then, depending on the outcome of that test, a second decision is made to follow it up with another more specific and possibly more costly test.

---

[12] The proof of this result for the general case requires a mathematical formulation that can be found in Chapter 9.

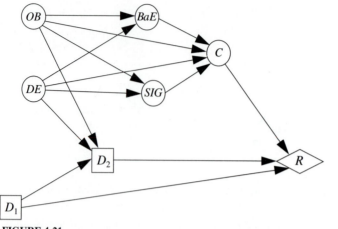

FIGURE 4-21
Sequential decision colon cancer problem.

Figure 4-21 shows an influence diagram for a problem where we consider sequential decisions to find the minimum expected cost per cancer found. We must first choose to perform the *DE* or the *OB* test, both of which are inexpensive and, depending on the outcome, make a second decision as to which if any of the more expensive tests *SIG* and *BaE* to perform.

The reader may be surprised at the large number of directed arcs in Figure 4-21. As we shall see, this is due to the conditional dependencies between test results; they cannot be avoided.

In the sections that follow we show how one constructs a decision tree from an influence diagram and the data associated with the problem; we then describe a standard solution procedure for finding optimal decisions that is easy to use with small decision trees. In Section 4.5 we return to the bank credit account, and in Section 4.6 to the colon cancer testing problems to illustrate these methods.

4.4   BUILDING AND SOLVING DECISION TREES

A decision tree is a well-known and popular way of displaying and analyzing sequential decisions and events. For example Keeney (1980) states

> *Often the complex problems are so involved that their structure is not well understood. A simple decision tree emphasizing the problem structure which illustrates the main alternatives, uncertainties, and consequences, can usually be drawn up in a day. Not only does this often help in defining a problem, but it promotes client and colleague confidence . . .*

Decision trees appeal to those interested in graphing the details of how chance outcomes and decision alternatives unfold over time. Unfortunately there is a trade-off between the details provided in a graphical representation of a tree and the size of the problems that can be practically represented. In many real-world problems it is not uncommon to find that tens of thousands of paths are required to characterize the possible sequences of choices and outcomes. The onerous task of drawing or labeling paths in the tree is somewhat simplified by the ability of modern computers to list and enumerate all possible paths and select the "best" ones, but the magnitude of the task of assessing probabilities and formatting and assembling payoff or cost data should never be underestimated.

A path in a decision tree is a series of decisions and random outcomes that lead from a single initial chance or decision node to a distinct end point or terminal node. The terminal nodes are often called result or value nodes whose numeric value is a payoff or loss or other scalar measure of the result of the decision problem. This value usually depends on the particular path that leads to the result node, as it depends on the particular decisions and chance outcomes that make up the path. Associated with each branch emanating from a chance node is a (conditional) probability that in general also depends on the decisions and random outcomes that precede it on a particular path. If the decision tree faithfully models the sequence of events and decisions that take place in a decision-making problem, each path represents a particular scenario that could unfold over time. Unfortunately, a decision tree has one fundamental weakness in that it does not make evident to the user the dependencies among branches on each path that arise in real decision problems. These dependencies are shown explicitly in an influence diagram. Our goal in this section is to show how to build a decision tree from an influence diagram, then use it together with a simple procedure to solve the decision problem.

4.4.1 Node Outcome and Alternative Sets

The key to being able to construct a decision tree from an influence diagram is to be able to specify every possible outcome of every node: chance outcomes from each chance node, decision alternatives from each decision node, and every possible result for the result or value node. Recall our notation that for a given influence diagram,

C = The set of chance (circular) nodes,
$\mathcal{D}$ = The set of decision (square) nodes.

In all our examples there is only one result node. Each chance node must have associated with it a set of possible outcomes of the random event, each decision node a set of possible decisions that can be made, and the result node a set of possible results (e.g., payoffs or losses). Let

$C(i)$ = The set of possible outcomes at chance node i, $i \in C$,
$\mathcal{D}(j)$ = The set of possible decisions at decision node j, $j \in \mathcal{D}$,
$\mathcal{R}$ = The set of possible results at the result node.

These sets may also be conditional on random events that have occurred or alternatives that were taken at earlier nodes. The notations $C(i|\mathbf{h})$ and $\mathcal{D}(j|\mathbf{h})$ are used when necessary to indicate a history vector $\mathbf{h}$ of chance events and decisions at nodes preceding the given node. We illustrate these sets using the engine maintenance example introduced in Section 4.2.1. The influence diagram for this problem is shown in Figure 4-1 (p. 139). It contains four nodes with $C = \{1,3\}$, $\mathcal{D} = \{2\}$, and the result node is 4.

To construct the sets associated with each node, we must elicit information from the decision maker as to alternative decisions he or she is willing to make and the specific level of detail of random outcomes that are needed to make a decision or measure a result. This is often an iterative process, as we illustrate in the next section. In general, when using letters for notation, we use lowercase for random outcomes and uppercase for decision alternatives. For the engine maintenance problem, at node 1 we denote the outcome of the test by the letters

> a: The engine meets performance specifications.
> b: It does not.

Thus $C(1) = \{a, b\}$. Decision node 2 has sets that depend on whether the outcome at node 1 is a or b. If it is a (meets specifications), the decision alternatives at node 2 are

> A: Accept As Is,
> M: Perform Maintenance.

If the test outcome is b (fails to meets specifications), the only decision alternative is

> RE: Replace Engine.

Thus $\mathcal{D}(2 \mid a) = \{A,M\}$, and $\mathcal{D}(2 \mid b) = \{RE\}$. Recall that node 3 represents the uncertain performance of the engine following the decision and corresponding action at node 2. Using numeric notation, possible outcomes are

> 1: completely satisfactory performance,
> 0: reliable operation but minor failures,
> −1: catastrophic failure.

It is assumed that the probability that any of these three outcomes occurs depends on the decision taken at node 2. In particular, we assume that the only outcomes that can occur with positive probability are

$$C(3 \mid A) = \{1,0,-1\}, \; C(3 \mid M) = \{1,0\}, \text{ and } C(3 \mid RE) = \{1\}.$$

We assume that the result outcome only depends on engine performance with $\mathcal{R} = \{0,1,10\}$.

These sets[13] contain all the information required to construct a decision tree. However, they do not yet contain the information required for analysis. In order to solve the
problem depicted by the tree, the following sets are also required:

$p(i|h)$ = The set of probabilities defined over the set $C(i \mid h)$ for each chance node
$i \in C$ and each history **h**,

$R(i|h)$ = The set of returns (payoffs or costs) defined over the $C(i \mid h)$ for each
chance node $i \in C$ and $D(i \mid h)$ for each decision node $i \in D$ and each history **h**.

Again we must allow for possible dependency of these sets on the path history **h** to reach
the particular node. The reader must be careful to distinguish between our use of R without parentheses defined earlier for the single result node with the sets $R(i \mid h)$ as defined
for chance and decision nodes.

The simple engine maintenance problem in Figure 4-1 (p. 139) is used as an illustration. Recall that possible outcomes of the test are given by $C(1) = \{a,b\}$. Because this
is the first node, we can drop the **h**. Let $p(1) = \{0.9, 0.1\}$, stating that 90% of all engines
tested meet specifications. Let $p(3 \mid a,A) = \{0.5, 0.4, 0.1\}$ and $p(3 \mid a,M) = \{0.85, 0.15\}$;
if an engine test meets specifications and the decision is to accept it as is, there is a 50%
chance it will perform completely satisfactorily, a 40% chance it will be reliable but with
minor failures, and a 10% chance of a catastrophic failure. If it meets specifications
when tested and the decision is to perform maintenance, there is an 85% chance it will
perform completely satisfactorily, a 15% chance it will be reliable but with minor failures, and no chance of a catastrophic failure. Recall that if the engine fails to pass the
test, it will be replaced. We assume that a replacement engine always performs completely satisfactorily, so $p(3 \mid b,RE) = \{1\}$. As we shall see when we draw the decision
tree this last set is not required.

In most published descriptions of decision trees it is assumed that all branch values encountered on a given path are summed and included in the value at the terminal
node. It may be preferable in some problems to place returns on the appropriate branches
of the decision tree rather than accumulate them on the terminal node. It will become
clear in our examples why it is sometimes important to measure branch values as they
occur rather than identifying them only when a terminal node is reached. For example,
in many construction projects and multistage acquisitions of complex systems, costs
may occur well before the completion of the project. Thus, the cost of making an early
decision may affect the decision choice; at a later decision point that same cost may be
irrelevant to future decisions because it is a sunk cost or it may come from different fund
sources that are not linked. For example, budgets for construction may be separate from

---

[13] Some texts distinguish ordered sets using parentheses ( ), from unordered sets using curly brackets { }. We
use them interchangeably for sets, ordered or unordered. The only reason for requiring ordering of chance outcomes is to ensure clarity in the assignment of probabilities to these outcomes. However, we believe the ordering will always be clear from the context and thereby avoid the additional formality that this distinction
requires.

those for operations; the latter may be derived from revenues that are partly due to the presence of the new building. Thus, timing of costs and returns is important.

For decision node 2, $\mathcal{R}(2 \mid a) = \{0, 0.1\}$ and $\mathcal{R}(2 \mid b) = \{2\}$. If the engine test meets specifications, there is no cost for accepting the engine as is, and a cost of 0.1 for performing maintenance. If it fails to meet specifications, it costs 2 to replace the engine. For the result node 4, $\mathcal{R} = \{0, 1, 10\}$; there is no cost for a perfectly satisfactory engine performance, a cost of 1 if it experiences minor failures, and a cost of 10 for a catastrophic failure.

4.4.2   Drawing Consistent Decision Trees

Having the sets of chance event outcomes and decision alternatives defined together with their sets of probabilities and costs/payoffs, respectively, we are now in a position to construct a decision tree. The tree is constructed in chronological order from left to right, consistent with the timing convention of the influence diagram with decision and chance nodes in the order in which they actually occur, and connected by branches. Each decision node j has one branch for each alternative in that node's set $\mathcal{D}(j)$. Each chance node i has one branch for each possible outcome in that node's set $C(i)$. This is sometimes called *Nature's Tree*. The steps for drawing a discrete decision tree are as follows:

1. Start with the first (leftmost) node in the influence diagram. If it is a decision node, draw a branch for each decision alternative. If it is a chance node, draw a branch for each possible outcome.

2. Label each branch emanating from a chance node i with a unique member of the set $C(i \mid \mathbf{h})$ (a chance outcome), the corresponding probability that it will occur from the set $\mathbf{p}(i \mid \mathbf{h})$, and the return or loss from that branch from the set $\mathcal{R}(i \mid \mathbf{h})$. In general, both the probability and return are dependent on events and decisions that have *preceded* it (hence the notation $\mathbf{h}$ for history). Probabilities must be coherent, satisfying the laws of probability.

3. Label each branch emanating from decision node j with a unique member of the set $\mathcal{D}(j \mid \mathbf{h})$ (decision alternative) and the return or loss from making this decision from the set $\mathcal{R}(j \mid \mathbf{h})$.

4. Place values (payoffs or costs) from the set $\mathcal{R}$ on the terminal nodes at the end of each path. In general these are path dependent, which is to say there are as many possible distinct values on terminal nodes as there are distinct paths in the decision tree. However, in many problems of interest, different paths with common sequences of nodes will have identical terminal values.

The reader must be careful to clearly distinguish the very different meanings of directed arcs in an influence diagram and branches in a decision tree. *The former represent influence or relevance of one event to another, while the latter identify specific random outcomes or decision alternatives.*

We illustrate this procedure using the engine maintenance example. In Figure 4-1 (p. 139) the first node is the test node numbered 1 with set $C(1) = \{a, b\}$. Figure 4-22(a)

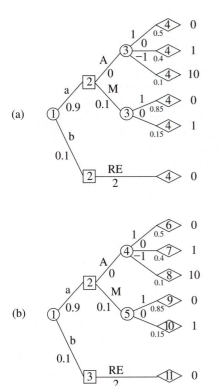

FIGURE 4-22
Engine maintenance decision trees.

starts on the left with a chance node and two branches labeled a and b to which we assign
the probabilities 0.9 and 0.1 respectively from the set $\mathbf{p}(1)$. In this example the set $\mathcal{R}(1)$
is empty as there are no returns or losses for these branches. The next node in Figure 4-1
is decision node 2, so each branch from node 1 in Figure 4-22 terminates with a decision
node numbered 2. Recall that $\mathcal{D}(2\,|\,a) = \{A,M\}$ and $\mathcal{D}(2\,|\,b) = \{RE\}$. From the upper
node 2 there are two branches labeled A and M, and from the lower node 2 a single
branch named RE. The appropriate costs have been added from the sets $\mathcal{R}(2\,|\,a)$ and
$\mathcal{R}(2\,|\,b)$. Each of these branches should terminate in a chance node 3, the next node
to occur in Figure 4-1. Recall that $C(3\,|\,A) = \{1,0,-1\}$, $C(3\,|\,M) = \{1,0\}$, and
$C(3\,|\,RE) = \{1\}$. Because the last of these sets has only a single element, we could place
a node 3 at the end of branch RE followed by a single branch labeled 1, which in turn
ends with a terminal node 4. But a little reflection should convince the reader that this
branch 1 is unnecessary; we lose nothing in structure by terminating the branch RE directly with a terminal node 4. This will always be the case whenever a chance node set
has only a single element. From the upper node 3 we add three branches labeled 1, 0, and
-1, with probabilities $\mathbf{p}(3\,|\,a,A) = \{0.5\ 0.4\ 0.1\}$. From the lower node 3 we add two
branches labeled 1 and 0 with probabilities $\mathbf{p}(3\,|\,a,M) = \{0.85, 0.15\}$. Again for this ex-
ample there are no returns for any of these chance branches. The only node not yet con-

sidered in Figure 4-1 is result node 4, so we add a terminal node 4 to each branch. Finally we assign the appropriate value to each terminal node from the set $\mathcal{R}$.

A decision tree could use a different node numbering convention from that in the influence diagram. An example of this is illustrated in Figure 4-22(b) where each chance, decision, and value node has been assigned an arbitrary unique number independent of the numbering system used for the nodes in the influence diagram of Figure 4-1. Such a system is commonly used in drawing small decision trees for visual inspection and hand calculations. If we wish to have a computer construct the tree from the node sets and perform subsequent calculations to determine optimal policies, there are great advantages to using the numbering system in Figure 4-22(a). When such a system is used, a given sequence of node numbers on a path may not define a unique path. For example, there are five paths with the node sequence $(1, 2, 3, 4)$ in Figure 4-22(a). Although leading to computational efficiency on a computer as well as using a node numbers that are consistent with the influence diagram, this feature significantly complicates the design of algorithms to find optimal decisions, as we shall see in Chapter 9. For small trees that can be solved by hand this complication is not required, and in the remainder of this chapter we use a simple algorithm that is based on trees with unique node numbers.

Both decision trees show the detailed outcomes and alternatives as events and decisions unfold over time. Both have three chance nodes, two decision nodes, six value nodes, and ten branches that must be identified and labeled. Recall, however, that in the influence diagram in Figure 4-1 there are only two chance nodes, one decision node, one value node, and five directed arcs. Note that nothing in the structure of the decision tree helps determine any dependencies among the random events. These dependencies can be seen explicitly in the influence diagram.

4.4.3   Perfect Information

We reconsider the engine maintenance problem encountered in Figure 4-1 (p. 139) where the decision affects the engine performance. This problem was described in Section 4.2.4 (p. 141). Because there are three possible outcomes for engine performance, we drew three different influence diagrams shown in Figure 4-6 (p. 143). In Figure 4-23 we draw the three corresponding decision trees. Again, we exclude the diagnostic test node from the decision tree since we have perfect knowledge about performance and there is no need to include a prediction made by a test.

If decision A is taken, there are three possible outcomes, if M is taken there are two, and if RE is taken there is one. But we are attempting to draw a tree where the decision is not made until the outcome of node 3 is observed. Observe that although the top three terminal nodes have equal cost (corresponding to the outcome completely satisfactory), the least costly decision at node 2 is to select the A alternative because the last two branches incur a cost of 0.1 and 2, respectively. Similarly, the best alternative to select from the middle two branches, when it is known that the engine performance is reliable operation, is again the A alternative. Finally, we see there is no choice in the bottom path. In this case we must use alternative A because we know that this decision is the *only* one that can lead to catastrophic failure.

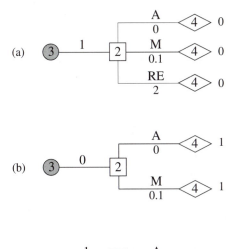

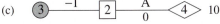

FIGURE 4-23
Perfect information decision trees for Figure 4-6.

Recall that in the original formulation {a, b} denotes whether the test outcomes states that the engine meets or does not meet performance specifications. There are three possible decisions {A, M, RE} corresponding to Accept As Is, Perform Maintenance, and Replace Engine. Under normal conditions the third decision is only used if test outcome b is obtained.[14]

It is not possible to assign meaningful prior distributions to the branches emanating from node 3 in Figure 4-23. Thus, expected value of perfect information is not a meaningful calculation. This demonstrates the difficulty one encounters in drawing and using decision trees to evaluate the expected value of perfect information about outcomes of decision-dependent random quantities.

4.4.4   Decision Tree Solutions

In solving decision problems represented on trees where there are at least two decisions to be made sequentially, we are always faced with the important *Decision Tree Paradox*:

> *One cannot decide what decision to follow today until one has decided what decision to make tomorrow that may be influenced by today's decisions.*

If we are to compute the best way to proceed, we must start at the terminal node of a path rather than at the starting node and compute answers in the reverse order in which events

---

[14] The reader should not confuse outcomes a and b for the Test node with the labels (a) and (b) attached to two of the three influence diagrams used in Figure 4-6 and Figure 4-23.

and decisions actually occur. The difficulties can be visualized as follows. Place yourself at a given intermediate node on a particular path of the decision tree. Look to the left (backward in time) noting chance or decision nodes that have already occurred, and look to the right to see which chance or decision nodes have not yet occurred but must end in a particular value node. The former path will probably influence the probability of future outcomes and the value of future decisions. The difficulty in solving the problem lies in the fact that the decisions from this point forward in general depend on how you got to where you are.

Let us assume that each node in the decision tree is given a unique number as demonstrated in Figure 4-22(b). The usual solution method is to proceed backward from the value nodes, taking expectations at random nodes and maximizing expected payoffs or minimizing expected losses at decision nodes, thereby determining the best decisions from each point forward. At a decision node we compare the expected values that would result from going down every possible branch and pick the best one. At a chance node we compute the expected values over all branches that depart from the node of interest. We emphasize that the particular calculation depends on whether you are at a chance, decision, or value node. If you are at a decision node i with the set of decision alternative $\mathcal{D}(i)$, the next node on a path from i to the terminal node may be either another decision node or a chance node. A branch emanating from chance node i corresponds to a member of the set $C(i)$, and one from a decision node i to an alternative belonging to $\mathcal{D}(i)$. With the numbering system that gives each node a unique number, the sets $C(i)$ and $\mathcal{D}(i)$ are just the sets of node numbers connected by branches emanating from i. In this node numbering system the set $\mathcal{V}(i) = \{v_i\}$ is the set containing the single value at the end of the path that ends in terminal node i.

Suppose we assign a label v_i to each node i that represents the *optimal expected return proceeding from that node to a terminal node*. Starting at the terminal nodes, our objective is to label all other nodes until the starting node is labeled. For any branch (i, j) that connects decision node i to a node j let $v(i, j)$ be the payoff or loss accumulated in traversing this branch. At each decision node i we attach a second label d_i^* that represents the optimal decision to be made at node i.

In the description that follows we assume that the objective is to maximize expected return. If the objective were to minimize expected loss we simply replace the *Max* operation described in the algorithm by *Min*. The following procedure is used to find these labels:

The Rollback Algorithm

1. Start with no labels on any nodes.
2. To each result or terminal node i, assign a label that equals the payoff or loss at that node, so that $v_i = r_i$, $r_i \in \mathcal{R}$.
3. For any unlabeled decision node i where all nodes j connected to i by a branch (i,j) are labeled, set

$$v_i = \underset{j \in \mathcal{D}(i)}{Max} \{r(i, j) + v_j\},$$

and set $d_i{}^*$ equal to the decision that yields this maximum value. If the object is to minimize loss, replace *Max* by *Min*, and set $d_i{}^*$ equal to the decision that yields this minimum.

4. For any unlabeled chance node i, where all nodes j connected to i by a branch (i,j) are labeled, set

$$v_i = \sum_{j \in C(i)} p_{ij} (r(i, j) + v_j) .$$

Steps 3 and 4 are repeated until the starting node is labeled. This starting node label will give the maximum expected payoff r^* (or minimum expected loss l^*) for the tree. The $d_i{}^*$'s on each decision node identify the optimal decisions.

This algorithm is easily demonstrated for the tree shown in Figure 4-22(b). We need to compute $\{v_i; i = 1,2, \ldots ,10,11\}$ and $\{d_i{}^*; i = 2,3\}$. We start by setting $v_i = r_i$, $i = 6,7, \ldots ,10,11$. From these recursions we have:

$$v_4 = 1.4,$$

$$v_5 = 0.15,$$

$$v_2 = Min\{r(2,4) + v_4, v(2,5) + v_5\} = 0.25,$$

with $d_2{}^* = M$. Because there is only one decision choice at node 3, no minimization is required so that

$$v_3 = r(3,11) + v_{11} = 2.0,$$

with $d_3{}^* = RE$. Finally, the minimum expected loss is

$$l^* = v_1 = 0.425.$$

The optimal policy that results in this minimum expected loss is: (1) if an engine meets test specifications, perform maintenance, and (2), if it does not, replace the engine.

In the next two sections we return to some of the examples described in Section 4.3. Our aim is to demonstrate building a decision tree that is consistent with an influence diagram, show how the decision tree depends on the conditional probability structure, and then demonstrate the rollback algorithm. The reader is encouraged to perform the rollback procedure on decision trees used in the text.

4.5   THE BANK CREDIT PROBLEM

Consider the problem depicted by the influence diagram in Figure 4-11 (p. 149) where an offer is to be made to all bank customers with the objective of preventing the most desirable customers from moving their accounts to a competitor's bank. At decision node D the possible decisions comprise the set $\mathcal{D} = \{\text{Make Offer, No Offer}\}$; we assume that there are negligible costs associated with making the offer. At chance node Y, $\mathcal{Y} = \{G, B\}$, where G denotes good and B denotes bad, it is estimated that 90% of all customers are good; thus, the probability set $\mathbf{p}(Y)$ is $\{0.9, 0.1\}$. At chance node X,

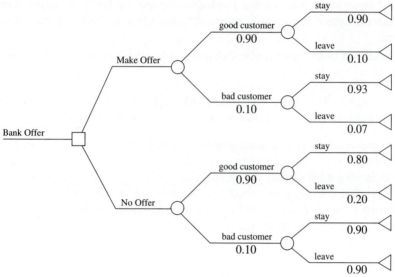

FIGURE 4-24
A decision tree for a single offer to all customers.

$X = \{\text{stay, leave}\}$. The influence diagram indicates that the distribution of X may depend on both D and Y.

We estimate that if a good customer is made an offer there is a 90% chance he or she will stay, whereas without an offer only 80% will stay. We estimate that 93% of the bad customers who receive an offer will stay, and 90% if they do not. This information leads to the following probability sets:

$\mathbf{p}(X \mid \text{Make Offer, good customer}) = \{0.9, 0.1\}$,
$\mathbf{p}(X \mid \text{No Offer, good customer}) = \{0.8, 0.2\}$,
$\mathbf{p}(X \mid \text{Make Offer, bad customer}) = \{0.93, 0.07\}$,
$\mathbf{p}(X \mid \text{No Offer, bad customer}) = \{0.9, 0.1\}$.

These probabilities are shown on the appropriate branches of the decision tree in Figure 4-24. However, notice that the ordering of the chance nodes does not agree with the timing of events as they actually occur in Figure 4-11 (p. 149), and we are therefore led to consider the decision tree in Figure 4-25 that shows the various paths that are obtained when the stay/leave chance node precedes the good/bad node; that is, events are displayed as they actually occur over time. We use Bayes' rule to determine the following probability sets when we condition on X rather than Y and obtain the new conditional probabilities

$\mathbf{p}(Y \mid \text{Make Offer, stay}) = \{0.897, 0.103\}$,
$\mathbf{p}(Y \mid \text{No Offer, stay}) = \{0.889, 0.111\}$,

$\mathbf{p}(Y\,|\,\text{Make Offer, leave}) = \{0.928, 0.072\},$

$\mathbf{p}(Y\,|\,\text{No Offer, leave}) = \{0.947, 0.053\}.$

A customer who stays and is a good produces a net profit to the bank of $120, whereas a bad results in a loss of $1,000. It is assumed that the bank gains or loses nothing if a customer leaves. Thus, the result set for the terminal nodes is $\mathcal{R} = \{-1,000, 0, 120\}$. Note that we do not need to add random X nodes to the leave branches because the payoffs for goods and bads are the same in this case, namely, zero. In this example there are no sets of payoffs or losses for the decision and random nodes.

Using the rollback algorithm one can calculate the optimal expected profit and the optimal decisions. The optimal expected profits are shown in the boxed labels. The expected profit per customer if an offer is made is $4.20; if it is not, a loss of $3.60 is expected. Thus, the optimal policy is to make the offer.

The reader is encouraged to check the meaning and structure of the decision trees and the conditional probabilities on the branches. The optimal solutions to the two trees must be the same, and solving the tree in Figure 4-24 is left as an exercise (see Problem 4.3).

4.5.1 The Economic Value of a Performance Forecast

How much better can the bank do if it makes a forecast of the performance of each customer account and then makes offers to those customers that it believes are likely to be goods. We use the influence diagram in Figure 4-12 (p. 151) as our model and again assume that the assessed conditional probabilities are for X given Y as shown on the branches of Figure 4-26. The first node is F, the forecast of customer performance. The decision node D occurs next with the same alternative set as before, followed by node Y and then X. The forecast indicates a good customer 83% of the time. Of all customers forecast to be good 98% turn out to be good, whereas 49% of those forecast to be bad turn out as bad. The net profit or loss of each customer type remain the same. The reader should add the payoffs at the terminal nodes, apply the rollback algorithm, and verify

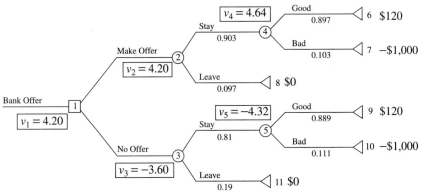

FIGURE 4-25
Alternative tree for the single offer to all customers.

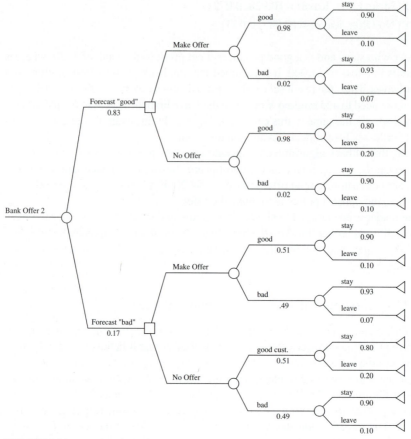

FIGURE 4-26
Bank offer based on performance forecast.

that the optimal policy is to make the offer only to those customers forecast to be goods (see Problem 4.4). The overall expected net profit per customer is $5.76 per customer. Note that the most the bank should be prepared to spend on screening customers of the offer is $1.56 per customer, which is equal to the difference between $5.76 and the result we obtained with the baseline forecast.

4.5.2   Perfect Information about Performance

The bank credit example provides us with an excellent opportunity to understand the many uses of perfect information. Recall that Figure 4-10 (p. 148) illustrates an influence diagram that applies when good/bad performance is assumed independent of the individual's decision to stay or leave. If we had perfect information about the eventual performance of the individual, but not about the individual's decision to stay or leave, we would have an influence diagram such as that shown in Figure 4-27 where the

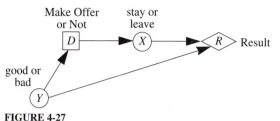

FIGURE 4-27
Perfect information on performance.

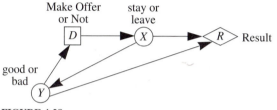

FIGURE 4-28
Infeasible model for perfect information.

good/bad performance node precedes the decision. The same branch probabilities used in the decision tree that represents the influence diagram in Figure 4-10 can be used to calculate the value of perfect information.

Let us now consider a slightly more complicated and realistic case associated with the decision tree in Figure 4-25. To calculate the economic value of perfect information on performance, we would have to solve a decision tree in which (1) Y is conditional on X but in which (2) Y precedes D. The corresponding influence diagram is shown in Figure 4-28 in which there is an infeasible directed cycle of nodes D, X, and Y.

There is an easy way to resolve this difficulty as we already have the assessments for $p(x \mid y)$ in Figure 4-24 (p. 170). Note that this would correspond to a reversal of arc (X, Y), in which case we would have a feasible and meaningful influence diagram whose associated decision tree could be easily solved. The result is shown in Figure 4-29 in which performance precedes the decision. In fact, this decision tree makes use of the original conditional probability assessments that were given in the problem formulation on page 170.

4.6 COLON CANCER DECISION PROBLEMS

In Section 4.3.4 we constructed a number of influence diagrams to model some decision problems associated with the diagnosis of colon cancer. In this section we construct the decision trees associated with some of these diagnostic tests to illustrate how the tests play a role in solving both the detection and cost problems associated with testing.

We start with the problem depicted in the influence diagram of Figure 4-18 (p. 158) where the objective is to minimize expected number of cancers found. Recall

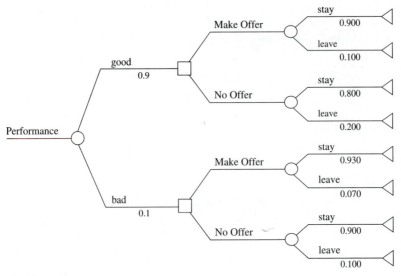

FIGURE 4-29
Decision tree with perfect information on performance.

that we considered the following three possible test sequences in order of increasing cost per patient:

1. *DE* test only,
2. *OB* followed by *BaE* if *OB* test positive,
3. *DE* followed by *SIG* if *DE* test negative.

These test sequences make up the decision set $\mathcal{D}$. In this example we assign costs to the terminal nodes because in two out of the three choices in $\mathcal{D}$ the cost incurred depends on the uncertain outcome of future events and cannot be determined when the decision is made.

Using the same convention as that of Section 3.5.2 (p. 111) we let the categorical forecast take on the value "c" if the test indicates cancer, and "nc" if it does not, so the forecast outcome set is {"c", "nc"}. For the *C* node the outcome set is {c, nc} to denote presence or absence of a real cancer. Before a solution can be found, we need to determine the probability sets for each chance node. The data that are available for a problem of this type are often in the form of a probability that a test will be positive or negative, given that cancer is present or not present. Following Weinstein and Fineberg (1980) we assume (as we did on page 112 in Chapter 3) that:

1. Colon cancer is present in 1 of every 1,000 in the particular target population,
2. *DE* finds all of the 10% of cancers in the very low end of the colon,
3. *SIG* finds all of the 85% of cancers in the first 25 to 30 cms,

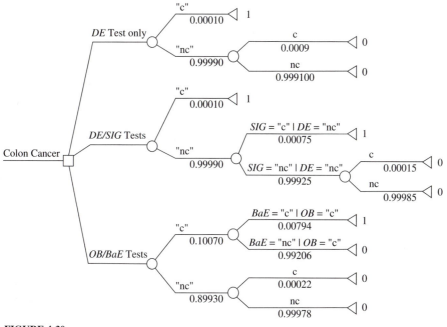

FIGURE 4-30
The decision tree for Figure 4-18.

4. *OB* will find 80% of any cancers present but will indicate that 10% of those with no colon cancer have the disease,

5. *BaE* finds all cancers with no false indications.

Figure 3-10 (p. 112) shows the structure of an influence diagram consistent with these assumptions. It was used in that section to help in finding the required decision probabilities.

The decision tree for the problem of maximizing the expected number of cancers detected (as depicted in the influence diagram of Figure 4-18 (p. 158)) is shown in Figure 4-30. Notice that some test nodes are directly followed by result nodes with no intermediate *C* node. In these cases the tree has been simplified because a positive result on these tests (*DE*, *SIG*, and *BaE*) indicates cancer with certainty; this is not the case for a single *OB* test. The payoff values at the result nodes are 1 on those paths with a positive *DE*, *OB*, *SIG*, or *BaE* test result, and 0 otherwise. The reader is encouraged to apply the rollback algorithm and show that if each decision alternative were applied to 100,000 patients, *DE* alone would find on average 10%, *DE/SIG* 85%, and *OB/BaE* 80% of the 100 cancers expected to be present. Thus, of the three test sequences considered in this single decision problem, the optimal policy is to administer a *DE* test and, if negative, to follow it with a *SIG* test. Before leaving this example, note that the tree could be further simplified by the removal of the remaining cancer (*C*) nodes if we use total cancers found as a measure of effectiveness. Why? Look at the payoffs at the ends of these branches.

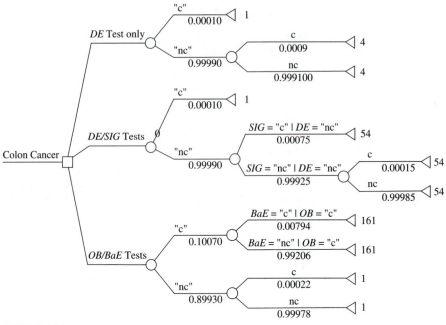

FIGURE 4-31
The decision tree for the ID in Figure 4-19.

We turn next to the minimum expected cost formulation shown in Figure 4-19 (p. 158). The decision tree for this problem is shown in Figure 4-31. As stated in Section 4.3.4, the costs of the *OB*, *DE*, *SIG*, and *BaE* tests are $1, $4, $50, and $160, respectively. The costs of each path are shown at the terminal nodes. The rollback algorithm shows that the expected costs for screening 100,000 patients are (1) *DE* only – $400,000, (2) *DE* with *SIG* if *DE* = "nc" – $5,399,500, and (3) *OB* with *BaE* if *OB* = "c" – $1,711,200. Clearly, the optimal expected cost policy is *DE* only.

Suppose we construct a decision tree from Figure 4-20 (p. 159) for the minimum expected cost problem. This is shown in Figure 4-32. Notice how much smaller and less complex it is than the one in Figure 4-31 and yet it gives precisely the same answer.

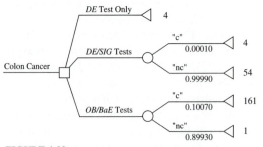

FIGURE 4-32
The decision tree for the reduced ID in Figure 4-20.

Great savings can often be achieved by properly and efficiently constructing the influence diagram before attempting to draw the decision tree.

The expected cost per patient, the expected number of cancers found, and the expected cost per cancer found in 100,000 patients, are shown in Table 4.1 for each of the three decision alternatives. Obviously, we have only included a few of the many possible sequences of tests, and more could be added to the list. With any list of test sequences, one must estimate the number of expected true cancers detected, calculate the expected cost of such a program, and select a preferred policy. In doing so, one must be extremely careful to estimate correctly the conditional probabilities that a particular test can identify a new cancer given the outcomes of previous tests in the sequence. There is a difficulty with the tree in Figure 4-30 that is inherently linked with the making of lists. At only seven of the eleven terminal nodes can we be certain that a correct diagnosis has actually been made, be it detecting a cancer or concluding that one is not there. In the second, third, sixth, and seventh terminal nodes from the top the reports of the test are not conclusive; there is a chance of false negatives, that is a report of no cancer when one may indeed be present. Thus, we are comparing apples and oranges if we compare the detections at each terminal node. From the point of view of the patient it would make sense to continue testing in those cases where a correct diagnosis is not ensured. From the point of view of an insurance company proposing a procedure for a comprehensive health plan, extended testing may not be affordable for every patient. Nevertheless, every decision maker should realize that seven of the eleven terminal nodes represent correct diagnoses while in four we must remain uncertain about the presence or absence of colon cancer.

In Table 4.1 we see (perhaps not surprisingly) that the ordering of test sequences from cheapest to most expensive cost per patient tested does not coincide with the one obtained from policies that order from most to least expected number of cancers found. The average cost per cancer found increases about $20,000 in going from Policy 2 to 1 and again in going from 1 to 3. Thus, a commonly used test pair, *DE* followed by *SIG*, is actually the most expensive (in terms of cost per cancer found) of the three sequences considered. Again, we emphasize the phrase *true cancer found* because we are interested in measuring costs of true cancers, not costs of forecasts or reported detections that may be false.

4.6.1   Sequential Decisions for Colon Cancer Detection

Figure 4-21 (p. 160) shows an influence diagram for a colon cancer problem with two decisions, where the outcome of the first decision can affect the choice for the second

TABLE 4.1
Results for single decision model on colon cancer tests

Test sequence	Expected cost per patient (dollars)	Expected number colon cancers found (per 100,000 patients)	Expected cost per colon cancer found (dollars)
1. *DE* only	4.00	10	40,000
2. *OB,BaE* | *OB* = "c"	17.11	80	21,390
3. *DE,SIG* | *DE* = "nc"	54.00	85	63,523

decision. As pointed out, there are many possible sequences of the four tests, but not all of them make sense economically. Because a *BaE* test is assumed to always find any cancer that is present with no false alarms, it does not make sense to follow it with any of the other three tests. Here, we assume that the first test decision D_1 is either *OB* or *DE*, the two least expensive tests. There are a total of 32 possible test sequences that start with these. The number of test sequences starting with *OB* is $1 + 3 + 6 + 6 = 16$, counting the *OB* test alone, all pairs starting with *OB*, all triplets, and all four tests; there are the same number starting with *DE*. Eight of the test sequences initiated with *OB* are worthy of consideration, and eight are not. As just pointed out, the sequence {*OB*, *BaE*, *SIG*} is dominated by the less expensive sequence {*OB*, *BaE*}, and a *SIG* test should never be followed by a *DE*. How does one go about analyzing this problem with so many different and difficult choices? Which tests should be performed and in what order based on earlier results? The answers to these questions depend on the assessed probabilities, the incidence of cancer in the population being tested, the costs of each test, and the expert opinions of doctors familiar with patient care and the interpretation of the tests. One must not only provide a general strategy and a topology for the influence diagram and decision tree that makes medical and economic sense but one must also be able to estimate correctly coherent decision probabilities on branches and paths of the tree that recognize the outcomes of earlier tests. As in the single decision colon cancer examples, we assume that these have been calculated. Details of how this is done are given in Chapter 3.

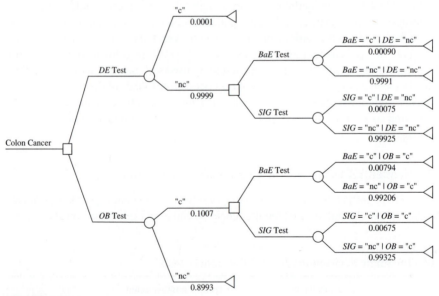

FIGURE 4-33
Two sequential decisions for colon cancer detection.

One approach to the problem is to make the first decision a selection of either a *DE* test or an *OB* test, as each of these, though not perfect discriminators are reasonably inexpensive compared to the other two. The next decision is based on the results of the first test; in each case we have the new option of performing more expensive *BaE* or *SIG* tests, but we *do not specify a priori* that one testing sequence is preferred to another. The conditional probabilities and trade-offs between number of cancers detected and costs of detection will help us decide which are the preferred testing combinations. The reader is encouraged to write out the node sets using these assumptions and to verify the structure of the tree in Figure 4-33 using the probabilities found in Section 3.5.2 (p. 111).

By approaching the testing problem in this way one can "let the decision tree speak for itself" by using a rollback procedure to suggest good combinations of tests (i.e., the number, order, and type of test). We have already found that the least discriminatory tests are the least expensive so it probably makes sense to consider them first.[15] As mentioned, a decision tree formulation requires a Bayesian analysis of the likelihood data provided by medical experts that can then be used by the decision maker (clinical doctor or public health official) to analyze optimal strategies for effective use of funds.

The reader should apply the rollback algorithm to Figure 4-33 using first the costs given earlier and then the appropriate values to find the optimal policy that maximizes the expected number of cancers found. Using cost as our measure of effectiveness, the optimal policy is

1. Administer an *OB* test initially,
2. a. If the result is positive, follow up with a *SIG*,
   b. If the result is negative, end the testing.

The resulting expected cost is $6.00 per patient, but only 68% of cancers are found. Using cancers found as our measure of effectiveness, the optimal policy is

1. Administer a *DE* test initially,
2. a. If the result is positive, end the testing,
   b. If the result is negative, follow up with a *BaE*.

This policy finds every cancer for an average cost per patient of $163.98.

4.7 IRRELEVANT DECISIONS AND A GAME-SHOW PROBLEM

Consider the first decision node in Figure 4-3 (p. 141), and note that its direct successor node set consists of D_2 and R. Contrast this with the first decision in the influence dia-

---

[15] We do not assume that any of the tests provide additional risk to the patient when the test is performed, at least by comparison with the cost of an undetected cancer in the colon; if they did, this feature could also be included.

grams in Figure 4-16 (p. 155). In the latter figure the only direct successor node to the first decision is another decision. We illustrate by example that in such cases the first decision is irrelevant and can be omitted from the diagram. The general proof of this result is mathematical and left until Chapter 9.

The example chosen to illustrate this result involves a game show where there are three doors on a stage clearly marked 1, 2, and 3. You as the player in this game are told that behind each of two of the doors is a goat, and behind the other is a luxury car. The car is equally likely to be located behind each door and its location is known to the game-show host who waits on stage. The rules of the game are as follows in order of timing:

1. The player picks and announces a door number (decision D_1 with $\mathcal{D}_1 = \{1, 2, 3\}$,
2. The host opens one of the *other* two doors to reveal a goat (a random choice is made in the case where the player picked the door with the car in step 1),[16]
3. The player must decide whether or not to switch from the original choice to the other unopened door (decision D_2 with $\mathcal{D}_2 = \{$Stay, Switch$\}$,
4. The player wins the object behind the final door choice.

This problem has received considerable national publicity recently, as many professionals with a background in probability failed to come up with the correct solution (see vos Savant (1991)). An initial influence diagram for this problem is shown in Figure 4-34, where the chance node X can take on "1st Choice Correct" or "1st Choice Wrong" where by correct we mean that the player's first choice was in fact the door with the car behind it. Assuming that the player has no prior knowledge of the car's location, the probability distribution of X is not affected by either decision, and so there are no directed arcs from either D_1 or D_2 into X. Clearly, the event "1st Choice Correct" has probability 1/3 and "1st Choice Wrong" 2/3. Note that X is shown to the right of D_2. This is done to empha-

Initial Door
Choice

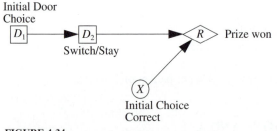
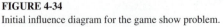

FIGURE 4-34
Initial influence diagram for the game show problem.

---

[16] Note that the strategy of the host is preannounced and does not allow the possibility of revealing a goat behind the first door selected by the player.

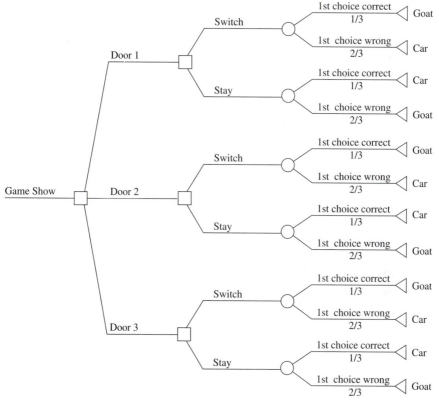

FIGURE 4-35
Initial decision tree for the game show problem.

size that X is not known when D_2 is made. The decision tree for this problem is shown in Figure 4-35.

Note first that the optimal second decision is always to switch since this doubles the probability of winning a car from 1/3 to 2/3. But also note that this second decision is the same no matter what choice is made for the first decision. This should be obvious to readers after a little thought because of the symmetry of the problem. But it could have been determined directly from Figure 4-34 because the only direct successor node to D_1 is D_2. We should have removed node D_1 from the influence diagram *before* constructing the decision tree. If we had, we would have drawn the much smaller and simpler tree shown in Figure 4-36. Not only has the tree size been reduced by two-thirds but also the optimal solution is much clearer to see and understand. One should always look for decision nodes in an influence diagram that have only decision nodes as direct successors and eliminate them before constructing the decision tree.

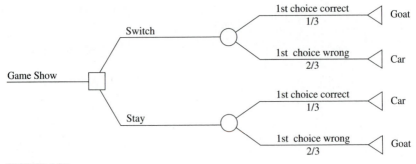

FIGURE 4-36
A simpler game show decision tree.

4.8   A BUDGET PLANNING PROBLEM

We end this chapter with the formulation of a typical two-stage budget allocation problem that arises in many federal activities. It is used to further illustrate model formulation through the development of an influence diagram and its associated decision tree. In a real situation there may be more alternatives to consider, but the basic structure we assume of timing of information and decisions is quite common. We start by describing the problem as it is perceived by the decision maker.

It is close to the start of a new fiscal year for the federal government. The Navy Department has an overall plan that includes spending $54 (units are in millions of dollars) in the coming year at one of its activities (for example, the Naval Postgraduate School). Although this figure is in a plan, the actual available budget may be quite different from this. The planning figure, as well as the actual expenditure of $50 in the year that is just ending, are known to the person in charge of the activity whom we refer to as DM (Decision Maker). DM must approve an internal financial plan that authorizes expenditures for the coming year at a decided-upon rate (in $millions/year). This rate will determine the fraction of the activity's mission that will be accomplished during the coming year. DM's basic dilemma is that a starting rate at the beginning of the year sufficient to accomplish the planned work could lead to overspending the (unknown) available budget; a starting rate lower than this would not exceed the budget but would fail to accomplish the complete mission.

The two figures given are the only hard data available at the time the initial expenditure rate decision must be made. It is also known that Congress has not yet passed a budget for the department and probably will not do so until some 4 to 5 months after the start of the budget year. Experience has shown that the activity will receive a "firm" yearly budget figure at mid-year, at which time the activity can review and adjust its financial plan. In past years, additions or reductions in available budgets have often occurred in the last few weeks of the year. Such adjustments can be expected for the upcoming year. DM's objective is to choose an initial spending rate that maximizes the efficiency of the activity, knowing that an adjustment to the rate can be made at mid-year and without overspending (a federal offense) or underspending (resulting in an almost certain cut in the Activity's budget for the following year). In a federal department,

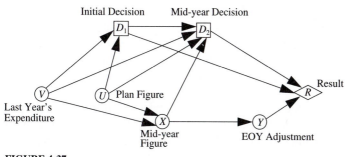

FIGURE 4-37
Initial influence diagram for the budget problem.

unlike the private sector, it is almost always the case (there are exceptions, but we assume they do not apply here) that money left unspent at midnight on 30 September of a given year disappears. It cannot be banked and carried into the next year.

The DM's advisors at the activity meet to formulate a plan and recommendation. They discuss their options and decide that two decisions are to be made in the year, (1) the annual spending rate D_1 at the start of the year, and (2) a revised figure D_2 at mid-year. Because the budget figure will not be known to the activity for some months, they decide to treat it as a random variable X. They also assume that the year-end adjustment is a random variable Y whose value will probably depend on X and the planning figure $U = 54$, and last year's figure $V = 50$. Although at this point they are uncertain how to measure their performance, it is clear that their planning will give some result they call R. They call in their analyst to help formulate and solve the problem.

The analyst's first task is to draw an influence diagram that accurately represents the structure of the problem. This is done by soliciting the following agreed-upon information and dependencies from the advisors:

1. Both U and V are known before the first decision D_1 is made.
2. The mid-year figure will depend on both U and V, but not on the first decision D_1.[17]
3. The mid-year decision D_2 will depend on X which will be known when D_2 is made and may depend on D_1.
4. The end-of-year (EOY) adjustment Y will depend on X but not on D_2.
5. The result will probably depend on both the initial and mid-year decisions as well as any EOY adjustment.

These five agreements lead to an initial influence diagram in Figure 4-37. Note that a no-forgetting arc has been included from U to D_2.

---

[17] One can conceive of dependencies between D_1 and X, for example, cutbacks in X if the people who decide its level know that the activity has chosen an initial spending rate D_1 that is low. If this were possible, we would need a directed arc from D_1 to X.

On discussing the implications of this diagram with the DM's advisors, the analyst is particularly interested in the roles of U and V. It is important to determine if these vary at all during the period of the decision problem; if not they can be viewed as given data and thus do not have to be represented by uncertain chance nodes in the influence diagram. They all agree that this is the case, so the analyst simplifies the diagram to that in Figure 4-38.

In order to proceed with the analysis the analyst must solicit some agreed-upon measures for the uncertainties, and the alternatives they will assume at each decision node. Using their corporate intelligence together with known data, the advisors predict that there is a 60% chance that the mid-year figure X will be $54 (same as the plan), 10% that it will be $50 (same as last year's expenditure), and 30% that it will be $52. They assume that at mid-year, once X is known, they will have three options for the mid-year decision D_2 so that $\mathcal{D}_2 = \{A,B,C\}$, where:

A: Adjust the expenditure rate to produce an annual rate X, so $D_2 = X$,
B: Plan on receiving EOY funds equal to 54–X, so $D_2 = 54$,
C: Plan on receiving EOY funds equal to 5% of X, so $D_2 = 1.05X$.

They now use their collective intelligence and past experience to quantify as best they can what will happen at year end. They agree on the following:

$$Pr\{Y=0 \mid X=54\} = 0.8,$$
$$Pr\{Y=0.05X \mid X=54\} = 0.2,$$
$$Pr\{Y=0 \mid X=52\} = 0.6,$$
$$Pr\{Y=2 \mid X=52\} = 0.3,$$
$$Pr\{Y=-1 \mid X=52\} = 0.1,$$
$$Pr\{Y=0 \mid X=50\} = 0.4,$$
$$Pr\{Y=2 \mid X=50\} = 0.4,$$
$$Pr\{Y=4 \mid X=50\} = 0.2.$$

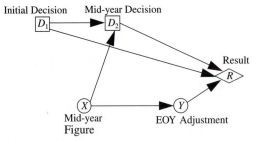

FIGURE 4-38
Second influence diagram for the budget problem.

Considering practical constraints on the decision alternatives for D_1 they agree to consider only three possibilities, $\mathcal{D}_1 = \{A,B,C\}$, where:

A: Start spending at a rate of $54,
B: Start spending at a rate of $52,
C: Start spending at a rate of $50.

The analyst asks the advisors how they wish to measure performance. They consider three possible measures, minimize (1) the expected overspending for the year, (2) the expected underspending, and (3) the disruption from their planned $54 program. They agree that all three measures are important. In Chapter 7 we show how to solve the problem considering all three simultaneously as a multiattribute problem. At this point, seeing that the Comptroller is both vociferous and responsible for ensuring that the activity does not overspend, they agree to use measure (1). Because the total yearly expenditure will be D_2 and the amount received $X + Y$, the analyst realizes that by using measure (1) the result is given by

$$R(X, Y, d) = Max\{0, d - (X + Y)\},$$

where d must be one of the elements of $\mathcal{D}_2 = \{X, 1.05X, 54\}$. Thus, the result does not depend on D_1, so the arc from D_1 to R can be removed. When this is done, the only direct successor node to D_1 is D_2, so D_1 can be removed, leading to the influence diagram in Figure 4-39.

The analyst now has enough structure to draw the decision tree shown in Figure 4-40. The rollback procedure, described and illustrated in earlier sections, is used to determine the minimum expected overexpenditure. It is straightforward to determine the value at each of the 22 terminal nodes, then find expected values at the Y nodes, and find minima at the D_2 nodes and expected value at the X node.

The reader can check that the optimal decision is: At mid-year when X is known, readjust the spending rate to give an annual rate of X.

This is the mid-year decision the advisors plan to give to DM who will make the actual decision. Note that the initial spending rate is not determined because D_1 is not a factor in the model when the objective is to minimize overexpenditure. The analyst

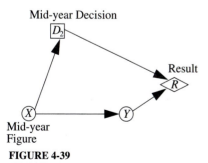

Mid-year Decision

Result

Mid-year
Figure

FIGURE 4-39
Third influence diagram for the budget problem.

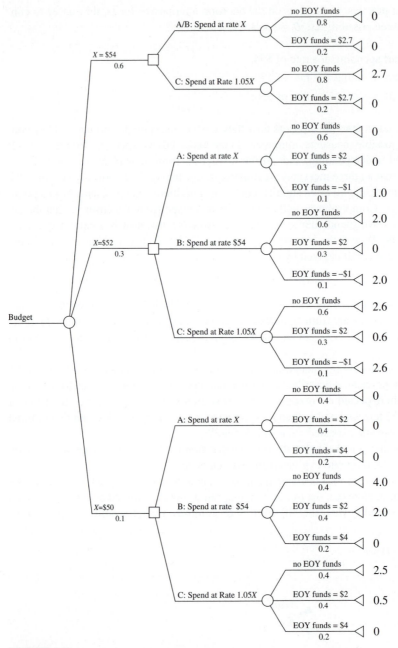

FIGURE 4-40
Budget problem decision tree with overspending.

points out that the same optimal mid-year decision applies no matter what original spending rate is chosen. As the mission plan calls for an expenditure rate of $54, they plan to recommend this to DM as the starting rate.

From past experience the advisors know they can expect some probing questions as to how they came up with this answer, what assumptions they made, and what are the risks to the activity if their advice is taken. Some easily calculated measures of risk that the reader can check are:

$Pr\{$Activity Underspends$\} = 0.27,$

Expected amount underspent at year end = \$664,000,

$Pr\{$Activity Overspends$\} = 0.03,$

Expected amount overspent at year end = \$30,000.

Probably the most important result the advisors should tell DM is that *all three alternatives for the first decision give the same expected loss. The answer is independent of the distribution assumed for X.* It is often this type of result that decision makers are surprised to hear. Whenever future uncertainties can be shown to have no effect on the outcome of a decision, this is extremely valuable information.

Many variations of this budgeting problem arise in the real world. Some are given as problems at the end of the chapter, and the problem is revisited in later chapters. In Section 7.9 (p. 288) we solve it for the three (conflicting) measures of effectiveness we mentioned.

4.9   SUMMARY AND INSIGHTS

Influence diagrams and decision trees play an important *joint* role in the interactive process of building and testing decision models and deriving solutions to decision problems. Influence diagrams capture and illustrate the important influences and relationship of decisions with uncertainties; they also help us understand the probability structure of the problem without getting into the details of specific alternatives and outcomes that can occur, and they help us see the ordering and timing of events. For these reasons an influence diagram is an excellent tool for discussing the formulation of a problem with decision makers who may not have and may not be interested in quantitative skills or formulations; it is possible to use the influence diagram as a framework in which to discuss different formulations and ensure that the correct problem is being modeled before one initiates the more detailed and elaborate assessment of costs and probabilities and the specification of actual outcomes and alternatives.

Figures 4-1 through 4-21 demonstrate a number of ways to convert an initial understanding of a problem into models suitable for analysis. Although it is difficult to portray the interactive process of model building in a text, we hope that the reader can at least get a flavor of this important aspect. To the greatest extent possible the model-building phase should be carried out in close cooperation with the person(s) who will make decisions as well as those people who have knowledge about and can comment sensibly on the probabilistic structure of uncertain events and the design of forecasts. In general, these are different people, possibly even different institutions; the model builder must have access to a great diversity of expertise.

TABLE 4.2
A comparison of influence diagrams and decision trees

Model features	Influence Diagram	Decision tree
Modeling and Representation		
Timing	Shows timing of all decisions and uncertain events.	Shows timing of all decisions and uncertain events.
Conditional Independence	Shows dependence among uncertain events and decisions.	Dependence among uncertain events and decisions not shown.
Size	Number of nodes grows linearly with the number of variables.	Number of terminal nodes and paths grows exponentially with the number of variables.
Data	Identifies dependencies of variables without need for data.	Decision, probability and result data shown explicitly.
Variable Type	Both continuous and discrete decisions and probabilities.	Adequate only for discrete variables.
Asymmetry	Scenarios with different event sequences not distinguished.	Shows asymmetric structure of problem.
Modeling Usefulness	Most useful in initial stages of modeling. Captures interaction between decision maker and analyst	Useful in depicting detailed uncertain event outcomes and decisions, and model solution. Difficult to display large problems.
Solution Process		
Bayes' Rule	Indicated by arc reversal, but calculation not show.	Indicated by node reversal (a separate event tree may be used as an aid for calculation).
Method	Reduction by a set of reduction operations possible using advanced methods (see Chapter 9).	Uses simple rollback algorithm.

Decision trees show quite different aspects of decision models than do influence diagrams. Whereas the latter stress internal model structure, assumptions and dependencies, and the interrelationships between the variables and their timing, the former show the details of all possible sequences of decisions and uncertain outcomes. Thus, they portray the multiplicity of paths or scenarios that lead to possible outcomes. Table 4.2 summarizes the differences, strengths, and weaknesses of the two graphical methods.

The rollback procedure we have described in this chapter can be used for hand calculations on small trees, can be easily implemented on a spreadsheet for larger problems, and forms the basis for many solution methods. Efficient solution of very large problems requires refinements to the basic rollback procedure that recognize asymmetry, avoids repetitive calculations, efficiently stores information or results that are used repeatedly, and recognizes more complicated path dependencies than the ones we have worked with thus far. Influence diagrams can also be used directly for model solution, without constructing decision trees. The methods for these more advanced techniques are discussed in Chapter 9.

In proceeding from an influence diagram to a decision tree it is important to avoid an overly complex structure. Even simple conditional independence assumptions can go a long way toward explaining the structure that is thought to be an important aspect of the decision problem. Once an initial formulation has been obtained, one should take the additional step to search for the presence of irrelevant decisions or barren nodes that remain in the formulation. The size of a decision tree can be significantly reduced by removing irrelevant decision nodes and, wherever possible, by eliminating barren nodes in the influence diagram.

Once we have a model and initial solution using given data, an important next step is to be able to demonstrate the sensitivity of optimal decisions to changes in important model parameters. In the colon cancer case, for example, it was assumed that 1 in 1,000 of the target population have the disease. We can never know the true value of this parameter without testing every individual with a perfect test. Would the same sequence of tests be optimal if the frequency of occurrence were 1 in 500, or 1 in 2,000, or 1 in 10,000? The analysis of the sensitivity of an optimal decision policy to important parameter changes is the subject of the next chapter.

There are many excellent books and papers on decision *theory* (see, for example, French (1993), Lee (1989), Lindley (1985), and Smith (1988a)). Little attention is given to model-building aspects as the authors are primarily interested in the theoretical underpinnings of decision making. Thus, there is little need to confront the complexities that arise when one or more decisions and many chance events are interrelated and how one might go about untangling the complicated web of dependencies.

There is much discussion in the original cloud seeding paper by Howard, Matheson, and North (1972) on how to estimate the conditional probability distribution of V, given U and D; even though we have not emphasized that aspect of the problem in this chapter it is obviously a most important aspect of finding a solution to the problem.

The colon cancer testing problem we describe in Section 4.3.4 is based on material in Chapter 8 of Weinstein and Fineberg (1980); they discuss five different test sequences as admissible strategies for screening out of the sixty-four possible sequences. Although their analysis makes use of the probabilities we label as forecast likelihoods, they do not discuss the conditional dependence and independence structure implied by the different tests in Figure 3-10 (p. 112); thus, their calculations and optimal test strategies differ from ours.

The simple tree-labeling procedure in the rollback algorithm (Raiffa (1968)) is based on the well-known principle of optimality (Bellman and Dreyfus (1962)). Although the authors of tree rollback algorithms do not mention the connection, it should be remembered that they are nothing more than a simple application of the principle of optimality in dynamic programming.

The game show problem described in this chapter is only one of a number of versions. At least two have been popularized by vos Savant (1992), one of which coincides with our formulation where the decisions and actions of the game show host are announced prior to the selection of the first door. For further details and interesting reading, see also Gillman (1991, 1992), Klein (1993), and Tierney (1991).

PROBLEMS

4.1 Select a problem from your own experience or from a newspaper that suggests decision making under uncertainty.

   a. Make a list that names and clearly identifies (1) the decision variable(s), (2) quantities that are known at the time the decision(s) are made, (3) quantities that are uncertain at the time decisions are made, (4) different possible decision makers, and (5) the appropriate objectives and value function(s) for each decision maker.

   b. Draw an influence diagram that illustrates the sequence of decisions and events. (1) Identify and write out the sets of decision alternatives for each decision node, (2) uncertain outcomes for each random node, and (3) the list of all conditional independencies.

   c. Draw the corresponding decision tree.

   d. Redraw the influence diagram and decision tree when one of the uncertain quantities is known perfectly before you made the decision.

4.2 Use Bayes' rule to check the probabilities for the bank credit problem used in Figure 4-26 (p. 172).

4.3 Calculate the optimal solution to the bank credit problem in Figure 4-24 (p. 170).

4.4 Complete the calculation of the optimal solution to the bank credit problem in Figure 4-26 (p. 172).

4.5 Draw a new decision tree where the X and Y nodes are reversed in Figure 4-26 (p. 172); calculate the appropriate new branch probabilities, and obtain the optimal policies and solutions. Compare the solutions with the results obtained in the previous problem.

4.6 For the bank credit problem calculate the optimal policies and expected returns when we have perfect information on performance as shown in Figure 4-29 (p. 174).

4.7 In Figure 4-11 (p. 149) discuss the difficulties that arise when we attempt to move node X to the left of node D, that is when we have perfect information on the stay/leave node.

4.8 Assume that two forecasts are available in the bank credit problem; one that predicts performance is now denoted by F_P (formerly F), and one that predicts customer behavior (stay or leave) is denoted by F_B. Modify the influence diagram shown in Figure 4-13 (p. 152) and draw the associated decision trees. Are F_P and F_B dependent?

4.9 Once the bank manager understands the formulation of the bank credit problem in Figure 4-13 (p. 152) she decides that she would like to see an analysis made where there are three alternatives: (1) make no offer, (2) make one with an incentive to keep goods, and (3) make an offer that encourages bads to leave. Under independence assumptions identical to those used in Figure 4-13, draw a decision tree that models this problem.

4.10 In the car and goats game problem, recall that the host never opens the door that reveals the car after the player's initial choice. Let q denote the probability that the host opens door 1 given a goat is behind door 1.
   a. Draw the influence diagram and decision tree for this problem.
   b. What is the probability that you (the player) win if you switch given that you selected door 1 and the game host opened door 3? *Hint:* You may want to use Bayes' rule.
   c. How does this solution compare with the solution already given in the text?

4.11 For the budget planning problem of Section 4.8, show that if R is the amount left unspent at the end of the year, $R = Max\{0, (X + Y) - D_2\}$. Find the optimal decisions using this result function, including any that are equivalent. Assuming that an optimal policy is used, calculate for each one the probability of overspending, the expected amount overspent, the probability of underspending, and the expected amount underspent. State how you would use these results to resolve ties.

4.12 Suppose in the budget problem that the planned program can only be completed by expending at a $54 rate all year. Any loss due to the program not being completed in the first half of the year cannot be made up in the second half. Show that the appropriate result function is

$$R = 27 - 0.5D_1 + Max\{0, 0.5D_1 + 27 - D_2\}.$$

Solve the problem using this result function.

4.13 Consider the influence diagram in the following figure. You have been told that $\mathcal{D} = \{0,1\}$, $X = \{a,b\}$, $\mathcal{Y} = \{s, m, g\}$ and that $R = R(y,d)$.

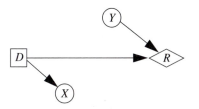

Draw the decision tree that includes all the nodes in and is consistent with the influence diagram where X occurs before Y. Discuss the implication of perfect information on X and separately on Y.

CHAPTER
5

MODEL
ANALYSIS

5.1 INTRODUCTION

Having built a decision model resulting in a decision tree and having found a solution using the rollback algorithm, in this chapter we show how one can determine the sensitivity of the solution to important parameter values. As we have stressed earlier in this book, in many decision problems judgment often plays a major role in assessing parameter values. If the consequences of a decision are significant, an important and perhaps crucial feature of a model is whether or not it allows one to measure the sensitivity of the optimal solution to this judgment. This idea was introduced in Chapter 1 using a decision sapling and is illustrated in Figure 1-8 (p. 29). The parameter used was the probability of success in the risky venture. The reader could see that it is always optimal to choose the risky venture when this probability is above a certain critical level that depends on the payoff values for the problem. Many of the examples cited in Table 1.2 (p. 30) are decision problems with serious consequences. In those cases it would be very important for a decision maker to know whether or not their estimate of the risky venture success probability is close to the critical level.

Our first example is a revisit to the football betting problem of Chapter 2. This is done in Section 5.2 and is the first of a number of examples that demonstrate an analysis when we vary two of the model parameters. The models discussed in Section 5.3 can be thought of as extensions to the simple decision sapling model. Rather than having simply a choice between the risky venture and the riskless alternative, a third option is available, that of seeking more information in the form of a forecast to reduce the uncertainty in the risky venture. Sections 5.4 and 5.5 contain detailed analysis of this model, first in terms of the decision probabilities, and second in terms of the forecast likelihoods. Section 5.6 looks at a model when more than one forecast is available before a single decision is made, and Section 5.7 considers models where obtaining the forecasts is decided sequentially. Both algebraic and numeric methods are demonstrated using a number of examples.

5.2 BETTING ON THE FOOTBALL GAME

In Chapter 2 we introduced a problem where two friends, Nancy and Ron, are in the stands watching a football game. They contemplate betting with each other on the outcome that will result after the coach has made his decision to have the team kick or attempt the two-point conversion. They agree on the following:

1. If the coach decides on the two-point conversion and the team scores, Ron pays Nancy $2. If they do not score, Nancy pays Ron $1.
2. If the coach decides on the kick and the team scores, Ron pays Nancy $v. If they do not score, Nancy pays Ron $1.

Unlike the previous presentation we assume they have yet to decide how big the amount v should be (it was 25 cents in Chapter 2). We want to compare the two bets for different values of v to see whether we might prefer one to the other or be indifferent to either. The result R is again the payoff to Nancy. It depends on both the coach's decision and the outcome of the play and is shown in Table 5.1. Recall that we model the coach's decision with a random variable. The entries in this table together with 0 constitute the elements of the set $\mathcal{R}$.

Let us assume that

$$Pr\{X = 1 \mid Y = k\} = p_k > Pr\{X = 1 \mid Y = t\} = p_t.$$

Almost everyone would agree that p_t is less than p_k and that v should be less than $2. Because the chance of kicking for one point is thought to be higher than the chance of scoring two points, it seems reasonable to place a smaller bet on the scoring outcome with the higher odds. We also know that each of the conditional probabilities must lie between 0 and 1 so that we have

$$0 \le p_t < p_k \le 1. \tag{5.1}$$

TABLE 5.1
The payoff to Nancy

	Play outcome	
Coach's decision	Score $X = 1$	No score $X = 0$
$Y = k$	v	-1
$Y = t$	2	-1

Our objective is to find Nancy's optimal betting strategy for all possible values of p_k, p_t, and v.

The decision tree for Nancy's problem, originally shown in Figure 2-5, is reproduced in Figure 5-1. Note that it consists of two decision saplings joined to a random node Y. The criterion we use to measure the result of the decision is expected value. Applying Equation (1.19) (p. 19), if the coach's decision is kick, the expected return is

$$E[R(Y,X)\,|\,Y=k] = vp_k + (-1)(1 - p_k) = (1 + v)p_k - 1,$$

whereas if the coach selects the two-point conversion it is

$$E[R(Y,X)\,|\,Y=t] = 2p_t + (-1)(1 - p_t) = 3p_t - 1.$$

Nancy also believes that $p_t < p_k$. Moreover, if she wants both bets to have nonnegative expected values, then the following two additional inequalities must also be satisfied:

$$p_k \geq \bar{p}_k = 1/(1 + v) \quad \text{and} \quad p_t \geq \bar{p}_t = 1/3. \tag{5.2}$$

We can think of $\bar{p}_k$ and $\bar{p}_t$ as critical values of the probabilities p_k and p_t that indicate when betting is as good as not betting. Probabilities larger than these make betting profitable for Nancy; probabilities just equal to these values make her indifferent to betting or not betting. Thus, they are referred to as *indifference* probabilities. If we plot the values of p_k and p_t that satisfy the inequalities (5.1) and (5.2), we obtain the shaded region I in Figure 5-2. Any combination of p_k and p_t inside this region gives a positive expected payoff to Nancy no matter whether the coach decides on k or t. Thus, if she is not sure

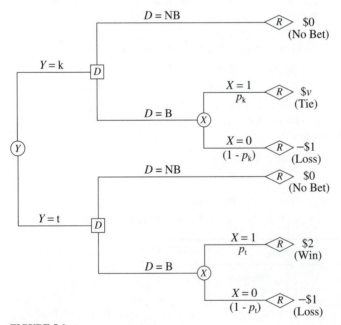

FIGURE 5-1
Nancy's decision tree, coach's decision known.

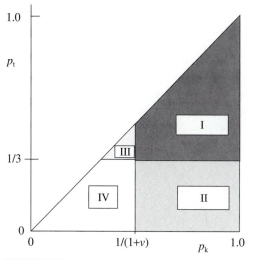

FIGURE 5-2
Policy space analysis for Nancy's betting problem.

how well the team will perform, as long as her estimates of p_k and p_t fall in region I, her expected payoff is positive. Another way to state this result is to note that in this region *the optimal decision is independent of the coach's decision.* If p_k and p_t fall in region II, the expected payoff will be positive only if the coach elects to kick, and in III only if the coach elects to try for the two points. Because we are assuming that she knows the coach's decision before placing the bet, she can combine area I with either II or III as appropriate. The remainder of the feasible region (i.e., where $0 < p_t < p_k$) indicated by the unshaded portion IV shows those values of p_k and p_t for which a bet has a negative expected payoff no matter what the coach decides. In summary the optimal policy is:

Case I: If $Y = k$ or $Y = t$, Bet ($d = B$),
Case II: If $Y = k$, Bet, and if $Y = t$, Do Not Bet ($d = NB$),
Case III: If $Y = k$, Do Not Bet, and if $Y = t$, Bet,
Case IV: If $Y = k$ or $Y = t$, Do Not Bet.

Notice that when $v > 2$ there is no area III; any combination of p_k and p_t with $0 < p_t < p_k$ that gives a positive expected payoff if the coach elects t also gives a positive expected payoff if the coach elects k.

The reader may find it easier to understand the structure of Figure 5-2 by viewing it as resulting from the superposition of two figures, each for a decision sapling as in Figure 1-9 (p. 33), one of them plotted horizontally and the other vertically.

We now look more closely at the regions of positive expected payoff to Nancy. Only in area I does she bet no matter what the coach decides. She will be indifferent to a kick or two-point conversion choice by the coach provided the expected gains from each bet are equal. This is true when

$$3p_t = (1 + v)p_k.$$

This "indifference" equation is shown in Figure 5-3 as a straight line with positive slope $(1 + v)/3$ passing through the southwest corner of region I. Points (p_k, p_t) lying on one of the solid lines in regions I, II, and III all have the same expected payoff for a given value of the probability the coach decides on the kick, $p_C = Pr\{Y = k\}$. These lines are referred to as *isocontours*. The optimal expected payoff is zero in area IV, a ramp parallel to the p_k axis and increasing as p_t increases in Area III, a ramp parallel to the p_t axis and increasing as p_k increases in Area II, and a ramp increasing toward the northeast corner in area I.

As in Chapter 2 we now turn to Nancy's betting decision when she does not know the coach's decision before placing the bet. The decision tree for this problem is reproduced in Figure 5-4. Unlike the previous case, if Nancy makes bets before she knows the coach's decision, she must assess the probability that the coach will make decision k or t. We assume that her decision will be to bet only if the expected return is nonnegative. Recall that $p_C = Pr\{Y = k\}$, Nancy's assessment of the probability that the coach will select k. Then

$$E[R(Y,X) \mid D = B] = p_C[(1 + v)p_k - 1] + (1 - p_C)(3p_t - 1),$$

(5.3)

$$E[R(Y,X) \mid D = NB] = 0.$$

Nancy would bet only if the first of these expressions is greater than zero; of course, the decision she selects depends not only on the p_C value but on p_k and p_t as well. The reader should find the boundaries of the new feasible region where Nancy is willing to enter a bet under the assumption that the coach's decision is not known before she must make her bet (see Problem 5.1).

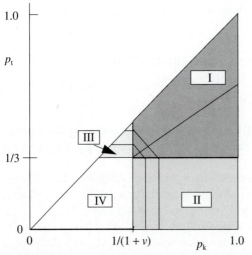

FIGURE 5-3
Analysis of the optimal expected gain.

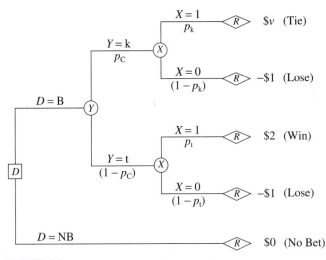

FIGURE 5-4
Nancy's decision tree, coach's decision unknown.

5.3   AN EXPERT OPINION MODEL

We now turn to the formulation and analysis of a decision model that describes many problems found in real-life situations; we call it an "Expert Opinion" model. Suppose we start with the simple decision sapling as shown in Figure 1-8 (p. 29). In addition to the choices of taking a Risky Venture or a Riskless Alternative given our state of knowledge, a third alternative is to seek more information from an expert as to the success probability of the risky venture. Look at the examples in Table 1.2 (p. 30). All except the lottery example are candidates for seeking a forecast from an expert. Such forecasts or expert opinions are sought in order to improve the predictions of outcomes and reduce uncertainties, typically at an additional cost. Whether or not it pays to obtain such a forecast depends on whether it contributes sufficient value to the solution of the original problem. Although variants of this problem can lead to complex probabilistic structures, we concentrate on some simple models.

Suppose that our current estimate p_X of a successful outcome of the Risky Venture can be modified by obtaining a forecast or an expert opinion. If our first decision is to obtain a forecast, we must then make a second decision, D_2, based on the forecast information, to choose either the Risky Venture or the Riskless Alternative. Formulation of this problem is illustrated in the influence diagram of Figure 5-5 and the corresponding decision tree in Figure 5-6.

We start by assuming that an initial decision D_1 must be chosen from the decision set $\mathcal{D}_1 = \{RA, RV, Forecast\}$, where RA and RV represent choosing the Riskless Alternative and Risky Venture, respectively. If RA is chosen, the result r_2 is obtained. If RV is chosen, one of the results r_1 or r_3 is determined once the outcome X of the Risky Venture is known. In either of these cases no second decision is made and we need not consider a second decision set $\mathcal{D}_2$. If Forecast is chosen, the forecast outcome F must first

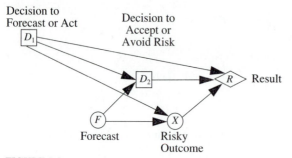

FIGURE 5-5
An influence diagram for the expert opinion problem.

be obtained, then a second decision D_2 must be chosen from the set $\mathcal{D}_2(\text{Forecast}) = \{RA, RV\}$; again the choice of RA results in r_2, whereas RV results in either r_1 or r_3. As was the case in Chapter 1 we assume that $r_1 > r_2 > r_3$, but in addition, in this chapter for simplicity of notation we assume that $r_3 = 0$.

Following our timing convention, the decision, random and result nodes are laid out from left to right in order of their occurrence as shown in Figure 5-5. The arcs connecting D_1 and F to D_2 indicate that both D_1 and F (when measured) are known before D_2 is made and that their values may influence the second decision D_2. The three arcs leading into R show that the result may depend on both decisions and on the uncertain quantity X. The absence of an arc from F to R illustrates the important concept and use of *conditional independence*. Given D_2 and X, the result R is conditionally independent of the forecast F; the effect of the forecast on R is seen through D_2 and X.

The probability distribution we use for the random quantity X depends on our first decision choice, and if this is Forecast, on the outcome of the forecast. This is indicated in Figure 5-5 by the two directed arcs from nodes D_1 and F to node X. Let $X = 1$ if the riskless alternative produces a result r_1, and $X = 0$ if it produces a result r_3. Similarly, $F = 1$ if the riskless alternative is forecast to produce a result r_1, and $F = 0$ if the forecast is for a result r_3. The following probabilities required are required:

$$Pr\{F = 1\} = p_F,$$

$$Pr\{X = 1 \mid D_1 = RV\} = p_X,$$

$$Pr\{X = 1 \mid D_1 = \text{Forecast}, F = 1\} = p_1,$$

$$Pr\{X = 1 \mid D_1 = \text{Forecast}, F = 0\} = p_0,$$

where we have used the notation introduced in Equation (3.22) (p. 100).

The decision tree for this problem is shown in Figure 5-6. In all the decision sapling problems shown in Table 1.2 (p. 30), with the exception of the nuclear power plant, the objective was to maximize an expected return. In many problems, results are stated in terms of *losses*, and our objective is to find decisions that minimize overall expected loss. In this section our examples are cast in this form so that the Riskless Alternative produces a loss of r_2; the Risky Venture results in either a *worst* outcome with a loss of r_1, or a *best* outcome with a loss of 0.

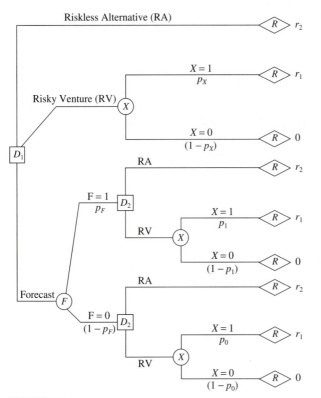

FIGURE 5-6
The decision tree for the expert opinion problem.

The remainder of this section is devoted to the analysis of examples of this simple but general expert opinion problem. The first is concerned with a decision on whether or not to test and replace an aircraft engine part. The second is concerned with protecting an agricultural crop against the danger of freezing weather. Both of these problems were mentioned in Chapter 3.

5.3.1   An Aircraft Part Decision Problem

Expert opinion problems find application in many diverse areas. We illustrate one such problem originally stated as a textbook problem by Lindley (1985):

> *A part of an aircraft engine can be given a test before installation. The test has only a 75% chance of revealing a defect if it is present, and the same chance of passing a sound part. Whether or not the part has been tested it may undergo an expensive rework operation which is certain to produce a part free from defects. If a defective part is installed in the*

engine the loss is [r]¹ (utiles). If the rework operation costs r/5 utiles and 1 in 8 parts are initially defective, calculate how much you could pay for the test and determine all the optimum decisions.

Let $X = 1$ if the (unobservable but true) condition of the part is defective, and $X = 0$ if it is not defective. This problem can be solved in several different ways; we make direct use of the expert opinion tree in Figure 5-6 by inserting specific names on the chance and decision nodes along with numerical values for the decision probabilities and the applicable costs. The first decision D_1 has alternatives $\mathcal{D}_1 = \{$Rework, Install, Test$\}$, where Rework is a riskless alternative that costs $r/5$ (r_2), and Install is the Risky Venture that will cost either r or 0 depending on whether the part fails $(X = 1)$ or not $(X = 0)$. The second decision D_2 has alternatives $\mathcal{D}_2 = \{$Rework, Install$\}$. The reader is encouraged to check the formulation of this problem shown on the tree in Figure 5-7.

Consider the top two alternatives in Figure 5-7. If we rework the part, the cost is $r/5$. If we install the part as is, we are risking r that the part will be defective with probability $1/8$ and risking zero with probability $7/8$ that the part is not defective; the Risky Venture has an expectation equal to $r/8$. If the Test alternative were not available to us, the problem would be a simple decision sapling with the least costly alternative being to select the Install alternative since $Min\{r/5; r/8\} = r/8$.

Up to this point we have a one-stage decision and there is no need to consider the effect that decisions made at a later time might have on the current decision. However, the situation changes when we consider the Test alternative on the bottom branch leading out of D_1 that includes a cost of c for the Test itself. Obviously, every path that contains the F node must include a fixed cost of c because every outcome or decision is now being made under the assumption that the test has been performed. We have chosen to label the Test branch c rather than each of the end nodes with the understanding that every path from this point on must include the cost c. The reason for this convention will become apparent later in the book.

What numerical values do we use for the decision probabilities p_0 and p_1, the probability that the part fails given that the test indicates "not defective" (i.e., $Pr\{X = 1| F = 0\}$) and "defective" $(Pr\{X = 1| F = 1\})$, respectively? These were calculated in Section 3.4.6 using Bayes' Rule and shown on event tree (b) in Figure 3-6 (p. 105). We found that $p_0 = 1/22$ and $p_1 = 3/10$.

Let us now focus attention on the two possible forecasts produced by the test. When the test indicates "not defective" and the Install alternative is chosen, the expected loss is

$$E[R| D_1 = \text{Test}, D_2 = \text{Install}, F = 0] = r/22.$$

Note that we have not yet added the cost of the test because the cost applies to all paths passing through the D_2 decision nodes. Reworking a part guarantees us that it will be in perfect condition, so

---

[1] The notation used by Lindley has been changed from u to r to agree with ours.

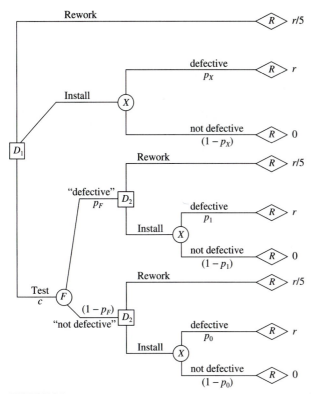

FIGURE 5-7
The aircraft engine part decision problem.

$$E[R| D_1 = \text{Test}, D_2 = \text{Rework}] = r/5,$$

independent of the result of the test. Thus, the minimum expected loss at the bottom D_2 decision node is

$$Min\{r/22;\ r/5\} = r/22,$$

so the best decision when the test result is "not defective" is Install.

We proceed in similar fashion to consider alternatives in the branches leading out of the upper D_2 node (when the test yields a forecast of "defective"). Again, we begin our analysis at the terminal end points of the decision tree and proceed backward (from right to left) using the rollback procedure. Even though we do not know with certainty that we shall ever reach a particular decision node in the decision tree, we must know the optimal policy to use at each such node in order to compute the best decisions at an earlier stage. Making use of our earlier calculations for the decision probabilities when the categorical forecast tells us that the part is "defective," we find

$$Min\{3r/10;\ r/5\} = r/5,$$

so the preferred alternative is Rework. If we follow the preferred alternatives at both D_2 decision nodes, we can label the topmost one $r/5$ and the bottom one $r/22$ because these are the minimal achievable losses at each of these nodes. Again we stress that at this stage in solving the problem we do not yet know whether we will ever choose Test at D_1 and so take a path through one of the lower D_2 nodes.

We also know from our calculations in Chapter 3 that the chance of the "defective" forecast ($F = 1$) is $5/16$ and the chance of the "not defective" forecast ($F = 0$) is $11/16$. Using these we can label the F node with an expected loss equal to

$$E[R|\text{ optimal } D_2 \text{ policy with either forecast}]$$

$$= 5/16(r/5) + 11/16\ (r/22) = 3r/22.$$

Of course, we must add the cost of testing, c, if we are to evaluate correctly the total expected loss when we select the Test alternative:

$$E[R|\ D_1 = \text{Test, optimal } D_2 \text{ policy with either forecast}] = c + 3r/32.$$

We are now in a position to compare the three alternatives at the first decision node and thereby obtain the unconditional expected loss under an optimal policy. Denoting this by l^*, we see that

$$l^* = Min\{r/5;\ r/8;\ c + 3r/32\} = Min\{r/8;\ c + 3r/32\},$$

where the three choices apply to the decisions Rework, Install, and Test, respectively. We see that depending on how large or small is c, the best choice may correspond to the Test alternative or to the Install alternative; the expected cost of choosing the Rework alternative at D_1 is always larger than that for the Install alternative so that it never pays to choose Rework at D_1. However, the Rework alternative may be the best choice if the test indicates "defective." The maximum we should pay for the test is $r/32$. For any c higher than this, $r/8 < c + 3r/32$. Thus, we can summarize the optimal policies and smallest expected costs as follows:

(1) For $c < r/32$: $d_1^* = $ Test; If $F=1$, $d_2^* = $ Rework, $l^* = c + r/5$,
        If $F=0$, $d_2^* = $ Install, $l^* = c + r/22$.

(2) For $r/32 \le c$: $d_1^* = $ Install, $l^* = r/8$.

We return to this problem in Section 5.5 when we discuss sensitivity analysis in the expert opinion problem.

5.3.2   A Crop Protection Problem

Many commercial crops are subject to damage when temperatures drop below a certain level; for our example we assume that damage occurs whenever temperatures fall to freezing or below. Methods are available in some cases to protect the crop (or its value) from such damage, but all cost money. A decision maker (farmer, producer, insurer) must make a decision D_1 whether or not to protect or not protect a crop, or obtain an expert's weather forecast. Let $D_1 = $ NP if the decision is to Not Protect, $D_1 = $ P if it is to Protect, and $D_1 = $ F if a forecast is obtained. Thus $\mathcal{D}_1 = \{P, NP, F\}$, where P is the riskless alternative and NP is the risky venture. Let the random variable F denote the forecast of weather. The set $\mathcal{F}$ will depend on the type and complexity of the forecast. In this chapter we consider that the forecast outcome is one of two categories, and set $F = 1$ if the forecast is "freezing" and $F = 0$ if it is "not freezing." Thus, $\mathcal{F} = \{0,1\}$. In the next chapter we revisit this problem to illustrate the use of two distinct types of forecasts, namely, probability and categorical. Let the random variable X represent the uncertain weather with $X = 1$ if it freezes, $X = 0$ if it does not, so $X = \{0,1\}$. If the initial decision is to obtain a forecast, the decision maker must then decide whether or not to protect the crop. Let r_1 be the loss incurred if freezing weather occurs and the crop was not protected, and r_2 the cost of protecting the crop where $r_1 > r_2$. There is no cost if the crop is not protected and it does not freeze. The structure of this expert opinion problem is shown in Figure 5-8.

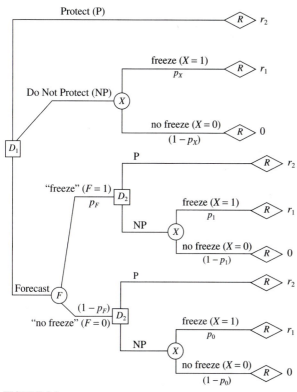

FIGURE 5-8
The crop protection decision problem.

In real crop protection problems the set of decisions $\mathcal{D}$ might include a large number of alternatives; one might consider a much larger number of possible weather forecasts in $\mathcal{F}$ and a very large number of weather outcomes in X. Although a real-world problem is more complex than the one considered here, it differs more in the number of weather states and decision alternatives available to the decision maker than in the underlying structure of the problem.

Many features might make real-world problems more difficult. These include: (1) the dynamic decision environment in which the decision to protect or not protect is made repeatedly, say daily or during certain months of the season; (2) losses that depend on decisions and random quantities occurring over many periods where there can be partial, rather than complete, crop losses in successive periods; (3) successive weather patterns that are not independent; (4) whether crop vulnerability (such as resistance of fruit buds to low temperatures) is stationary over time; and (5) whether different forecasting experts provide similar or very different forecasts to the decision maker. There are a number of publications on crop protection problems in the forecasting and meteorological literature that illustrate the important linkage between forecasting and decision making. A number of these can be found in the references at the end of the chapter.

The next two sections demonstrate how one can increase one's understanding of structure and implications of the expert opinion problem by performing sensitivity analyses, using the decision probabilities p_1 and p_0 in the next section, and the forecast likelihoods f_1 and f_0 in Section 5.5. Aircraft part replacement and crop protection are used to illustrate the ideas.

5.4 SENSITIVITY ANALYSIS USING DECISION PROBABILITIES

A close look at the decision tree in Figure 5-6 (p. 199) shows that the decisions we face with each forecast outcome have the same decision sapling structure as does the original decision problem if a decision to forecast is not considered. The crucial difference between the three subtrees is the value used for the probability of occurrence of the random event X. These probabilities are path-dependent. At the top X node, which is reached without having obtained a forecast (F not observed), the appropriate probability for the event $\{X = 1\}$ is p_X. The middle and lower X nodes are reached after observing F to be 1 and 0, respectively, so the appropriate probability for the branches where $X = 1$ are now p_1 and p_0.

Starting at the terminal R nodes in Figure 5-8 and looking back through the tree, the first nodes we see are the three random nodes designated X. Starting at the top of the tree and working toward the bottom these have expected losses $p_X r_1$, $p_1 r_1$, and $p_0 r_1$. As we move back to the two decision nodes marked D_2, we must choose the decision at each of these that minimizes the expected loss. Again, going from top to bottom of the tree, we choose $Min\{r_2, p_1 r_1\}$, and $Min\{r_2, p_0 r_1\}$. Moving back to the F node the expected loss is $p_F Min\{r_2, p_1 r_1\} + (1 - p_F) Min\{r_2, p_0 r_1\}$. Finally at the D_1 node, let l^* denote the *minimum expected loss* for the tree, so

$$l^* = Min\{r_2, \ p_X r_1, \ p_F Min\{r_2, p_1 r_1\} + (1 - p_F) Min\{r_2, p_0 r_1\}\}. \tag{5.4}$$

Note how we are invoking the principle of optimality. Within the large braces we have calculated the result of using optimal policies at each of the D_2 nodes. This expression may at first look quite formidable, but it can be simplified when we take into account the ordering of the probabilities involved. Although the results in the remainder of this section apply to the more general expert opinion problem, we use the terminology of the crop protection example in Figure 5-8 to help clarify the analysis for the reader.

From the law of total probability we know that the marginal probability of actual freezing weather is

$$p_X = p_F p_1 + (1 - p_F)p_0. \tag{5.5}$$

Because p_F lies between 0 and 1, it follows that p_X lies between p_0 and p_1. We say a forecast is discriminating if the actual event is more likely to occur when it is predicted than when it is not.[2] Mathematically, we say that the forecast is discriminating if $Pr\{X = 1 | F = 1\} > Pr\{X = 1 | F = 0\}$, or $p_1 > p_0$. This property is assumed to hold for all forecasts in the remainder of this chapter. This inequality together with Equation (5.5) imply that the three decision probabilities used in the expert opinion decision tree have the order

$$0 \le p_0 \le p_X \le p_1 \le 1. \tag{5.6}$$

Based on the ordering in (5.6) we can resolve the minimization functions in Equation (5.4), depending on the relative values of r_1 and r_2. We consider three cases:

CASE I. $p_1 r_1 \le r_2$.
It follows from (5.6) that $r_1 p_X \le r_2$ and $r_1 p_0 \le r_2$. This shows that at both D_2 nodes the optimal decision is not to protect, or $d_2^* = NP$. Equation (5.4) reduces to

$$l^* = Min\{r_2, p_X r_1, p_F p_1 r_1 + (1 - p_F)p_0 r_1\}.$$

But from Equation (5.5) we see that the second and last terms in this minimization are identical, and from (5.6) are both smaller than the first. Unlike the part testing problem of the previous section where the cost of a test was explicitly included, we have not considered the cost of a forecast. If we had, such a cost would be added to the second term, and the first term would be the minimum. Instead, our goal is to find the economic value of a forecast by measuring the difference in expected cost between using and not using the forecast. In this case it is zero, and we break the tie by choosing not to forecast. Thus $d_1^* = NP$ and

$$l^* = p_X r_1.$$

CASE II. $r_2 \le p_0 r_1$.
Using arguments similar to those used in Case I we see that, no matter what the result of the forecast is, the minimum expected loss is given by protecting. The optimal policy is $d_1^* = P$ and

---

$$l^* = r_2.$$

CASE III. $p_0 r_1 \leq r_2 \leq p_1 r_1$.

The right-hand inequality implies that if $F = 1$, $d_2^* = P$, and the left-hand one that if $F = 0$, $d_2^* = NP$. Thus, it is optimal to protect if the forecast is for freezing weather and not protect otherwise whenever the loss ratio r_2/r_1 is between p_0 and p_1. The minimum expected loss at node F is $p_F r_2 + (1 - p_F) p_0 r_1$, and Equation (5.4) reduces to

$$l^* = Min\{r_2, \ p_X r_1, \ p_F r_2 + (1 - p_F) p_0 r_1\}.$$

Substituting $r_1 p_0 \leq r_2$ into the last term in this expression shows that r_2 cannot be the minimum so it reduces to $l^* = Min\{p_X r_1, \ p_F r_2 + (1 - p_F) p_0 r_1\}$. Substituting $r_2 \leq p_1 r_1$ into the second term in this expression and using Equation (5.5) shows that

$$l^* = p_F r_2 + (1 - p_F) p_0 r_1.$$

These results are demonstrated graphically in Figure 5-9 to compare the minimum expected losses with forecast available, denoted l^*, to one with only a baseline forecast, denoted l_B^* (see Section 2.2); both are plotted as functions of the loss r_2 when $r_1 = 1$ (equivalent to measuring losses in units of r_1). The heavy line shows the minimum expected loss when a forecast is available, l^*, and the light line when only a baseline forecast is available; clearly $l_B^* = Min\{r_2, \ p_X r_1\}$. The shaded area shows the only situation where there is a positive reduction in expected loss due to forecasting and corresponds

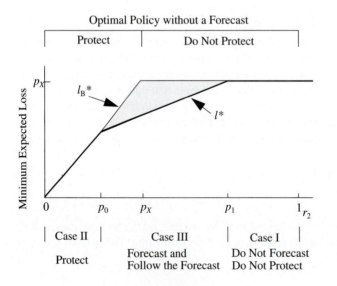

FIGURE 5-9
Expected minimum losses with and without a forecast.

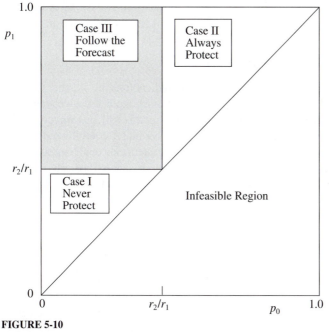

FIGURE 5-10
Optimal policies in the (p_0, p_1) plane.

to Case III; in Cases I and II *there is no advantage to forecasting*. We pursue this observation further with an analysis in the (p_0, p_1) plane as shown in Figure 5-10.

In Case I the optimal policy is never to protect; in Case II the optimal policy is always to protect. For these cases the optimal policies are independent of the outcome of the forecast and are identical to the policies we obtain when only the climatological forecast p_X is available to us. These optimal policies are indicated in the two unshaded triangles lying above the diagonal line in Figure 5-10. The shaded rectangle corresponds to Case III where $d^* = $ P if $F = 1$, and $d^* = $ NP if $F = 0$ (protect if the forecast predicts freezing weather and do not protect otherwise). We now restrict our analysis to this region where there is an economic advantage to forecasting.

5.4.1 The Economic Value of a Forecast

Let Δ_F be $l_B^* - l^*$, the difference between the two lines in Figure 5-9. This represents the reduction in expected loss that can be expected by choosing to forecast ($D_1 = $ Forecast). It can also be thought of as the expected economic gain from forecasting.

Recall that $l_B^* = Min\{r_2, p_X r_1\}$. For Cases I and II we find that $\Delta_F = 0$. In Case III, by using Equation (5.5) we see that

$$\Delta_F = (1 - p_F)(r_2 - r_1 p_0) \geq 0 \qquad \text{if } p_0 \leq r_2 \leq p_X r_1,$$

$$= (r_1 p_1 - r_2) p_F \geq 0 \qquad \text{if } p_X r_1 \leq r_2 \leq p_1. \tag{5.7}$$

We now look more closely at the shaded area in Figure 5-10, where Δ_F is positive; Figure 5-11 (p. 210) contains three representations of this region. We make an assumption that the forecast is calibrated in expectation,

$$E[F] = E[X]. \tag{5.8}$$

This assumption states that, on average, the forecast predicts freezing weather the same fraction of time that freezing weather actually occurs.[3] For example, if the weather freezes on average 10% of the time, then we require that the forecasts will, on average, predict freezing weather 10% percent of the time. Since X and F are both Bernoulli random variables, Equation (5.8) is equivalent to $p_F = p_X$. Some forecasters may tend to predict good weather more often than actually occurs. We could think of such a forecaster as being overly optimistic with a positive bias. Our assumption would not allow this.

The calibrated forecast assumption together with earlier results leads to some interesting observations that are illustrated in Figure 5-11. By using $p_F = p_X$ with Equation (5.5) we obtain

$$p_1 = 1 - p_0(1 - p_X)/p_X. \tag{5.9}$$

With p_X fixed, p_1 is a straight-line function of p_0 with origin at the point in the (p_0, p_1) plane where we have perfect forecast discrimination (i.e., $p_1 = 1$ and $p_0 = 0$), and slope $-(1 - p_X)/p_X$. Equation (5.9) is plotted in Figure 5-11(a), for two values of p_X; each gives a downward-sloping straight line emanating from the northwest corner. The reader should check that a line from the northwest to the southeast corners partitions the rectangle into cases where $r_2 < p_X r_1$ and $r_2 > p_X r_1$. One can think of each value of p_X as characterizing the weather of a particular geographical region and r_2/r_1 as characterizing a particular user of the forecasting service. The line for Equation (5.9) rotates counterclockwise with increasing p_X (poorer average weather conditions). We have indicated by $\bar{p}_0$ and $\bar{p}_1$ the values that correspond to the intersections of the constant p_X lines with the $p_1 = r_2/r_1$ and $p_0 = r_2/r_1$ boundaries, respectively. If the line segments with negative slope are extended beyond the boundary of the shaded region, they eventually intersect the p_0 axis at a value equal to the odds of inclement weather. We shall show that moving in a northwest direction on each line segment increases the economic value of the forecast for a constant p_X. This increase in value occurs because p_1 increases while p_0 decreases; the forecasts become more discriminating and hence more valuable. We shall also see that this corresponds to an increased correlation between F and X.

We now turn to the expected value of the forecast Δ_F. This will clearly vary with both p_0 and p_1 and should be plotted in a third dimension above the (p_0, p_1) plane. We illustrate the shape of the function with isocontours, that is, lines of constant Δ_F, but first we express p_1 as a function of p_0 and the normalized economic value of the forecast Δ_F/r_1. Using the expressions derived for Δ_F in Equations (5.7) together with Equation (5.5) and our assumption that $p_X = p_F$, we obtain

---

[3] See Chapter 8 for a more complete discussion of forecast calibration.

$$p_1 = \begin{cases} 1 - \dfrac{\dfrac{\Delta_F}{r_1} p_0}{\left(\dfrac{r_2}{r_1} - \dfrac{\Delta_F}{r_1}\right) - p_0} & \text{if} \quad p_0 \le \dfrac{r_2}{r_1} \le p_X \le p_1, \\[2em] \dfrac{\dfrac{\Delta_F}{r_1} + p_0\left(\dfrac{r_2}{r_1} + \dfrac{\Delta_F}{r_1}\right)}{\dfrac{\Delta_F}{r_1} + p_0} & \text{if} \quad p_0 \le p_X \le \dfrac{r_2}{r_1} \le p_1. \end{cases}$$

For feasible p_1 the denominator in the first fraction must be positive, which means that p_0 must be less than $(r_2/r_1 - \Delta_F/r_1)$, which, in turn, is less than r_2/r_1. The contours of constant Δ_F derived from these two equations are shown in Figure 5-11(b). The revealing insight from this figure is that, for a given value of p_0 and r_1, the same value of Δ_F is obtained from *two* values of p_1. Because one value corresponds to low and the other to high p_X, the same increase in economic value results from smaller discriminatory power in the case of low marginal probabilities p_X, than it does in the case of large p_X. To state it differently, forecasts must provide a greater difference between p_0 and p_1 when faced with large probabilities of freezing weather to gain the same economic value of a forecast.

To be useful a forecast must correlate with the actual weather. The usual way of measuring the degree to which this occurs is with the covariance or correlation coefficient. For the random variables F and X these are defined to be (see Section 1.5.7)

$$Cov[F, X] = E[FX] - E[F]E[X]$$

and

$$\rho_{FX} = \frac{Cov[F, X]}{\sqrt{Var[F]\,Var[X]}},$$

respectively. For the two-category forecast when F and X are both Bernoulli trials it is easy to show that

$$\rho_{FX} = (p_1 - p_0)\sqrt{\frac{p_F(1 - p_F)}{p_X(1 - p_X)}}.$$

Using the assumption that $p_X = p_F$, this reduces to

$$\rho_{FX} = (p_1 - p_0). \tag{5.10}$$

Isocontours of constant correlation are shown in Figure 5-11(c) using Equation (5.10), with $\rho_{FX} = 1$ in the northwest corner and 0 in the southeast corner. Lines of constant ρ_{FX} are parallel to the diagonal boundary between feasible and infeasible regions.

It is easy to show that the economic value of the forecast, Δ_F, remains zero until the correlation coefficient between forecasts and weather exceeds a critical value, which

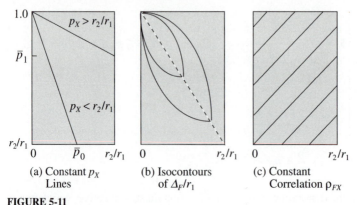

1.0

$\bar{p}_1$

r_2/r_1

$p_X > r_2/r_1$

$p_X < r_2/r_1$

0    $\bar{p}_0$ r_2/r_1 0 r_2/r_1 0 r_2/r_1

(a) Constant p_X (b) Isocontours   (c) Constant
    Lines   of Δ_F/r_1 Correlation ρ_{FX}

FIGURE 5-11
Analysis of case III in the (p_0,p_1) region

we denote by $\bar{\rho}$, and then increases with the correlation between forecast and actual weather. The critical correlation coefficient $\bar{\rho}$ is the value obtained when the line segment of constant p_X intersects the boundary of the shaded region, that is, either $(\bar{p}_0, r_2/r_1)$ or $(r_2/r_1, \bar{p}_1)$ in Figure 5-11(a) depending on whether $p_X < r_2/r_1$ or $p_X > r_2/r_1$. Beginning at any point on the $p_1 = p_0$ diagonal, a line of constant p_X or a line perpendicular to the contours of constant ρ_{FX} must obviously first pass through one of the unshaded triangles of Figure 5-10, where $\Delta_F = 0$ (Case I or Case II). Beyond the intersection of this line with the boundary of the shaded area, the economic gain increases with the correlation.

The reader should pay special attention to the fact that high correlation between X and F does not necessarily lead to economic benefit in forecasting. This is easily demonstrated by superimposing (b) on top of (c) in Figure 5-11. The correlation is represented by a 45° line in (c) that cuts across the contours in (b). As one travels from the northeast of a constant correlation line to the southwest, the value of the forecast increases to a maximum on the ridge represented by $p_X = r_2/r_1$ and then decreases. If it cuts the east and south boundaries of the shaded region, it has no economic value at these points no matter how high the correlation.

5.5   SENSITIVITY ANALYSIS USING FORECAST LIKELIHOODS

Recall that in the original statement of Lindley's aircraft part problem in Section 5.3.1 we are given the forecast likelihoods $p_{F|X}(f|x)$. If $F = 1$, the test shows "defective," and if $F = 0$, it shows "not defective." In the formulation of the problem we are told that the test correctly identifies the condition of the part 3/4 of the time. With our notation this means that $f_0 = 1/4$ and $f_1 = 3/4$.[4] For many forecasting systems, whether they be diag-

---

[4] The reader should be careful to note that in general f_0 and f_1 are not required to add to 1, even though they do in Lindley's numeric example.

nostic tests as in this problem, weather forecasts as in the crop protection problem, or any of a host of other possible predictions of an outcome of interest, one is often more likely to have data on how well the forecast performed under known conditions (i.e., $p_{F|X}$ are known) rather than the other way around ($p_{X|F}$ are known).

Analysis of the expert opinion problem could have been done directly in terms of the forecast likelihoods f_0 and f_1 without deriving the decision probabilities p_0 and p_1. The formulas for finding them are given by Bayes' rule,

$$p_1 = \frac{f_1 p_X}{f_0 (1 - p_X) + f_1 p_X},$$ (5.11)

$$p_0 = \frac{(1 - f_1) p_X}{1 - (f_0 (1 - p_X) + f_1 p_X)}.$$ (5.12)

Although these are not required for calculation if the problem is analyzed using f_0 and f_1, they play an important role in finding the boundaries between the regions where each of the decisions Protect, Do No Protect, and Forecast is optimal.

Our objective is to determine the optimal policy regions in the (f_0, f_1) plane corresponding to those shown for the (p_0, p_1) in Figure 5-10. To do so, we need to find the equations of the boundaries between these regions in terms of f_0 and f_1. We invert Equations (5.11) and (5.12) to find the expressions for f_1 in terms of f_0, p_X and the decision probabilities,

$$f_1 = \left(\frac{p_1}{(1 - p_1)} \right)\left(\frac{1 - p_X}{p_X} \right) f_0,$$

and

$$f_1 = \left(\frac{p_0}{1 - p_0} \right)\left(\frac{1 - p_X}{p_X} \right) f_0 + \left(\frac{p_X - p_0}{p_X (1 - p_0)} \right).$$

Recall that the boundary between Cases I and III is found when $p_1 = r_2/r_1$, and the one between Cases II and III when $p_0 = r_2/r_1$. Substituting these into the above equations yields the equivalent boundaries in the (f_0, f_1) plane,

$$f_1 = \left(\frac{r_2}{r_1 - r_2} \right)\left(\frac{1 - p_X}{p_X} \right) f_0,$$ (5.13)

and

$$f_1 = \left(\frac{r_2}{r_1 - r_2} \right)\left(\frac{1 - p_X}{p_X} \right) f_0 + \left(\frac{r_1 p_X - r_2}{p_X (r_1 - r_2)} \right).$$ (5.14)

These expressions contain only the assumed-known quantities, namely, the forecast likelihoods f_0 and f_1, the event probability p_X, and the loss values r_2 and r_1. They are shown plotted in Figure 5-12(a) and (b) together with the line $f_1 = f_0$ and should be com-

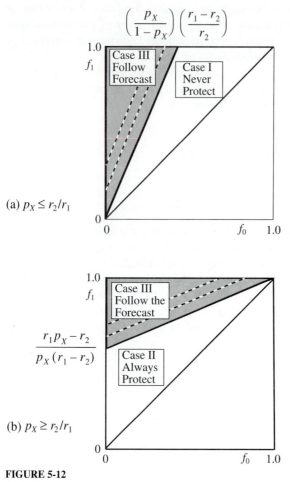

$$\left(\frac{p_X}{1-p_X}\right)\left(\frac{r_1-r_2}{r_2}\right)$$

(a) $p_X \le r_2/r_1$

(b) $p_X \ge r_2/r_1$

FIGURE 5-12
Optimal policies in the (f_0, f_1) plane

pared and contrasted with Figure 5-10. Using Figure 5-12, we can determine the optimal decisions for various feasible combinations of f_0 and f_1.

When $p_X = r_2/r_1$, Equations (5.13) and (5.14) both reduce to $f_1 = f_0$. When $p_X < r_2/r_1$, the line for (5.13) bisects the upper triangular area. Points below this line correspond to Case I where $d_2{}^* =$ DNP. Points above the line constitute Case III. As r_2/r_1 decreases from 1, the line for Equation (5.13) rotates clockwise about the origin from a vertical position, and the line for Equation (5.14) rotates clockwise about the point $(1,1)$. When $r_2/r_1 = p_X$, they both lie on the diagonal, in which case the conditions for Cases I and II are vacuous. For $r_2/r_1 < p_X$ the line for Equation (5.14) bisects the upper triangular area. Points below this line correspond to Case II where $d_2{}^* =$ P. As r_2/r_1 increases from 0, the line for Equation (5.14) rotates counterclockwise about $(1,1)$ from a horizontal position.

We now look at the economic value of the forecast as we did in Section 5.4.1. First, note that the denominator of Equation (5.11) gives p_F in terms of p_X and the forecast likelihoods, whereas that in (5.12) gives $1 - p_F$. Thus, we could write these equations as

$$p_1 = f_1 p_X / p_F, \qquad \text{and } p_0 = (1 - f_1) p_X / (1 - p_F). \qquad (5.15)$$

Substituting these into Equations (5.7) in Section 5.4.1 gives the expected value of the forecast in terms of the forecast likelihoods,

$$\Delta_F = (1 - p_F) r_2 - (1 - f_1) p_X r_1 \qquad \text{if } (1 - f_1) p_X r_1 / (1 - p_F) \le r_2 \le r_1 p_X,$$

$$\qquad (5.16)$$

$$= f_1 p_X r_1 - p_F r_2 \qquad \text{if } r_1 p_X \le r_2 \le r_1 f_1 p_X / p_F.$$

Figure 5-13 shows a plot of the normalized expected value of the forecast as a function of r_2 / r_1 (the reader should check that this curve represents the difference between the two curves in Figure 5-9). It is maximized when $r_2 / r_1 = p_X$. Whereas in the space of the decision probabilities the isocontours of constant Δ_F (shown in Figure 5-11(b)) are somewhat complex, in the space of the forecast likelihoods they are straight lines parallel to the boundaries of the feasible region (see the dashed lines in the shaded areas, Case III, in Figure 5-12).

5.6   PROBLEMS WITH ONE OR MORE FORECASTS

We now turn to a class of decision problems that occur in many fields of interest and in many different contexts; single or sequential decisions are made after one or more forecasts or expert opinions are made available to the decision maker.

The colon cancer diagnosis problem in Chapter 4 is one where up to four forecasts can be obtained. In more complex versions of the bank credit problem the decision to forecast whether a person is a good or bad credit risk may come after a prediction has

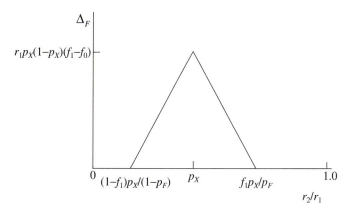

FIGURE 5-13
Expected payoff of a forecast as a function of (r_2/r_1)

been made of the individuals' likely acceptance of a new offer. In exploring an inaccessible mountainous region for mineral deposits it may be cost-effective to undertake a coarse and high-speed search of large areas from the air and, then, if the first forecast predicts the existence of large deposits, undertake a more precise but costly search on the ground in a localized area. The Coast Guard may want to identify boats in distress or potential contraband-carrying vessels in a busy shipping area. In military operations one may want to discriminate between warheads and decoys being carried by missiles in a hostile attack or hostile military units interspersed with civilians in terrorist-type actions or moving rather than stationary targets. In these applications the decision problem is one of identifying and classifying a subpopulation of targets or objects that warrant attention.

We start by assuming that a forecasting system (possibly including devices or sensors and software in combination with judgmental assessments) has been designed to discriminate among objects to identify or classify a certain object or characteristic that we call a target. The word *sensor* should be thought of broadly; it could be a radar or other electronic device that picks up electromagnetic radiation. For the insurance and banking examples described earlier it could be a scoring system used to assess the risk of an applicant for a loan or credit card.

We assume that the sensor or forecasting system predicts a target or nontarget. With this information, the decision maker must decide whether or not to take action. The sensor is not perfect, so that action can be taken against nontargets or no action can be taken against a real target. In the banking example, action against a nontarget would equate to approving a loan of an individual that turned out to be a bad credit risk, whereas nonaction against a target would be turning down a loan applicant who is a good credit risk. Given performance characteristics of the sensor or forecasting system together with information on the population being investigated, what is the structure of the decision problem and how is an optimal decision characterized?

The notation used is similar to that in the expert opinion problem. In the population of objects being considered, let $X = 1$ if an object is a target and $X = 0$ if not. Assume that an estimate of the fraction of objects that are targets, $p_X = Pr\{X = 1\}$, is available. If the sensor is used repeatedly to discriminate among objects in the population it will identify some as targets and others as nontargets. For an arbitrary object observed by the sensor, let $F = 1$ if the sensor identifies it as a target, $F = 0$ if not, and define $Pr\{F = 1\}$ to be p_F, the long-run fraction of objects in the population that are identified as targets by the sensor.

The influence diagram for the one-decision, one-forecast problem is very similar to the expert opinion problem. The decision taken will probably depend on the forecast, as will the conditional probability that the object is a target. The decision has no effect on whether or not the object is really a target and, typically, the result depends on both X and D. The sets X and $\mathcal{F}$ of possible outcomes of the random variables X and F each contain 0 and 1 with the interpretations as given. The decision set $\mathcal{D}$ is assumed to contain two elements:

$D = $ A if action is taken,

$\quad = $ DN (Do Nothing) otherwise.

The set $\mathcal{R}$ is assumed to contain the following three loss elements:

$R = 0$ if object is a target and action is taken, or object is not a target and no action is taken,

$= r_1$ if object is a target and no action is taken (loss due to a false negative or "leaker"),

$= r_2$ if object is not a target and action is taken (loss due to a false-positive or "false alarm").

Once these sets have been identified we can draw the decision tree. This is shown in Figure 5-14 where the decision probabilities $p_1 = Pr\{X = 1 | F = 1\}$ and $p_0 = Pr\{X = 1 | F = 0\}$, are shown on the appropriate branches. If the forecast likelihoods f_1 and f_0 are given as in the aircraft part problem in Section 5.3.1, we can use Equations (5.11) and (5.12) to find them. Note that if the sensor is calibrated, no calculations are required as $p_1 = f_1$ and $p_0 = f_0$.

5.6.1   Optimal Policies and Expected Returns

Let $R(X,D | F)$ be the result as a function of the object type X and the decision D, given the sensor output F. Also let $l^*(i)$ be the minimum expected loss when $F = i$ and l^* be the minimal expected loss for the tree. Starting at the result nodes, taking expected values and rolling back to D, at the upper node D

$$l^*(1) = E[R(X,d^* | F = 1)] = Min\{(1 - p_1)r_2, \; p_1 r_1\},$$

and at the lower node D

$$l^*(0) = E[R(X,d^* | F = 0)] = Min\{(1 - p_0)r_2, \; p_0 r_1\}.$$

The minimum expected loss for the tree is

$$l^* = p_F l^*(1) + (1 - p_F)l^*(0).$$

Our objective is to find the decisions that minimize this expected loss. The critical value in determining the optimal decision is the ratio

$$\frac{r_2}{r_1 + r_2},$$

and we look at the three cases where this ratio falls in each of the intervals $(0, p_0)$, (p_0, p_1), and $(p_1, 1)$.

CASE I: $p_1 \le (r_2 / r_1 + r_2)$.
This together with (5.6) implies that $(1 - p_1)r_2 \ge p_1 r_1$ and $(1 - p_0)r_2 \ge p_0 r_1$. From Figure 5-15 we see that the optimal policy is to Do Nothing ($d^* = $ DN) no matter whether the sensor indicates $F = 1$ or $F = 0$. The cost of a false alarm is so high that no action should be taken whether or not it is identified as real. For this case

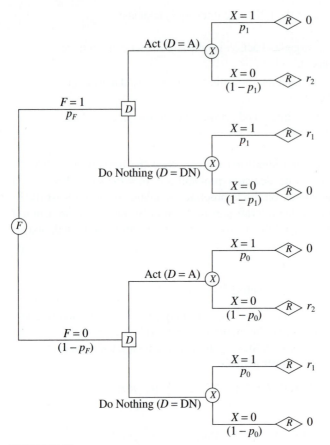

FIGURE 5-14
The decision tree for the single sensor target problem.

$$l^* = p_X r_1.$$

CASE II: $(r_2/(r_1 + r_2)) \leq p_0$.
For this case $(1 - p_0)r_2 \leq p_0 r_1$, which together with the inequalities in (5.6) gives $(1 - p_1)r_2 \leq p_1 r_1$. From Figure 5-15 we see that the optimal policy is to Act ($d^* = A$) no matter whether the sensor indicates $F = 1$ or $F = 0$. The cost of a false alarm is so small that action should be taken whether or not it is identified as false. For this case

$$l^* = (1 - p_X)r_2.$$

CASE III: $p_0 \leq (r_2/(r_1 + r_2)) \leq p_1$.
These inequalities imply that $(1 - p_0)r_2 \geq p_0 r_1$ and $(1 - p_1)r_2 \leq p_1 r_1$. From Figure 5-15 we see that $d^* = DN$ if the sensor indicates $F = 0$ and $d^* = A$ if $F = 1$. The costs are such that action is taken in accordance with the sensor indication. For this case

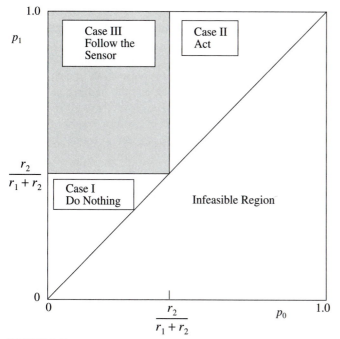

FIGURE 5-15
Optimal policy regions for the single sensor problem

$$l^*(0) = p_0 r_1, \; l^*(1) = (1 - p_1)r_2,$$

$$l^* = p_F(1 - p_1)r_2 + (1 - p_F)p_0 r_1$$

$$= f_0(1 - p_X)r_2 + (1 - f_1)p_X r_1.$$

Figure 5-15 shows the optimal policy regions in the (p_0, p_1) plane. The reader should compare this with Figure 5-10 for the crop protection problem and construct and interpret the appropriate analysis for Case III as was shown in Figure 5-11.

5.6.2 A Numeric Example

Suppose we have a sensor with $f_0 = 0.244$ and $f_1 = 0.800$ against a population where $p_X = 0.55$. These give $p_F = 0.55$, and we see that the sensor is calibrated. Thus, $p_1 = 0.800$ and $p_0 = 0.244$. Let $r_2 = 1.$[5] The minimal expected loss is

$$
\begin{aligned}
l^* &= 0.55r_1 & \text{if } 0 \le r_1 \le 0.25, \\
&= 0.11 + 0.11r_1 & \text{if } 0.25 \le r_1 \le 3.10, \\
&= 0.45 & \text{if } 3.10 \le r_1 .
\end{aligned}
$$

---

[5] This is equivalent to measuring the losses in units of the loss due to taking action against a false alarm.

This function is plotted in Figure 5-16 with a solid line.

Suppose the sensor could be improved, by which we mean that it can better discriminate targets from nontargets. We assume that f_1 is increased from 0.800 to 0.900 and that f_0 is reduced from 0.244 to 0.200. The reader can check that p_F is now 0.585 and that the sensor is no longer calibrated. It tends to overstate the presence of targets. Equations (5.11) and (5.12) now give $p_1 = 0.846$ and $p_0 = 0.133$, and the minimal expected loss is

$$l^* = 0.55r_1 \qquad \text{if } 0 \le r_1 \le 0.182,$$
$$= 0.09 + 0.055r_1 \qquad \text{if } 0.182 \le r_1 \le 6.54,$$
$$= 0.45 \qquad \text{if } 6.54 \le r_1.$$

This function is plotted with a dashed line in Figure 5-16. Note that the minimal expected loss is reduced when r_1 lies in the interval corresponding to Case III, namely, $[(1 - p_1)/p_1, \ (1 - p_0)/p_0]$, the reduction being a maximum when the loss due to not taking action against a leaker is 3.1 times the loss due to taking action with a false alarm. For r_1 below 0.182 or above 6.54 (extreme cost differential cases), there is no reduction in minimum expected loss when the improved sensor is used.

Like some of the previous examples this one serves to point out again that, although calibration might appear to be a desirable characteristic for a sensor or forecast, it is not necessarily the case when the forecast or sensor output is used as input to a decision problem. Finally, note that for this problem, if you knew the target type before the decision is made you could avoid all errors and the expected loss would be zero. Thus, in this case l^* gives the value of perfect information.

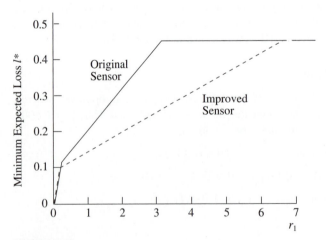

FIGURE 5-16
Optimal expected loss as a function of r_1.

5.6.3   A Single Decision with Two Forecasts

It is clear from Figure 5-16 that expected loss can be reduced for a large range of r_2 by replacing the original sensor with the improved one. The inquisitive reader might wonder whether there would be any advantage to using the original (poorer) sensor in addition to the improved version. For this reason we now construct a model using two sensors.

The formulation of the decision problem is shown in the influence diagram of Figure 5-17 in terms of the forecast likelihoods. Clearly, it cannot be used to construct a decision tree whose ordered decisions and events coincide with what we would observe in real time. For example, X would have to be observed before F and G could be observed, yet, X is the uncertainty on which the decision problem is based. But it does illustrate a very important assumption that is being made, that *the outputs of the two sensors are conditionally independent, given X*. That is, we assume

$$p_{F,G|X}(f, g \mid x) = p_{F|X}(f \mid x)p_{G|X}(g \mid x). \tag{5.17}$$

This *does not* mean that F and G are independent random variables! Indeed, one would hope that F and G are highly correlated. Consider the extreme case where the two sensors are perfect, so that $f_1 = g_1 = 1$ and $f_0 = g_0 = 0$. In this case they are both calibrated so that $p_F = p_G = p_X$. Also, because neither sensor makes any mistakes, $Pr\{G = 1 \mid F = 1\} = 1$, and the joint probability $Pr\{G = 1, F = 1\} = p_F \neq p_F p_G$ unless every object in the population is a target. They are clearly dependent and one can show that their correlation coefficient is 1.0.

Figure 5-18 shows the influence diagram with arcs drawn from F and G to X and an arc added from F to G. Using the same node sets as in the single sensor example and $G = \{0,1\}$, the decision tree is constructed and shown in Figure 5-19. The notation defined in Section 3.5.1, namely,

$$p_{i,j} = Pr\{X = 1 \mid F=i, G=j\}, \qquad i,j = 0, 1$$

is used for the decision probabilities. Recall that sensor 2 is said to be better than sensor 1 if the odds of the prediction of the object when it is known to be present relative to the odds when it is known to be absent are greater for sensor 2 than for sensor 1; that is,

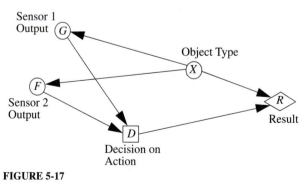

FIGURE 5-17
Influence diagram in terms of forecast likelihoods.

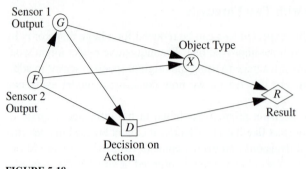

Sensor 1
Output

Object Type

F

Sensor 2
Output

R
Result

D

Decision on
Action

FIGURE 5-18
Influence diagram in terms of decision probabilities.

$$\frac{\dfrac{g_1}{(1-g_1)}}{\dfrac{g_0}{(1-g_0)}} \geq \frac{\dfrac{f_1}{(1-f_1)}}{\dfrac{f_0}{(1-f_0)}}. \tag{5.18}$$

When this is true, the inequalities

$$0 < p_{00} < p_{10} < p_{01} < p_{11} < 1 \tag{5.19}$$

hold. The following expressions for each of these were determined in Equation (3.34) (p. 110):

$$p_{11} = \frac{f_1 g_1 p_X}{f_1 g_1 p_X + f_0 g_0 (1-p_X)},$$

$$p_{01} = \frac{(1-f_1) g_1 p_X}{(1-f_1) g_1 p_X + (1-f_0) g_0 (1-p_X)},$$

$$\tag{5.20}$$

$$p_{10} = \frac{f_1 (1-g_1) p_X}{f_1 (1-g_1) p_X + f_0 (1-g_0) (1-p_X)},$$

$$p_{00} = \frac{(1-f_1) (1-g_1) p_X}{(1-f_1) (1-g_1) p_X + (1-f_0) (1-g_0) (1-p_X)}.$$

The earlier notation for the result is extended so that $R(X,D|\,F,G)$ is the loss as a function of the object type X and the decision D, given the output of the two sensors. Taking expected values and folding back to D, at the four nodes D in order from the top of Figure 5-19,

$$E[R(X,D|\,F=1,G=1)] = Min\{(1-p_{11})r_2,\ p_{11}r_1\},$$

$$E[R(X,D|\,F=1,G=0)] = Min\{(1-p_{10})r_2,\ p_{10}r_1\},$$

$$E[R(X,D|\,F=0,G=1)] = Min\{(1-p_{01})r_2,\ p_{01}r_1\},$$

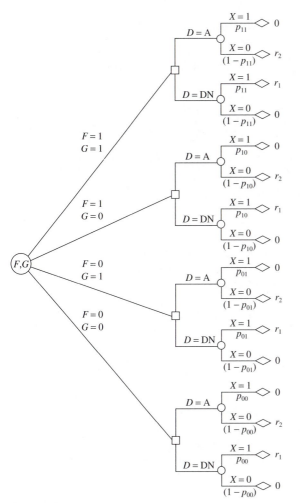

FIGURE 5-19
Decision tree with two sensors and single decision.

$$E[R(X,D| F = 0, G = 0)] = Min\{(1 - p_{00})r_2, \ p_{00}r_1\}.$$

The important measure in determining the optimal policy is again the ratio of costs $r_2/(r_1 + r_2)$. We need to consider five cases that correspond to this ratio being in each of the intervals

$$(0, p_{00}), (p_{00}, p_{10}), (p_{10}, p_{01}), (p_{01}, p_{11}), (p_{11}, 1).$$

The optimal decisions and minimal expected costs are summarized in Table 5.2. In Case I action should never be taken, and in Case II it should be taken no matter what the sensors indicate. The sensors add no value in these cases. In Case III action should be taken only when both sensors indicate a target, and in Case IV as long as at least one

TABLE 5.2
Optimal policies and expected costs with two sensors

Case No.	Interval for $r_2/(r_1+r_2)$	Sensor output (F, G)	Optimal decision d^*	Minimal expected loss l^*
I	$(p_{11}, 1)$	Any	DN	$r_1 p_X$
II	$(0, p_{00})$	Any	A	$r_2(1 - p_X)$
III	(p_{01}, p_{11})	(1, 1)	A	$r_2(1 - p_X)f_0 g_0$
		Any other	DN	$+ r_1 p_X(1 - f_1 g_1)$
IV	(p_{00}, p_{10})	(0, 0)	DN	$r_2(1 - p_X)(f_0 + g_0 - f_0 g_0)$
		Any other	A	$+ r_1 p_X(1 - f_1)(1 - g_1)$
V	(p_{10}, p_{01})	(0, 0) or (1, 0)	DN	$r_2(1 - p_X)g_0$
		(0, 1) or (1, 1)	A	$+ r_1 p_X(1 - g_1)$

does. Note that Case V will not exist when $p_{10} = p_{01}$, which occurs when $f_0 = g_0$ and $f_1 = g_1$ (identically performing sensors). When it does exist, the optimal policy is to act only when the better of the two indicates a target.

Consider a numerical example based on a population and the two sensors described in Section 5.6.2, with $p_X = 0.55$, $f_0 = 0.244$, $f_1 = 0.800$, $g_0 = 0.200$, and $g_1 = 0.900$. The reader should check that Inequality (5.18) holds. Using Equations (5.20),

$$p_{00} = 0.039, \qquad p_{10} = 0.334, \qquad p_{01} = 0.593, \qquad p_{11} = 0.947.$$

Figure 5-20 shows plots of the minimal expected loss using only the best sensor (solid line) and both sensors (dashed line). Note that for the three intervals representing cases I, II, and V there is no economic value to using both sensors compared to using only the best. However, there are two distinct ranges for r_1, $[(1 - p_{11})/p_{11}, (1 - p_{01})/p_{01}]$ and $[(1 - p_{10})/p_{10}, (1 - p_{00})/p_{00}]$, where adding an *inferior* sensor reduces the minimal expected loss.

5.7   SEQUENTIAL DECISIONS USING SEQUENTIAL FORECASTS

Let us consider a search problem that occurs in real practice. One obtains a forecast or takes a reading from a sensor, F, which, typically, is an inexpensive but coarse search or classification scheme. Based on the response from the coarse search, one then decides, (D_1), whether or not to immediately act or not act, or whether to obtain a second, more expensive forecast possibly derived from a second and different sensor. This second sensor has the feature that if an object is present it has a much higher probability of finding it than does the first forecast or sensor but at a much higher cost.

The colon cancer diagnosis problem in Chapter 4 is an example of this type of problem where the initial sensor is typically either a digital examination or occult blood test; the second sensor is barium enema or sigmoidoscopy. This type of problem is often met in mineral exploration where a large-scale coarse search looks for large ore deposits,

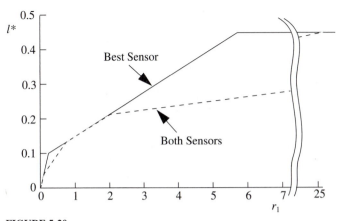

FIGURE 5-20
Comparison of best sensor alone, with both sensors.

possibly using a magnetometer in an airplane covering many thousands of square miles each day. There is a high probability of getting an operational "false alarm," that is, a prediction that a deposit is present when in fact it is not there. Based on the first prediction one might have to decide whether to launch a much more expensive but reliable "fine" search that involves landing a team of people and instruments in a remote site to drill and sample for actual ore deposits. Typically the forecasting system or sensing devices used in the second stage have a much larger probability of finding a target if one is there. Unfortunately, they are much more expensive and cannot cover as large areas as the first one.

The first event is a reading or forecast from the first sensor, denoted F, where $F = 1$ if it indicates that the object of interest is present, and $F = 0$ if not. Next comes the first decision where the set of alternatives is given by $\mathcal{D}_1 = \{A, DN, SS\}$ corresponding to the decisions to Act, Do Nothing, or use a Second Sensor, respectively. If the decision is made to use a second sensor, its output is denoted $G = 1$ if it indicates the object of interest is present and $G = 0$ if not. A second decision is then chosen from the set $\mathcal{D}_2 = \{A, DN\}$. The true type of object is denoted $X = 1$ if it is an object of interest and $X = 0$ if not. The influence diagram is shown in Figure 5-21 in terms of the forecast likelihoods. Note that as no directed arc connects nodes F and G, the outputs of the sensors are assumed to be conditionally independent of each other, given X. As in the two sensor examples in the preceding section, the arcs from X to F and X to G need to be reversed using Bayes' rule, and an arc added from F to G.

We assume that the second sensor is better than the first in the sense that the inequality in (5.18) holds; we also assume in what follows that the first sensor is less expensive than the first and that the objective is to minimize expected costs. The reader might wonder, given two forecasts, whether one can do better (lower the minimum expected cost) by making sequential decisions, and if so, should the better of the two sensors be used first or vice versa? We discuss this at the end of the section.

The reformulated influence diagram and associated decision tree are shown in Figures 5-22 and 5-23, respectively. Notice how large and bushy the tree becomes even

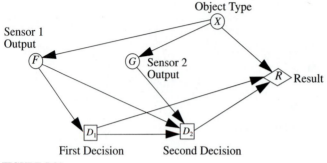

FIGURE 5-21
Sequential problem in terms of forecast likelihoods.

with this small problem. But notice also that it has as its main components two subtrees similar to those of Figure 5-14 (p. 216) for the single-sensor problem. Note that the probabilities have not been inserted on the figure. The reader is encouraged to do this as he or she works through the following material. We use the same numeric example as in the previous subsection, repeated here for convenience, but with added costs.

Population: $p_X = 0.55$,

Sensor 1: $f_0 = 0.244$, $f_1 = 0.800$, so $p_F = 0.55$ with this population,

Sensor 2: $g_0 = 0.200$, and $g_1 = 0.900$, so $p_G = 0.585$ with this population.

In addition to these, we need conditional probabilities for the decision tree (for each group we start at the top of the tree and work toward the bottom. The cases where the random variable X or G is 1 are shown; subtract these from 1 to get the cases when X or G are 0):

1. Terminal node branches:

$$Pr\{X = 1|F = 1\} = 0.800,$$

$$Pr\{X = 1|F = 1, G = 1\} = p_{11} = 0.947 \text{ (next two branches)},$$

$$Pr\{X = 1|F = 1, G = 0\} = p_{10} = 0.334 \text{ (next two branches)},$$

$$Pr\{X = 1|F = 1\} = 0.800,$$

$$Pr\{X = 1|F = 0\} = 0.244,$$

$$Pr\{X = 1|F = 0, G = 1\} = p_{01} = 0.593 \text{ (next two branches)},$$

$$Pr\{X = 1|F = 0, G = 0\} = p_{00} = 0.039 \text{ (next two branches)},$$

$$Pr\{X = 1|F = 0\} = 0.244.$$

2. Branches leading to D_2 (only $\{G = 1\}$ branches:

$$Pr\{G = 1|F = 1\} = [g_1 f_1 p_X + g_0 f_0 (1 - p_X)]/p_F = 0.760,$$

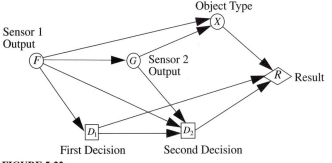

FIGURE 5-22

FIGURE 5-22
Sequential problem in terms of decision probabilities.

$$Pr\{G = 1|F = 0\} = [g_1(1 - f_1)p_X + g_0(1 - f_0)(1 - p_X)]/(1 - p_F) = 0.371.$$

In the preceding section the optimal policy was found for the loss ratio $r_2/(r_1 + r_2)$ in five distinct intervals (see Table 5.2 (p. 222)). In this section we assume that the following costs and losses apply. The costs of using Sensor 1 and Sensor 2 are $0.1 and $0.3 respectively. A missed object results in a loss of $10.0 ($r_1$), whereas a false alarm results in a loss of $1.0 ($r_2$). These are shown on the appropriate branches and terminal nodes in Figure 5-23. The objective is to minimize expected loss.

The tree is solved in the usual way using the rollback algorithm, the details being left as an exercise for the reader. The optimal policy is:

1. If Sensor 1 forecasts a real object, Act,
2. If Sensor 1 forecasts a false object, use Sensor 2, and
   a. If Sensor 2 forecasts a real object, Act,
   b. If Sensor 2 forecasts a false object, Do Nothing.

The overall minimum expected loss is $0.523. Some other measures of performance using this policy are:

$Pr\{$Do not act against a real object$\} = 0.011,$
$Pr\{$Act against a false object$\} = 0.178,$
$Pr\{$Correct action taken$\} = 0.811.$

Using the results in Table 5.2 (p. 222) for this example (Case IV holds) the reader should check that using both sensors as in Figure 5-19 (p. 221) gives a minimal expected loss of $0.688. Considerable savings are possible using the sensors sequentially.

Suppose we consider using the best sensor first. The structure of the problem remains the same, but the conditional probabilities change; the numeric values for f_1 and f_0 are interchanged with g_1 and g_0. The solution to this problem is left to the reader, but it is straightforward to show that when the best sensor is used first, the optimal policy is

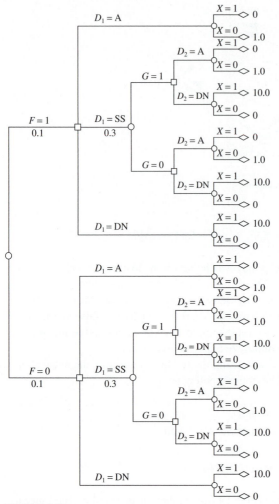

FIGURE 5-23
The sequential decision tree with two forecasts.

unchanged. However, the overall minimum expected loss increases to $0.629, but the other performance measures are unchanged. All that has been achieved by reversing the order of the sensors is to increase expected loss with no improvement in performance. Note however that the expected loss is still lower than when the sensors are not used sequentially.

Suppose that the fraction of real objects in the population is not known and that we wish to see how sensitive our "optimal solution" is to the fraction p_X. First, let us assume that, using the worst sensor first, it has indicated a real object. Thus, we are at the upper D_1 node in Figure 5-23. Instead of assuming that the population is composed of 55% real objects, let us assume that p_X can be any fraction between 0.01 and 0.22. We

Sensitivity Analysis on p_x

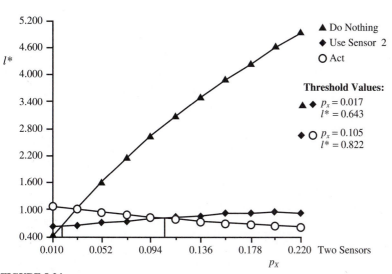

FIGURE 5-24
Sensitivity analysis when first sensor indicates "real."

plot the expected loss for each of the three decision alternatives for p_X in this range in Figure 5-24.[6] If $p_X < 0.017$ the optimal decision is to Do Nothing, whereas if $p_X > 0.105$, it is to Act. For any p_X in the range (0.017, 0.105), the second sensor should be used. If the initial sensor reading is "false," we are at the lower D_1 node in Figure 5-23. Now consider p_X between 0.10 and 0.70. The sensitivity analysis for this problem is shown in Figure 5-25. In this case the optimal decision is to Do Nothing when $p_X < 0.177$, Act when it is above 0.592, and use the Second Sensor when it is between these values.

5.8   SUMMARY AND INSIGHTS

In many real decision problems the data are often subjective, difficult to estimate and based on historical data that may not apply in a changing future. It is important that an analyst illustrate for the decision maker how sensitive the choice of optimal decisions is to such estimates or judgments. An in-depth sensitivity analysis forms an integral part of what we call a solution to the problem and is an essential aid for real decision-makers. We illustrated with a decision sapling where we found ranges of the risky venture success probability yielding the same optimal decision. We obtained a sensitivity analysis

---

[6] Figures 5-24 and 5-25 were produced using the one-way sensitivity analysis feature of DATA™ software by TreeAge.©

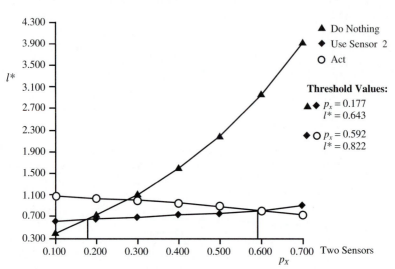

FIGURE 5-25
Sensitivity analysis when first sensor indicates "false."

for the football problem where the two-parameter policy space is described a plane rather than on a line.

The simple decision sapling model of Chapter 1 has been extended to include the possibility of obtaining a forecast to reduce the uncertainty in the risky venture. We called this model the expert opinion problem. Examples of an aircraft engine part and crop protection problem, each with a two-category forecast, were used to demonstrate how a sensitivity analysis could be carried out. The first analysis used the decision probabilities, and the second the forecast likelihoods. An important result is that even though a forecast can reduce our uncertainty, it may not always pay to use one. By plotting the optimal expected payoff or loss as a function of certain parameters such as the decision probabilities, forecast likelihoods, results, or the baseline probabilities, we demonstrate the sensitivity of an optimal policy to parameter value changes.

Modeling a decision problem under the assumption that you have perfect information on certain uncertain future events can be of some benefit in making decisions because it helps us place a bound on the value of optimal policies. The expected value of perfect information gives an upper bound on how much one can expect a forecast to increase the economic return of an optimal policy. It also suggests how resources might be invested in collecting and analyzing data, running experiments, or developing tests or prototype systems. One must be careful in describing and interpreting the meaning of perfect information models where decisions influence the probabilities of future outcomes.

When multiple forecasts of varying quality are available, it usually pays to include the results of the poorer ones along with the better ones. It also usually pays to consider sequential decisions on whether or not to obtain the forecasts, obtaining the cheaper and poorer ones before deciding whether or not to pay for the more expensive and better ones. We demonstrated these results in our sensor examples.

A number of publications on crop protection problems in the forecasting and meteorological literature illustrate the important linkage between forecasting and decision making. See for example Katz and Murphy (1987, 1990) and Oliver (1992).

PROBLEMS

5.1 Consider the betting problem faced by Nancy when she does *not* have any information on the coach's point-conversion decision but believes with probability p_C that the coach will choose the kicking play, that is, $Pr\{Y = k\} = p_C$. Draw the boundaries of the feasible region in the (p_k, p_t) plane that yield an expected nonnegative return for her risky venture. Carefully explain the meaning of any new subregions that do not coincide with the darkly shaded area I in Figure 5-2.

5.2 It is estimated that 15% of a particular part installed in aircraft engines are defective. Let $X = 1$ if a part is defective and 0 if not. If one of these parts fails in service, it results in a loss of $100. For a given part in an engine on an aircraft that is to receive maintenance, you can make an immediate decision D_1 to:

a. Leave it installed as is (at no cost). Call this action LI.

b. Rework it at a cost of $15. Call this action RW. A reworked part will be free from defects.

c. Test it at a cost $4. Call this action T1, and let $F = 1$ if the test indicates a defective part, 0 if not. The test has a 75% chance of revealing a defect if one is present; it also indicates a defective part 35% of the time when no defect is present.

If you decide to test, following your observation of the test result you can then make a second decision D_2 as follows:

If the first test indicates "no defect":

a. Leave it installed as is (at no cost), LI.

b. Rework it at a cost of $15, RW. A reworked part will be free from defects.

If the first test indicates "defective":

a. Leave it installed as is (at no cost), LI.

b. Rework it at a cost of $15, RW. A reworked part will be free from defects.

c. Make a second (independent) test that costs $6, denoted T2. Call the outcome of this test by G, similar to F for the first test. This second test has a 90% chance of revealing a defect if one is present; it also indicates a defective part 20% of the time when no defect is present. If this second test indicates a defect, the part must be reworked. If it does not, you leave it installed. The result of the decision problem will be measured by cost.

a. Draw the influence diagram for this problem, explain the meaning of each directed arc, and define each node set.

b. Draw the decision tree, and show all the costs.

c. Calculate all the conditional probabilities required to solve the tree.

d. Determine and state clearly the optimal policy for the given data.

e. Suppose you are uncertain as to the actual fraction of defective parts. Determine how the optimum policy changes as this fraction moves from 5% to 25%, and plot the minimum expected cost as a function of this fraction.

5.3 For the crop protection problem of Section 5.4.1, show that

a. $\rho_{FX} = (p_1 - p_0) \sqrt{\dfrac{p_F(1-p_F)}{p_X(1-p_X)}}$

b. $\rho_{FX} = (f_1 - f_0) \sqrt{\dfrac{p_X(1-p_X)}{p_F(1-p_F)}}$.

5.4 For the crop protection problem of Section 5.4.1, when $p_F = p_X$, the correlation coefficient $\rho_{FX} = p_1 - p_0$. Show in this case that the decision probabilities are given by

a. $p_0 = p_X(1 - \rho_{FX})$ and $p_1 = p_X + (1 - p_X)\rho_{FX}$.

b. Plot these as functions of ρ_{FX} on $(0, 1)$ and comment on the result.

5.5 For the crop protection problem of Section 5.4.1 with $p_F = p_X$, plot the normalized economic gain Δ_F/r_1 as a function of the correlation coefficient ρ_{FX} and show that

a. The economic gain is zero until the correlation exceeds

$\bar{\rho}_{FX} = (r_2/r_1 - p_X)/(1 - p_X)$, when $r_2/r_1 > p_X$, or

$\bar{\rho}_{FX} = (p_X - (r_2/r_1))/p_X$ when $r_2/r_1 < p_X$.

b. For $\rho > \bar{\rho}_{FX}$ and fixed p_X, the relative gain is linear in ρ_{FX} with slope $(1 - p_X)/p_X$ Explain. (*Hint*: Use the results in Problem 5.4)

5.6* For the crop protection problem of Section 5.4.1 with $p_0 < p_1$, plot the optimal protect decisions in the (p_X, p_1) plane; discuss and interpret the feasible and optimal regions

a. Under assumption that f_0 and f_1 are fixed,

b. Under both the assumptions that f_0 and f_1 change as p_X changes in such a way that $p_F = p_X$ on the range $(0, 1)$.

5.7 In comparing historical weather patterns with forecasts made by a commercial service it has been found that 70% of the time adverse weather was predicted when it actually occurred and that 8% of the time adverse weather was predicted when it did not occur. The weather is adverse 10% of the time. Find the conditional predictors $p(x|f)$ and the optimal expected gain when $r_2/r_1 = 0.2$.

5.8 Suppose that the condition of the part in the (Lindley) replacement problem in Section 5.3.1 depends on the manufacturing supplier Y = a or b. X denotes condition of part; assume that $Pr\{X = 1 | Y = a\} = 1/16$; $Pr\{X = 0 | Y = b\} = 3/4$, and that, as before, outcome of test is independent of supplier. Redraw the influence diagrams and decision trees, and solve the new decision problem when the subcontractor is known.

5.9 Suppose that, in the preceding problem, it is found that the shipping department mixed the parts together and placed them in one bin so that identification of the manufacturer is no longer possible. Solve the problem when it is believed that about two-thirds of the bin's contents are from subcontractor Y = a. Describe two ways to solve the problem. Explain.

5.10 You manufacture a device that is imperfect with probability p. You consider only two decisions: $D = 0$, ship without inspection, or $D = 1$, inspect the device with a defective measuring instrument that gives a faulty reading with probability q, that is, the report "imperfect" when the device is perfect and the report "perfect" when the device is imperfect. Whenever the inspection reports "imperfect," you destroy the device and do not ship, otherwise you ship. The cost of each device is $300, an inspection costs $100, and you collect $r from your client for each item shipped. Whenever your client receives an imperfect device, you lose the client at an estimated additional cost of $1,000. Your objective is to maximize expected profit.

   a. Draw an influence diagram and a decision tree for this problem in which the inspect node precedes the node describing the true state of the device. Label the branches of the tree with the appropriate probabilities.

   b. Draw the boundaries of the optimal policy region(s) in the (p,q) plane for two different "reasonable" values of r. Carefully label and explain the meaning of the optimal regions.

5.11 Check the formulas for the minimum expected loss function in column 5 of Table 5.2 (p. 222). Write out this function for the numeric example using both sensors. Check the plot shown by the dashed line in Figure 5-20 (p. 223).

5.12 For the sequential sensor problem in Section 5.7, check the stated minimum loss values for the cases where (a) the worst sensor is used first, (b) the best is used first.

CHAPTER
6

SUBJECTIVE
MEASURES
AND UTILITY

6.1 INTRODUCTION

In the models discussed and analyzed in preceding chapters we have assumed that the results of the decision problem could be represented by measurable observable quantities. Usually, the attribute chosen was money, and the payoff function was linear. In the airline seat allocation and hotel room pricing problems the results were measured by expected revenues. In the crop protection problem expected monetary losses were used. In the cancer diagnosis problem the results were measured by the number of cancers found and by the cost of testing a given number of patients. In the football example introduced in Section 2.3 (p. 48), the initial problem posed was that of the coach who must decide which play the team should attempt. Because the appropriate unit of measure to use for the result was not clear, we analyzed a related problem of two spectators betting on the game. In this chapter our objective is to present a procedure to quantify attributes that have no obvious and widely accepted unit of measure. This would allow the coach to quantify his or her preference for the three possible outcomes, "win," "tie," or "loss." Being able to quantify such attributes is the first of two important reasons we study utility theory. The second is the measurement of risk.

Some readers may question the appropriateness of using linear return functions even when we can measure results in terms of monetary payoffs or losses. There are many situations where the maximum size of the loss in a given decision problem is small relative to one's total assets. In such cases we say that the risk is low. But if potential losses are large compared to one's total assets we say the decision has high risk, and in such cases the use of a linear payoff function is probably not appropriate. If we look at the business of gambling itself, large casinos have assets far in excess of the usual maximum return from any of the games they offer. They have very low probability of serious losses because the many small gambles they enter into with clients (players at the tables and slot machines) all have a positive expected return to the casino. Linear loss or return functions are appropriate for the casino management to use. But suppose you are considering visiting a casino. How much of *your* net worth are you willing to gamble? Even if you could find a game that has a positive expected value (for you the player), you would probably not be willing to risk losing a large fraction of your net worth on a single game. This of course is the reason why both individuals and institutions try to reduce their risk by diversifying assets. If linear return functions were appropriate, their use would lead one to invest all one's assets in the single venture with the largest expected return. The linear payoff function does not take into account the risk-averse behavior of many people. The second reason for studying utility theory is that it presents a coherent method for modeling risk.

Another way to illustrate that linear result or payoff functions may not be appropriate is to consider an individual's income. An extra \$5,000 a year means far more to someone earning \$20,000 than it does to someone earning \$100,000. Our sense is that money over such a range has decreasing returns as the amount one earns increases. This means that the value of money is nonlinear. It is usually more appropriate to use nonlinear payoff functions when faced with decision problems where the potential losses or returns are significant compared to one's total assets.

It may be easy to decide that the use of a nonlinear payoff function is appropriate in a given decision problem. It is much harder to determine what nonlinear form should be used. The problem of deciding on the most appropriate form by taking into account the preferences of the decision maker is discussed in the following sections.

Sections 6.2 and 6.2.1 summarize the important topics of subjective probability and certainty equivalence, both of which are central to modeling with utility. Section 6.2.2 presents the theoretical assumptions that underlie utility theory, and 6.3 demonstrates methods for finding a decision maker's utility function. Examples of how families of utility functions can be used are presented in Section 6.4, and these are used in 6.5 to illustrate and quantify the concept of risk. Section 6.6 gives some words of caution concerning the use of utility functions along with some important properties.

6.2   BASICS OF UTILITY THEORY

Subjective probabilities play a central role in determining utility functions. In many of our previous examples it has been assumed that the probability distributions of the random variables were known. In practice this would usually mean that historical data would have been collected, some stationarity assumptions made (demand, performance, or other random outcome, has not changed over time), and that the distributions have been estimated from this historical data. But in many decision models, especially those that incorporate executive or senior management level decisions, historical data may not be relevant; the judgment or beliefs of the decision makers may be most important in estimating relevant probabilities.

A method often used to elicit subjective probabilities is with what is often called a *standard device*, usually an urn filled with various mixtures of different colored balls (either real or conceptual). More will be said about this in Section 6.3.1, but whatever method is used, their determination must be done in a way that is (1) consistent with the rules of probability, and (2) done in a *coherent* fashion. For example, suppose that the following two events can occur in the next year: *A*—You receive a pay raise, *B*—You receive an annual bonus. If you believe that *A* is more likely to occur than *B*, the subjective probabilities reflecting these beliefs must satisfy $Pr\{A\} > Pr\{B\}$. The laws of probability tell us that the probability at least one of them will occur, $Pr\{A \cup B\}$, must satisfy

$$Pr\{A \cup B\} = Pr\{A\} + Pr\{B\} - Pr\{A \cap B\}, \tag{6.1}$$

where the last term on the right-hand side is the probability that both occur. You may believe there is an 80% chance of a pay raise, and a 50% chance of a bonus. If you are sure that you will receive at least one of them, then you must believe there is a 30% chance you will receive both. Any other assessment is inconsistent with Equation (6.1). You must also believe that, given you receive a pay raise, there is a 37.5% chance you will also receive a bonus, whereas if you do not receive a raise the chances of a bonus are 100%. It is important that the implied joint and conditional probabilities be checked for both reasonableness and consistency. Suppose that a third event *C* can occur, that you move to a new job. You may believe that this is less likely to occur than receiving an annual bonus. You cannot also believe that taking a new job is more likely than receiving a pay raise as this would imply $Pr\{C\} > Pr\{A\} > Pr\{B\} > Pr\{C\}$! One must always be careful when applying subject judgment to assess probabilities that it be done coherently.

6.2.1 Indifference Probabilities and Certainty Equivalents

When we first introduced the decision sapling in Section 1.8 we assumed that the payoffs for the results were given; r_1 if the risky venture is taken and it succeeds, r_3 if it is taken and it fails, and r_2 if the riskless alternative is taken. By equating expected payoffs for the riskless alternative and the risky venture we can determine the *indifference probability*:

> The *indifference probability* for a decision problem between a risky venture and a riskless alternative with given known results is that probability of success in the risky venture for which the decision maker is indifferent to the two alternatives.

We now look at the decision sapling in a different way; assume that we know the payoffs for success and failure in the risky venture, r_1 and r_3, respectively, together with the probability of success *p*, and are asked for that value of r_2 we would need for certain in order to be indifferent between the two alternatives. Instead of being asked for our indifference probability, we are now asked for *certainty equivalent*:

> The *certainty equivalent* to a risky venture is the least amount the decision maker would have to obtain for certain by choosing the riskless alternative.

In many situations neither the indifference probability nor the certainty equivalent is some fixed value that is the same for everyone, but is a measure that reflects the amount of risk that the particular decision maker is willing to take. Both concepts play a central role in utility theory. Because of their importance, we illustrate with an example.

Recall the civil lawsuit example described in Section 1.8 (p. 28), where, if you let the jury decide the case you could obtain a net return of $1,950,000 or lose $50,000. Your problem is to decide whether or not to accept a net settlement of $650,000 ahead of the jury decision. Whether or not you accept the offer depends on a number of factors, two of which are your net worth and your assessment of the probability that the jury will decide the case in your favor. After comparing the possible outcomes to your current assets, you may decide that you would continue with the case and wait for the jury's verdict only if you judge the probability of finding in your favor to be at least 0.85, otherwise you would accept the settlement offer. In this case your *indifference probability* is 0.85 (note that the expected payoffs are equal when the probability is 0.35).

A second way to view this decision problem is to first decide on the probability of winning the case, and based on this, determine the level of offer you would accept in settlement. Most people would not be willing to accept a small settlement offer, but would accept any offer above some value. This value is the *certainty equivalent* of the risky venture and will again usually depend on your financial worth and your propensity to assume risk. Suppose that the opinion of your legal council, who has considerable courtroom experience, is that the chances of jury finding in your favor are 2 to 1. If you accept this probability 2/3 of winning the case, your expected payoff from the jury decision is $1,283,333. If you think your chances of a favorable verdict are only even, the expected payoff is $950,000. Although both these figures are greater than the settlement offer, you still may decide to settle out of court for the $650,000. If you do, your certainty equivalent is some number no larger than $650,000. You may never have to decide on its value in a particular decision problem, but as we shall see, its determination, often obtained from abstract unreal decision problems, is central to the application of utility theory to decision making.

6.2.2 Assumptions of Utility Theory

Let us look again at some of our earlier examples where it is not obvious how to quantify the results. Consider the football example in Chapter 2 where the coach must decide which play the team should make. In order to make a decision, the coach needs to evaluate the result "game tied" relative to "game won" or "game lost." In the medical example of elective surgery in Section 1.8, the patient needs to evaluate the relative results "successful surgery/great improvement in quality of life," "continue in current state without surgery," and "unsuccessful surgery/deterioration from current condition, perhaps even death." In this section we present a coherent way of placing measures on such outputs that (1) consistently reflect the decision maker's preferences for results, and (2) can be used together with decision trees to analyze complex decision problems. This measure is called a *utility*.

In both of the examples just given there are three possible results. One result is clearly the "best," another the "worst," and the third is at some level in between these extremes. In general, let $\mathcal{R}$ be the set of possible results, and denote the *best* (most preferred) result by $\bar{r}$ and the *worst* (least preferred) result $\underline{r}$. The words *best* and *worst* are used because for cases where results are quantifiable, they encompass both cases where (1) $\underline{r}$ is smaller than $\bar{r}$ (a maximizing problem), and (2) $\underline{r}$ is larger than $\bar{r}$ (a minimizing problem). To ensure that we can find a measure of the utility of any element of $\mathcal{R}$, we make the following two assumptions:

A1. Every pair of elements in $\mathcal{R}$ can be compared, and there is a well-defined *preference ordering* on the elements of $\mathcal{R}$. If r_1 and r_2 are any two elements of $\mathcal{R}$, we write

$r_1 > r_2$ if result r_1 is preferred to result r_2,

$r_1 \sim r_2$ if one is indifferent between results r_1 and r_2,

$r_1 < r_2$ if result r_2 is preferred to result r_1.

A2. The preference ordering is assumed to be *transitive*; that is, if you prefer r_1 to r_2 and r_2 to r_3, then you must prefer r_1 to r_3.

These two assumptions guarantee consistency in comparing results. Our aim now is to associate numeric values with the results in such a way that (1) these numeric values are ordered consistently with our preferences, and (2) we can determine these values by some procedure. An additional assumption needed for both of these to occur is

A3. Let r_1, r_2, r_3 be any three elements of $\mathcal{R}$ (any three possible results) such that $r_1 > r_2 > r_3$. In the simple decision sapling in Figure 6-1 we show a choice of a risky venture between r_1 and r_3 and the certainty of getting the intermediate result r_2. Assume that we can find a probability p of winning the risky venture such that we are indifferent between choosing the risky venture or the certain result. This third assumption is called *continuity*.

The set $\mathcal{R}$ is assumed to contain both the worst and best as well as all possible intermediate results,[1] so that a continuous utility function can be defined on $\mathcal{R}$.

Let $u(r)$ denote the utility of result r. We assign utility values $u(\bar{r}) = 1$ to the *best* result, and $u(\underline{r}) = 0$ to the *worst*. If A3 holds, we can assign a number to the utility $u(r)$ of any intermediate result r. By replacing r_1 with $\bar{r}$, r_3 with $\underline{r}$, and r_2 with r in Figure 6-1, equating expected values shows that $u(r)$ must equal p, the indifference probability between a certain choice of r and a risky venture between the best and worst results. A1 and A2 guarantee the uniqueness of this number. *Thus, we have demonstrated a constructive way of determining the utility function.*

---

[1] To be mathematically precise, we assume that $\mathcal{R}$ contains all convex combinations of $\underline{r}$ and $\bar{r}$ and is closed.

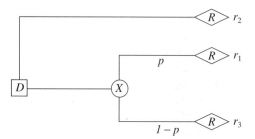

FIGURE 6-1
Decision sapling with indifference probability.

In order to ensure that this function can be used consistently in decision tree analysis we need to make three more assumptions:

A4: When comparing risky ventures among different results we assume that the laws of probability apply. Let $\{r_i; i = 1, 2, \ldots, m\}$ be any subset of results in $\mathcal{R}$. Let $\{p_i\}$ and $\{q_i\}$ be two sets of probabilities of achieving results $\{r_i\}$. Suppose that a risky venture using the p_i's will occur with probability s and one using the q_i's with probability $(1 - s)$. Now consider the single risky venture where the probability of achieving r_i is $(sp_i + (1 - s)q_i)$. It is assumed that you are indifferent between these alternatives. This assumption is referred to as *reduction of mixtures*.

Figure 6-2 illustrates these risky ventures with $m = 3$ in a decision tree with two decisions *alternatives* d_1 and d_2. Since it is assumed that there is no change in utility by choosing the single risky venture (d_1) over the double risky venture (d_2), this assumption is sometimes referred to as "no fun in gambling."

A5: It is assumed that if a decision maker is willing to substitute the risky venture between the worst and best results (shown in Figure 6-1) for the certainty of getting the intermediate result, then he or she will be willing to do this in every context. This assumptions is called *substitutability*.

A6: A decision maker, when confronted with two risky ventures, each on the best and worst results, always prefers the one with the largest probability of obtaining the best result. This assumption is referred to as *monotonicity*.

These assumptions guarantee that the constructed utility function can be applied to decision analysis problems in a coherent manner.

6.3   DETERMINATION OF UTILITY FUNCTIONS

A theory of utility would have little practical use if we could not determine a decision maker's utility function. Our assumptions and theory lead to constructive ways to elicit such a function. We present two methods, both based on the decision sapling.

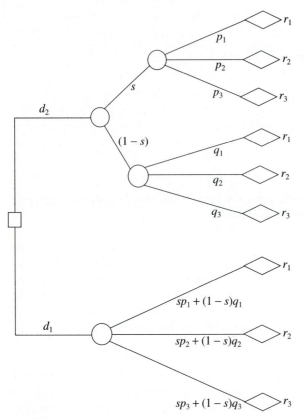

FIGURE 6-2
Simplification of complex risky ventures.

6.3.1   Utilities as Indifference Probabilities

Our first approach to determining utility functions is by successively finding indifference probabilities for a sequence of simple risky ventures using a decision sapling. We illustrate with the football example of Chapter 2.

Suppose in Figure 6-1 that r_1 is {win the game}, r_3 is {lose the game}, and r_2 is {tie the game}. The problem is to find $u(r_2)$ $(= p)$. If p were very close to 1, you would prefer the risky venture to the certain event. If p were close to 0, the reverse would be true. How about $p = 0.95$, or 0.5, or 0.90, and so on? The interval containing your indifference probability is narrowed by closing in from both ends. In this particular example it is not clear what meaning one could place on any convex combination of members of the result set $\mathcal{R}$, so the *mixture set* would have little meaning and we would not attempt to draw a continuous utility on the "interval" (win, lose). But we shall see an example of a plaintiff's decision problem in a court case where results are measured in terms of money, and convex combinations of best and worst results have real meaning.

For a student of probability who is used to working with the concept, finding an indifference probability would be a straightforward, though not necessarily easy, procedure. For people unfamiliar with probabilities it can be a daunting process. A *standard device* is usually proposed to help in such cases, the one most commonly cited being the "balls-in-an-urn" model. In the football example, the decision maker would be asked to compare going for the conversion kick with picking a ball randomly out of an urn that contains 100 balls of which n are black and $(100 - n)$ red. Picking a black ball means winning the game. As n is varied, the decision maker can home in on his or her indifference point between the kick and the risky venture, say $\bar{n}$. On a scale where winning the game has utility 1 and losing 0, their utility for a drawn game is $\bar{n}/100$.

6.3.2   Utilities from Certainty Equivalents

We now illustrate a method of determining a utility function associated with money that avoids thinking of probabilities except for 50/50 risky ventures. For example, suppose that you are the plaintiff in the civil lawsuit described in Section 6.2.1 where you are seeking $2,000,000 damages against the defendant. Ignoring your $50,000 expenses, if you win you get $2,000,000, if you lose you get nothing. Assume that the case has been heard, the jury is deliberating, and the defendant offers to settle for $700,000. You must decide before the jury returns whether or not to accept the offer. Let us assume that you believe the odds of the jury deciding the case in your favor are even. If you use the dollar value directly as a result measure, the expected winnings are $1,000,000. But how many people of modest means would give up the *certainty* of $700,000 for a 50/50 risky venture of getting $2,000,000 or nothing? Perhaps a nonlinear utility value of money is more appropriate to use when making the decision.

Although the result set in this example is {0, 700,000, 2,000,000}, let $\mathcal{R}$ be the set $\{r; 0 \leq r \leq 2{,}000{,}000\}$, so that $\underline{r} = 0$ and $\bar{r} = 2{,}000{,}000$. First, we assign utilities of 0 and 1, respectively, to these; these points are plotted in Figure 6-3. The decision maker is asked what amount he or she would accept for certain versus a 50/50 risky venture with $0 and $2 million (the certainty equivalent described in Section 6.2.1 with $p = 0.5$). The technique of bracketing the answer from both ends of the range can be used. The answer will vary from one person to the next, but let us assume your answer is $500,000. From Figure 6-1 it must be that

$$u(500{,}000) = 0.5u(0) + 0.5u(2{,}000{,}000) = 0.5.$$

Thus, the point (500,000, 0.5) falls on the utility curve and is plotted in Figure 6-3. Next, the decision maker is asked to find the certainty equivalent of a 50/50 risky venture between 0 and $500,000 and the certainty equivalent of a 50/50 risky venture between $500,000 and $2 million. Let these be $100,000 and $1,200,000, respectively, so that

$$u(100{,}000) = 0.5u(0) + 0.5u(500{,}000) = 0.25,$$

$$u(1{,}200{,}000) = 0.5u(500{,}000) + 0.5u(2{,}000{,}000) = 0.75.$$

The two points (100,000, 0.25) and (1,200,000, 0.75) are plotted in Figure 6-3. We would continue to evaluate points on the curve in each subinterval until convinced we

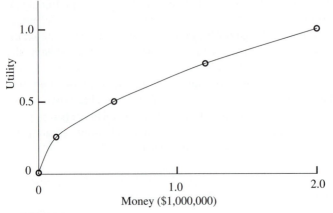

FIGURE 6-3
Example of the utility value of money.

have enough points through which we can reasonably draw the decision maker's utility function. Figure 6-3 shows such a utility function drawn through the five points.

It cannot be overemphasized that the points you have obtained on your utility curve are obtained using *your* subjective judgment. It is very important that you make consistency checks as you proceed. For example, having obtained the five points in Figure 6-3, you should now consider a 50/50 risky venture where a win gives you $1,200,000 and a loss gives you $100,000. Your certainty equivalent for this must be $500,000, otherwise your judgments are not consistent (see Problem 6.5 for a famous historical example where many people exhibit inconsistent choices).

The above method of using 50/50 risky ventures is a widely accepted method of assessing continuous utility functions. It is by no means the only one, but it is considered by many to result in a curve with the least bias. Note that this method cannot be easily applied to problems such as the football example where there is no continuum of results. What would be your certainty equivalent of a 50/50 chance of winning or losing the game?

6.3.3 Cautionary Comments

It is widely recognized that eliciting a decision maker's utility function is one of the most difficult tasks in decision analysis. To quote from Bunn (1984):

> *The assessment of the utility function is not a trivial task and requires considerable sensitivity to the cognitive processes of the subjects and the sorts of biases to which they may be prone. Expected utility will only be the appropriate criterion if the decision maker accepts its assumptions. It will not always be the correct way to model certainty equivalence. Nevertheless, even if it is not appropriate, the process of discovering why this is the case can provide considerable insight into the decision maker.*

To give a more graphic idea of the difficulties involved, picture the following scene. You have been hired as an analyst to help the president of a large soft drink company in a major investment decision that involves marketing a new product. You decide that your first task is to find the president's utility function and will attempt to do this by finding indifference probabilities using the standard device. You arrive at the president's office with your urn full of balls. After some quizzical looks at your paraphernalia this serious decision maker explains the problem. A soft drink with a new taste has been tested in selected markets, and although it has not been universally accepted in all test areas, it shows potential for strong national sales. A decision must be made on whether or not to introduce it nationwide and move to full production. You cannot help in this decision without some measure of the "utility" to the company of a successful new product. For example, is success to be measured in increased profits, or increased market share, or both? What is a 10% increase in profits equal to, and is it worth the risk involved to obtain it? To start you might assume a utility of 1 if the new drink is successful on a national scale and 0 if it fails (of course it is not obvious how to know when success or failure has occurred, but we ignore this problem). To measure the utility of the current product on this scale you might start as follows: "This urn contains 50 red balls and 50 white balls, and if you draw a white one your new product will be successful. Which would you prefer, drawing a ball from the urn or continuing with your current product?" We suspect that a majority of presidents would tell you to get the h--- out of the office and what to do with your urn and balls! You may have more success using the second method with 50/50 risky ventures, but this method is not without its difficulties either.

Soliciting utility functions from decision makers using artificially constructed gambles is difficult. Another approach is to decide ahead of time on a particular nonlinear functional form for the utility that has desirable characteristics and whose parameters can be determined from the decision maker's answers to a few questions that fit naturally into the world in which he or she operates. This is the approach taken in the next section.

6.4 EXAMPLES OF UTILITY FUNCTIONS

Although sometimes very difficult to extract from a decision maker, the general shape of most decision makers' utility functions is similar to that shown in Figure 6-3. The precise mathematical function used is not so important as is the shape of a concave function that exhibits marginally decreasing utility as the return increases. In our example, the utility obtained from the first $500,000 is much greater than the additional utility obtained from the next $500,000. Two families of functions fit this general shape and have other desirable properties as we shall see later in the chapter. These are certain members of the exponential and logarithmic families. This section illustrates methods of fitting these families of curves that avoid solicitation of subjective probabilities or certainty equivalents from the decision maker through fictitious risky ventures. They require instead an estimate of the relative rates of change in the utility function near the extreme points of the range, that is, $\underline{r}$ and $\bar{r}$.

In the following sections we assumed that the attributes are well defined and measurable and that the worst (least preferred) level is smaller than the best (most preferred); that is, $\underline{r} < \bar{r}$. It is left to the reader to show that the exponential and logarithmic utility functions shown in Equations (6.3) and (6.5), respectively, can be used directly for the reverse case where $r > \bar{r}$. This will occur when lower values of an attribute are preferred to higher ones. As the name suggests, utility is defined in such a way that in decision problems it is always maximized.

6.4.1   An Exponential Utility Function

Our first example of a family of utility functions is an exponential utility function

$$u(r) = a\left(1 - e^{-b\left(\frac{r-\underline{r}}{\bar{r}-\underline{r}}\right)}\right), \qquad \underline{r} \le r \le \bar{r},$$

where a and b are to be determined for the particular decision maker in a particular setting. By setting $r = \underline{r}$, we see that $u(\underline{r}) = 0$ for any values of a and b. By setting $r = \bar{r}$, for $u(\bar{r}) = 1$, the following relationship must hold,

$$a = \frac{1}{1 - e^{-b}}.$$

Thus, the parameter a is completely determined by the parameter b.

To determine a suitable value for b we look at the *slopes* of the curve at the worst and best values. Straightforward differentiation gives

$$\frac{d}{dr}u(r) = \frac{ab}{\bar{r}-\underline{r}}e^{-b\left(\frac{r-\underline{r}}{\bar{r}-\underline{r}}\right)}. \qquad (6.2)$$

Suppose now that we solicit from the decision maker the ratio of the marginal returns from an extra unit if r near $\underline{r}$ and near $\bar{r}$. This ratio measures the relative slopes of the utility function at $\underline{r}$ and $\bar{r}$. For the money example shown in Figure 6-3 a dollar return near zero increases the utility about ten times as fast as a dollar near two million. Define β to be the ratio of the slope of the utility function at the worst level to the slope at the best level. Using Equation (6.2) it follows that

$$\beta = e^{b}$$

and our utility function becomes

$$u(r) = \frac{\beta}{\beta - 1}\left(1 - \left(\frac{1}{\beta}\right)^{\left(\frac{r-\underline{r}}{\bar{r}-\underline{r}}\right)}\right), \quad \underline{r} \le r \le \bar{r}. \qquad (6.3)$$

All we require for a practical specification of this utility function are the values of $\underline{r}$, $\bar{r}$, and the decision maker's value for β.[2]

---

[2] This is by no means the only exponential form of a concave utility function. Others can be found in the references, as pointed out at the end of the chapter.

6.4.2   A Logarithmic Utility Function

Our second example of a family of utility functions is

$$u(r) = a\log\left(1 + b\left(\frac{r - \underline{r}}{\bar{r} - \underline{r}}\right)\right), \qquad \underline{r} \le r \le \bar{r}.$$

For positive a and b this also has the general shape shown in Figure 6-3. By setting $r = \underline{r}$ we again see that $u(\underline{r}) = 0$ for any values of a and b. By setting $r = \bar{r}$ we see that the following relationship must hold,

$$a = \frac{1}{\log(1 + b)}.$$

As in the exponential case the parameter a is completely determined by the parameter b. To determine a suitable value for b we again look at the ratio of marginal returns as described. Straightforward differentiation gives

$$\frac{d}{dr}u(r) = \frac{ab}{(\bar{r} - \underline{r}) + b(r - \underline{r})}. \qquad (6.4)$$

Using the parameter β to be the ratio of the slopes at the worst and best value as in the exponential case, Equation (6.4) leads to

$$\beta = b + 1,$$

and the utility function becomes

$$u(r) = \frac{\log\left(1 + (\beta - 1)\left(\frac{r - \underline{r}}{\bar{r} - \underline{r}}\right)\right)}{\log\beta}, \quad \underline{r} \le r \le \bar{r}. \qquad (6.5)$$

Note again that $u(r)$ is completely determined by $\underline{r}$, $\bar{r}$, and β.

Both the exponential and logarithmic utility functions are illustrated using the earlier money example, where $\underline{r} = 0$ and $\bar{r} = 2$. Figure 6-4 shows Equations (6.3) and (6.5) plotted using these values with $\beta = 10$. The points derived using indifference probabilities are also shown.

The reader can see that the logarithmic curve starts above the exponential curve but finishes below it. However, they are quite close to each other and are quite good approximations to the points derived from indifference probabilities. We must be careful when using the word *approximation* here. The reader must remember that, although we continue to use mathematical expressions for utilities, they are inherently subjective. *Their whole purpose is to reflect the relative preferences of a decision maker.* The (relatively small) differences between utilities found by fitting functional forms such as those in Equations (6.3) and (6.5) or found by the much more difficult procedures using either indifference probabilities or certainty equivalents may not be significant when compared to the inherent difficulty and uncertainty a decision maker typically faces in determining the utility function.

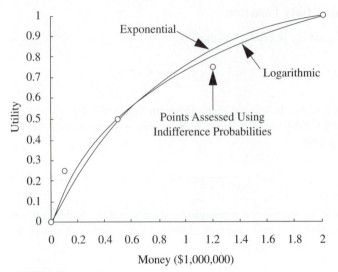

FIGURE 6-4
Exponential and logarithmic utilities for money.

6.5   MEASURES OF RISK

The purpose of this section is to explore and quantify what we mean by risk. Consider a possible risky venture with equal chances of success or failure, where success brings $(r + 1)$ and failure results in a loss of r. The expected gain is 50 cents, independent of r. Many people would take the venture if r were 1 (a possibility of doubling a bet of $1 with the toss of a coin), but few would take it for large values of r such as 50,000 or 1,000,000. A person's willingness to take risks will usually depend in part on the size of r relative to their total assets. People unwilling to risk a substantial part of their assets even for positive expected return are said to be *risk averse*; those willing to take a risky venture for a negative expected return are said to be *risk prone*. As we shall see, a concave utility function indicates a risk-averse person, and a convex function a risk-prone person. We explore ways of measuring the degree to which a person avoids or seeks risk. Because far more individuals appear to be risk averse rather than risk prone, in the following subsections we concentrate on the risk-averse case. But before proceeding to investigate measurements of risk, we first introduce the important concept known as *strategic equivalence*:

> Two utility functions are said to be strategically equivalent if and only if, when each is applied to two arbitrary lotteries, the same preference ordering is obtained.

It is easy to show that utility functions $u_1(r)$ and $u_2(r)$ are strategically equivalent if and only if they can be written in the form

$$u_2(r) = a + bu_1(r) \tag{6.6}$$

for some constants a and b with b positive. The use of either utility function would lead to the same optimal decisions for any decision problem. For example, consider using $u_1(r)$ in a decision problem represented by the decision sapling in Figure 6-1. Using the rollback algorithm, expectation is taken at the random node and the maximum taken at the decision node. At the random node the expected value is $pu_1(r_1) + (1 - p)u_1(r_3)$, and at the decision node we find

$$Max\{u_1(r_2), \ pu_1(r_1) + (1 - p)u_1(r_3)\}. \tag{6.7}$$

Solving the problem using $u_2(r)$, at the random node the expected value is $pu_2(r_1) + (1 - p)u_2(r_3)$, which after substituting Equation (6.6) is found to be $a + b[pu_1(r_1) + (1 - p)u_1(r_3)]$. Now at the decision node we find

$$Max\{a + bu_1(r_2), \ a + b[pu_1(r_1) + (1 - p)u_1(r_3)]\}. \tag{6.8}$$

The first term in Equation (6.8) is the maximum if and only if the first term in (6.7) is maximum. So the same decision is optimal no matter whether $u_1(r)$ or $u_2(r)$ is used as long as b is positive.

6.5.1   Risk Premium

Let $r_1 > r_3$ and assume you are considering a risky venture with equal chances of winning these two payoffs (see Figure 6-5). If a person's certainty equivalent (the minimum you are willing to accept for certain to avoid the risky venture) is $r_m = (r_1 + r_3)/2$, this person is indifferent between a risky venture and the certainty of obtaining its expected value.[3] Their utility curve on $(\underline{r}, \bar{r})$ would be the linear function $u(r) = (r - \underline{r})/(\bar{r} - \underline{r})$.

If a person's utility curve is the solid concave curve shown in Figure 6-5, their certainty equivalent is r_2, a number less than r_m, the midpoint between $u(r_1)$ and $u(r_3)$. The quantity $(r_m - r_2)$ is called the *risk premium* and is the difference between the expected value and the amount one would accept for certain rather than take a 50/50 gamble for r_1 or r_3; the higher this number for a given risky venture, the more risk averse is the individual, because that individual is willing to take a lower return from a riskless alternative.

If a person's utility curve is the dashed convex curve shown in Figure 6-5, their certainty equivalent r_2 is greater than r_m. In this case the *risk premium* $(r_m - r_2)$ is negative; the individual will take the riskless alternative only if the payoff is higher than the expected value. In general, the utility functions of risk-averse and risk-prone individuals are concave and convex, respectively.

6.5.2   A Risk Aversion Function

Suppose we have two utility functions $u_1(r)$ and $u_2(r)$ that are both concave and thus both indicate risk aversion. If they are strategically equivalent (they satisfy Equation (6.6))

---

[3] In the parlance of decision analysis, the expected value is often called the *expected monetary value* (EMV), and a person with a linear utility function is called an EMVer. Such persons or organizations are risk neutral.

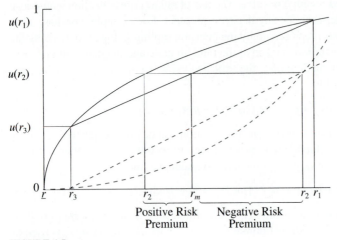

FIGURE 6-5
Utility functions and risk premiums.

one would like a measure of risk aversion to give the same value for both, as using either one in a decision problem will give the same optimal decision policy. Note that the first and second derivatives of u_1 and u_2 are related by $u_2'(r) = bu_1'(r)$ and $u_2''(r) = bu_1''(r)$. Dividing one of these equations by the other shows that the ratios of the first and second derivatives of strategically equivalent utility functions are equal. This leads us to define a risk aversion function $f(r)$ by

$$f(r) = -\frac{u''(r)}{u'(r)}, \qquad \underline{r} \le r \le \bar{r}. \tag{6.9}$$

In order to gain an understanding of what this function tells us about a utility function, we illustrate using the exponential and logarithmic examples shown in Figure 6-4 with $\beta = 10$.

We start by looking at the first derivatives, shown plotted in Figure 6-6. The start of each curve is at ten times the value of the end point in both cases. Clearly, they both show a decreasing marginal utility as wealth r increases. The logarithmic curve shows a higher marginal return for low and high values of r, whereas the exponential is higher in mid-range. In both cases their slopes (the second derivatives of $u(r)$) are negative. For a given wealth r suppose we look at the absolute value of the slope of the marginal utility curve relative to its value. This is what we have defined to be the risk aversion function $f(r)$. The two examples are shown plotted in Figure 6-7 for the exponential and logarithmic cases. Here we see a striking difference between the two forms of the utility that was not apparent in Figure 6-4.

The importance of the differences in the two utility functions depends on the size of the payoffs in a given decision problem relative to one's wealth. If we are comparing risky ventures with payoffs on the order of $1 million, we would use utilities over a large range in Figure 6-4, and small differences between the exponential and logarithmic forms of utility would probably not be important. But often one is faced with making decisions

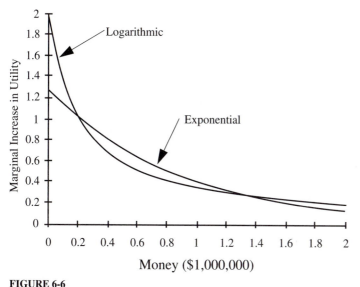

FIGURE 6-6
First derivatives of exponential and log utilities.

where the risk is small relative to one's total wealth. Figure 6-7 shows that for the exponential case, the fractional rate at which marginal utility is changing is constant. For the logarithmic case, the rate is high for low wealth and decreases as wealth increases. This appears to agree more with most people's behavior than does a constant rate. Suppose that your wealth level is r and that to increase it you are considering investing in a risky venture whose outcome is measured by the random variable X. Suppose that the expected outcome is zero but ranges over positive and negative values that are small relative to $(\bar{r} - \underline{r})$. One can show that for this case the risk premium is approximately proportional to the variance of X times the risk aversion function, or $Var(X)f(r)$. If one used the exponential form of the utility, the risk premium would not change with wealth level r, whereas with the logarithmic form it would decrease as wealth increases. This latter case would seem to agree better with actual behavior. For example, with $(\bar{r} - \underline{r}) = \$2,000,000$ you may be considering investing $10,000 in a stock fund that would increase or decrease with market fluctuations. If your liquid assets are only $50,000, you will probably show a much higher aversion to risk for this investment than if they are $500,000.

Not everyone's utility curve for money is necessarily concave, linear, or convex. People tend to be risk averse for uncertain results when faced with a possible loss on the order of their total wealth. For possible losses much smaller than this, it is not unusual for people to show some measure of risk proneness. The popularity of lotteries and casinos must be due to the fact that many people are willing to risk a few dollars on a game where the expected payoff is considerably less than the cost of playing but where the maximum prize for winning is considerably higher than both the cost of playing and the individual's worth. One explanation is that by spending a dollar, say, the player has a positive (often extremely small but positive) chance of obtaining considerable wealth with-

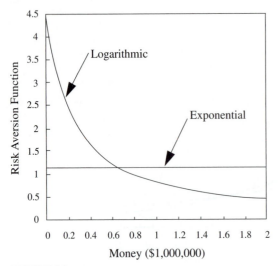

FIGURE 6-7
Logarithmic and exponential risk aversion functions.

out risking a significant part of his or her net worth. For most people the amount they are willing to risk is quite small. Individual utility curves for money might start near r being convex but usually change quite quickly to being concave for larger monetary values.

6.6   SOME PROPERTIES OF UTILITY FUNCTIONS

The power and usefulness of utilities come from their derivation as either (1) a numeric ordering of preferences for results of attributes that have no natural or obvious dimensional unit (win/lose/draw in the football game; ranking color preferences, satisfaction, etc.) or (2) measures of the decision maker's risk aversion or risk proneness toward given measurable levels of the attribute. But the reader should understand that utilities can only measure *order* of preferences, *not* their relative values. We illustrate this important point in a number of ways.

   If the utility of some result r_1 is assessed to be twice that of some result r_2, it does *not* follow that r_1 is preferred twice as much as r_2. Suppose you are soliciting information from a decision maker on the color of the result, say the purchase of a new car. People often have strong feelings on such outputs, but there is no obvious numeric way to measure these. One method of soliciting strengths of preferences is to ask for ratio weights. Suppose that the decision maker, in trying to decide on the color of a new car where the alternatives are blue, red, and white, states that blue is preferred twice as much as red and white preferred 2.5 times over blue. Suppose we arbitrarily assign r_r points $(r_r > 0)$ to red. Then we would assign $r_b = 2r_r$ points to blue, and $r_w = 5r_r$ points to white. It is easy to convert these points to a scale on [0, 1] and interpret the three numbers obtained as utilities. One way to do this is to define

$$u(c) = (r_c - r_r)/(r_w - r_r) \quad \text{for any color } c. \tag{6.10}$$

This gives $u(\text{white}) = 1$, $u(\text{red}) = 0$, and $u(\text{blue}) = 0.25$ independent of r_r. But suppose we used a decision sapling to find these utilities directly. As the most preferred and least preferred colors we would assign $u(\text{white}) = 1$ and $u(\text{red}) = 0$, respectively. Assume that $u(\text{blue})$ was assigned the value 0.25 as the indifference probability between getting blue for sure and a gamble between white and red. We should ask ourselves if these three utilities values always imply the relative weights 2.0 and 2.5 given above when these are derived using Equation (6.10). If r_r, r_b, and r_w are positive weights assigned to red, blue, and white, respectively, any values that satisfy

$$(r_b - r_r)/(r_w - r_r) = 0.25$$

would lead to this utility for blue. This equation can be written

$$\left(\frac{r_w}{r_r}\right) = 4\left(\frac{r_b}{r_r}\right) - 3 .$$

This shows that an infinite number of relative weights will lead to the same utility for blue of 0.25. For example $r_r = 1$, $r_b = 2$, and $r_w = 5$ gives the ratio weights 2 and 2.5, but $r_r = 1$, $r_b = 10$ and $r_w = 37$ also satisfy the equation but give ratio weights of 10 and 3.7. This demonstrates that, although utilities order preferences, they do not measure relative strengths of preferences.

An equivalent way to view these important observations is that, if the difference in utilities of two results r_1 and r_2 is greater than the difference between two other results r_3 and r_4, this does *not* imply that the difference between r_1 and r_2 is preferred to the difference between r_3 and r_4.

A final caution to the user of utilities is that they are almost always not additive. This should be clear from Figure 6-5 where additivity holds only for the linear case.

6.7   SUMMARY AND INSIGHTS

Subjective probabilities play a central role in decision modeling and decision making. It is through them that the decision maker quantifies his or her uncertainty about future events that affect the result of the decision. In addition to their use in estimating model parameters, they are also used to solicit a decision maker's utility function using a decision sapling.

Utilities are useful in at least two important areas, (1) to quantify attributes that have no obvious unit of measurement, and (2) to quantify and model a decision maker's propensity to accept or avoid risk. Experience has shown that obtaining a utility function can be difficult in practice because it requires involving the decision maker in fictitious, often unrealistic, risky ventures on a decision sapling. But this does not negate the usefulness of this procedure in decision modeling. The solicitation process itself often brings greater insight and clarity to the problem.

There are two common ways of proceeding when determining a utility function, one where an indifference probability is found for a fixed risky venture, and one where the riskless alternative payoff is found that is equivalent to the expected payoff of a given risky venture with even odds of success. The second approach, when it can be used, seems easier to use for many people than the first. But it can only be used when there is a continuum of possible payoffs to consider, as is the case of finding a utility function for money. To ask someone how much they would be willing to accept with no risk in order to avoid obtaining zero or $10,000 each with probability of 0.5 may be difficult but is not impossible to answer. To ask a football coach what outcome of a game would he accept for certain rather than take even chances on winning or losing the game is to ask an impossible question.

Most people exhibit risk-averse preferences when faced with uncertainties. This translates mathematically into their utility function being concave. Two typical families of concave functions that are often used to model utilities are the exponential and logarithmic curves. Each has properties that are characteristic of that family. One measure of risk called risk aversion function shows that the logarithmic family has characteristics that are more in keeping with most people's behavior than does the exponential family. If we have a vector of different possible results specified in a vector **r**, that of necessity includes the worst and best results, the only solicitation we need from the decision maker to compute the exponential or logarithmic function is β, the increase in utility at the worst level relative to that at the best level. A two-attribute example using costs and accidental deaths with exponential utility functions in a nuclear reactor decision problem can be found in Section 9.5.6 (p. 379).

The theoretical underpinnings of utility theory are based on the early work of von Neumann and Morgenstern (1953). See also Luce and Raiffa (1957). A more accessible account can be found in chapter 3 of Chankong and Haimes (1983), and also French (1993). Further reading into subjective probability can be found in O'Hagan (1988). An excellent discussion on coherence can be found in Lindley (1985). Utility assessment methods have been widely studied and published. A concise summary of the main issues is contained in Bunn (1984). The risk aversion function in Equation (6.9) is due to Pratt (1964); for a more accessible version see Keeney and Raiffa (1976).

6.8 PROBLEMS

6.1 Try to imagine that you are *really* faced with each of the following four decision problems; this is not easy to do, so do whatever you can to make the situation realistic. Determine your choice of decision in each case, and compare it to what you would do if you simply used expected values in each case:

a. Get $45 for certain, or toss a coin where a head wins you $100 but a tail results in nothing.

b. Pay $5 to participate in the coin tossing in (a), or do not play.

c. Pay $45 to participate in the coin tossing in (a), or do not play.

d. Pay $5,000 to toss a coin where a head wins you $100,000 and a tail zero, or do not play.

6.2 Rank in order of decreasing preference the following three games:

a. You are equally likely to win $2,000 or nothing.

b. You win $10,000 with probability 0.1, and nothing with probability 0.9.

c. You win $2,000 with probability 0.9 and lose $8,000 with probability 0.1.

Compute the expected value and the variance of each game, and comment on the results together with your preference ordering.

6.3 Consider choosing between the following six simple games where in each one you roll a fair die once. The payoffs for each game are as follows:

a. Show that the expected payoff for all six games is the same. Observe that, when comparing game G1 with G2, in five out of 6 possible results game G2 pays more than game G1.

b. Compare successive games, and show that using this criterion as a preference ordering violates the transitivity assumption of utility theory and leads to incoherence.

Game	Result					
	1	2	3	4	5	6
G1	1	2	3	4	5	6
G2	2	3	4	5	6	1
G3	3	4	5	6	1	2
G4	4	5	6	1	2	3
G5	5	6	1	2	3	4
G6	6	1	2	3	4	5

6.4 Show that

a. For the utility of a risk-averse person,

$$u(r_1 + r_2) < u(r_1) + u(r_2),$$

b. The reverse is true for a risk-seeking person.

c. If a is any real number and $b > 0$, $a + bu(r)$ is also a valid utility function.

6.5 Consider the following two separate decision problems (the Allais paradox):

(1) Choose between

d_1: Take $1 million with no risk

d_2: Win $5m, $1m, 0, with probabilities 0.10, 0.89, and 0.01, respectively.

(2) Choose between

d_1: Win $1m or 0 with probabilities 0.11, and 0.89, respectively

d_2: Win $5m or 0 with probabilities 0.10 and 0.90, respectively.

a. Draw the decision trees for each decision problem, and by inspection, make your personal choice of d_1 or d_2 in each case.

b. Show that choosing d_1 over d_2 in (1) and d_2 over d_1 in (2) cannot be consistent with the assumptions of utility.

6.6 Suppose you have two opportunities to invest $1m. The first will increase the amount invested by 50% with probability 0.6 or decrease it by 50% with probability 0.4. The second will increase the amount invested by 5% for certain. You wish to split the $1m between the two opportunities. Let x be the amount invested in the first opportunity, with $(1 - x)$ invested in the second. Find the optimal value of x

a. Using expected value as the criterion (linear utility)

b. Using the logarithmic utility function in Equation (6.5) with $\bar{r} = 2$, $\underline{r} = 0$, and $\beta = 10$. Comment on your results.

6.7 Show that Equations (6.3) and (6.5) can be used as utility functions when $\underline{r} > \bar{r}$.

CHAPTER
7

MULTIATTRIBUTE PROBLEMS

7.1 INTRODUCTION

In preceding chapters a single attribute, usually money, was used to measure the result of a decision problem. In the crop protection and the aircraft part rework problems we minimized expected monetary loss. In the airline ticket pricing problem we maximized expected revenue. In this chapter we show ways of modeling problems where there are multiple attributes for measuring the result of a decision. Often, these attributes conflict with each other in that optimizing on one results in a suboptimal solution for another. Thus, there is a need to think of trade-offs to resolve conflicts.

We use a number of examples to illustrate the ideas in this chapter, starting in Section 7.2 with a decision sapling problem. In Section 7.3 we revisit the crop protection problem in which the result is measured by both a yield and a crop quality attribute. Extending the sensitivity analysis results of Chapter 5, we demonstrate graphically that whether or not conflicts occur in determining optimal decisions depends on the values of important parameters. Section 7.4 considers a problem of comparing and ranking cars and extends this to a car replacement decision problem. Given a range of new cars from which to choose, how does one measure the relative merit of each one? Some attributes are price, performance, reliability, style, color, prestige, and the like. We concentrate on attributes that are measurable such as price and performance; the problem of how to measure the relative worth of items such as style, color, prestige, and so on is left to Section 7.8 and 7.10. Section 7.5 introduces the concept of fractional trade-off weights among attributes, and Section 7.6 demonstrates a method of finding these through preferences for risky ventures on a decision sapling. In Section 7.7 we analyze a hierarchical multiattribute model with emphasis on a cost–benefit analysis. Section 7.8 is devoted to an overview and analysis of what is known as the Analytic Hierarchy Process, or AHP.

The budget problem discussed and formulated in Chapter 4 is revisited in Section 7.9 and analyzed using three conflicting attributes. In Section 7.10 we extend the results of Chapter 6 on utility theory to multiattribute problems.

In many decision problems it is not always obvious how one should measure the criteria. This is particularly true as one moves from an operational level to a strategic or policy level. At an operational level the criteria are often measured in terms of income, profit, production, sales, inventories, and so on, all of which have well-understood and agreed-upon units of measurement. At a higher level the effectiveness of a decision problem may be measured by such criteria as quality, efficiency, relevance, prestige, creativity, and the like in an educational setting; stability, cooperation, trade, ambassa-dorial level, and so on in foreign diplomatic relations; or readiness, sustainability, force projection, forward presence, and similar qualities for the military. Some attempts to model these latter problems have an inherent, but usually unstated, assumption that, be-cause no measurement units exist, the attributes can be represented by dimensionless quantities. Using this approach can lead to disturbing inconsistencies in model output when attributes are combined to give an overall performance measure, as we see later in the chapter. To prevent such inconsistencies, we state and use the following principle as a fundamental guide in our model building:

> For a multi-attribute decision model to be consistent it should apply the same rules for com-bining attributes that cannot be measured directly as it does for those that can. If the problem under consideration has performance attributes for which there are no obvious measure-ment units, one should not assume that the weights assigned to these attributes are dimen-sionless and hence can be normalized in an arbitrary manner.

7.2   A DECISION SAPLING WITH TWO ATTRIBUTES

In Section 1.8 (p. 28) we considered a simple decision problem where the choice was either (1) select the riskless alternative, or (2) select a risky venture where the probabil-ity of winning was p. A number of illustrative examples were discussed. The decision problem of whether or not to accept a new job is revisited in this section.

Suppose that you are faced with the same general problem modeled by a decision sapling but now have two different ways to measure the result of your decision. Let p be the probability of success if the risky venture is chosen, and let r_{ij} be the payoff in at-tribute j of result i. Thus, for each attribute j the payoff is:

- r_{1j}:   If the risky venture is chosen and is a success,
- r_{2j}:   If riskless alternative is chosen,
- r_{3j}:   If the risky venture is chosen and is a failure,

and we assume that the objective is to maximize expected payoff. We assume that r_{2j} lies between r_{1j} and r_{3j} for every j, but it may not be true that $r_{1j} > r_{2j} > r_{3j}$. Figure 7-1 shows our decision problem.

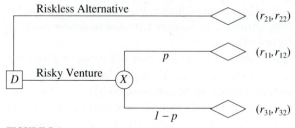

FIGURE 7-1
A decision sapling with two attributes.

A common approach to solving a multiattribute problem is to combine the different measures into a single numeric measure for each result and treat it as a single-attribute problem. In order to do this, we need to know how to combine the attribute measures. This is often a difficult problem, but as we shall see, we may never have to carry it out. Suppose we take a naive approach and solve the problem for each attribute separately. If our objective is to maximize the expected payoff the optimal policy when using the attribute j is:

If $r_{2j} \geq pr_{1j} + (1 - p)r_{3j}$, choose the Riskless Alternative.
If $r_{2j} \leq pr_{1j} + (1 - p)r_{3j}$, choose the Risky Venture.

As an example, suppose that you are considering either accepting a new job opportunity or staying in your current position. You would like your decision to be based on both promotion potential (attribute 1) and job satisfaction (attribute 2), and measure each of these on a scale from 1 to 10, 1 being the worst level, and 10 being the best. You estimate there is little or no risk in staying in your current position and rate it a (5, 5) job. The new job cannot be estimated with certainty, but you think there is a 55% chance it will be an (8, 3) job (better chance of promotion but lower job satisfaction), and a 45% chance it will be a (3, 8) job. Should you switch jobs?

For this job example,

$$(r_{21}, r_{22}) = (5, 5), (r_{11}, r_{12}) = (8, 3), \text{ and } (r_{31}, r_{32}) = (3, 8).$$

The above inequalities tell us that:

1. Using the promotion potential criterion,

   If $p > 0.4$, take the new job.
   If $p \leq 0.4$, keep current job.

2. Using the job satisfaction criterion,

   If $p \leq 0.6$, take the new job.
   If $p > 0.6$, keep current job.

Recall that the value of p that satisfies $r_{2j} = pr_{1j} + (1 - p)r_{3j}$ is called the *indifference probability,* and there is one for each attribute j. We denote them with $\bar{p}_1$ and $\bar{p}_2$ for $j = 1$ and 2, respectively, so that

$$\bar{p}_1 = (r_{21} - r_{31})/(r_{11} - r_{31}) \text{ and } \bar{p}_2 = (r_{22} - r_{32})/(r_{12} - r_{32}).$$

For our example these are $\bar{p}_1 = 0.4$ and $\bar{p}_2 = 0.6$.

   Figure 7-2 shows the optimal policy regions for each attribute in the job problem in terms of the probability p. Note that if p is between 0.4 and 0.6 both criteria lead to the same optimal decision—take the new job; the optimal decision is to take the new job *independent of which attribute is used*! If you stay with your estimate of 0.55 for p, the problem is solved. There is a conflict only if you believe that p lies in either of the intervals (0, 0.4) or (0.6, 1). Clearly, it pays to calculate $\bar{p}_1$ and $\bar{p}_2$ first and compare them to p in a given problem. Only if a conflict exists will we need to investigate possible trade-offs between the criteria.

   This simple example illustrates the ideas explored in more detail in the remaining sections of this chapter. The approach is to first examine the policy space to find regions where conflicts do not arise. If in the decision problem conflicts cannot be avoided, trade-offs must be made to resolve them. The reader will see that the decision sapling is important in determining these trade-offs but should keep clear its very different uses (1) as a way to model a simple single-stage decision problem, and (2) as a tool to determine trade-offs between attributes by using it to model fictitious decision problems.

   In the following sections we return to some of our earlier examples and introduce some new ones. All have at least two attributes to measure results so that to a result node i in the decision tree is assigned a vector $\mathbf{r}_i$ rather than a scalar r_i as in earlier chapters. The result set $\mathcal{R}$ is now a collection of vectors.

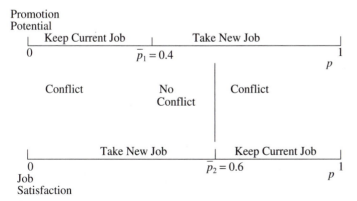

FIGURE 7-2
Optimal policy regions for the job decision problem.

7.3   THE CROP PROTECTION PROBLEM
WITH TWO ATTRIBUTES

In the crop protection problem analyzed in Chapter 5 there was a cost of protection that guaranteed the crop regardless of the weather outcome and a crop loss if the crop was unprotected and freezing weather resulted. The two-stage decision problem was analyzed to see whether it was cost-effective to invest in a weather forecast that might yield better predictions for the decision maker at the second stage.

We now consider the same problem, but measure results using two attributes: the first is the yield per acre using a volume measure such as bushels, the second is the quality of the produce measured on some standard industry scale such as Grade A, Grade B, and so on. Because these naturally measure return rather than loss, we modify the problem slightly by maximizing expected return rather than minimizing expected losses. The notation we use is:

$\mathbf{r}_2 = (r_{21}, r_{22})$ = Yield and Quality if crop is protected,

$\mathbf{r}_1 = (r_{11}, r_{12})$ = Yield and Quality if crop is unprotected and freezing weather occurs,

$\mathbf{r}_3 = (r_{31}, r_{32})$ = Yield and Quality if crop is unprotected and weather is not freezing.

For $j = 1, 2$ let $s_j = (r_{3j} - r_{2j})/(r_{3j} - r_{1j})$, the net gain in each attribute under a protect policy divided by the net gain if the crop is unprotected and freezing weather occurs. Because we can expect both lower yield and quality for an unprotected crop in bad weather than one that is protected, we assume that $r_{1j} < r_{2j}$, so that $0 < s_j < 1$ for $j = 1, 2$. In Figure 7-3 we assume that $s_1 > s_2$. If we solve the problem for each attribute alone, the plot of the optimal policies in the (p_0, p_1) plane is as shown in Figure 7-3. This follows immediately from Figure 5-10 (p. 207).

Note that there are six distinct areas in the feasible region above the line $p_1 = p_0$. We see that in areas 1, 2, and 3 the same policy is optimal using either attribute and that these agree with the interpretations made in Chapter 5. In areas 4 and 5, there is partial conflict. In 4, the optimal policy using the second attribute is "Protect," whereas using the first attribute we would "Forecast and protect only if bad weather is predicted." In this region both attributes could result in the same or different policies depending on the result of the forecast. This is also true in area 5. In area 6, the attributes always lead to conflicting policies, which is to say that for one the optimal policy is to always protect, and for the other it is always to not protect. Recall that more discriminating forecasts lead us toward the northwest corner where $p_1 = 1$ and $p_0 = 0$. In our example the reader can see that a better forecast leads to conflict resolution. It is true that better forecasts always lead to resolving any partial conflicts, but resolution can be in the direction of either decreased conflict as in this example, or increased conflict as in the example in Figure 7-2.

We now revisit the optimal policies in the (f_0, f_1) plane that we considered in Section 5.5 (p. 210). In this case different structures are obtained depending on the relative values of the s_j's and p_X. Figure 7-4 shows the three cases where (a) both s_j's exceed p_X, and (b) both s_j's are less than p_X, and (c) the case where s_1 is larger than, and s_2 smaller than p_X.

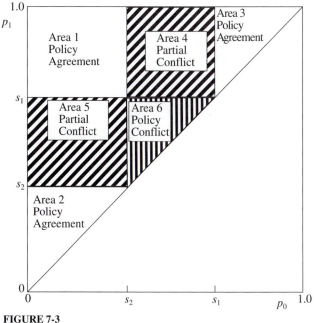

FIGURE 7-3
Two-attribute protection decisions in (p_0, p_1) plane.

In the northwest corner of Figure 7-4(a) there is policy agreement because the optimal policy is to follow the forecast using either attribute. In the southeast section above the diagonal there is also policy agreement, because the optimal policy is to never protect using either attribute. In the diagonally hatched region the optimal policy is to never protect if attribute 1 is used and follow the forecast if attribute 2 is used. Thus, there is disagreement on what action to take in this region only if the forecast is for non-freezing weather. In Figure 7-4(b), in the northwest and southeast sections there is no policy conflict; in the diagonally hatched area, using the first attribute the optimal policy is to follow the forecast, and using the second attribute it is to always protect.

The horizontally hatched area in Figure 7-4(c) is the region where the two attributes always conflict. The unshaded region in the northwest corner and above the diagonal yield the same optimal policy (follow the forecast). The diagonally hatched areas are those that may or may not result in the same decision to protect or not protect, depending on the outcome of the forecast. If we know the track record of the forecaster as measured by f_0 and f_1, the values of s_1 and s_2 determined from the return functions, and p_X (our estimate of the probability of freezing weather before obtaining a forecast), we can use this graph to determine whether or not a conflict exists.

We have shown that if the parameter values for a given multiattribute problem lie in certain ranges it is possible that a solution can be found that is the best for all attributes. Clearly, this will not always be the case, so we must investigate methods to resolve conflicts when they occur. The rest of the chapter is devoted to this topic.

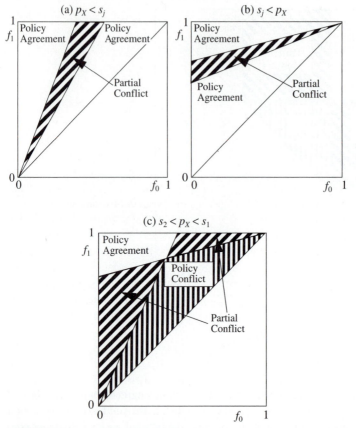

FIGURE 7-4
Two-attribute protection decisions in (f_0, f_1) plane.

7.4   CAR RANKING AND REPLACEMENT

In this section we consider two decision problems that many readers will at some time face, (1) ranking a number of cars by order of decreasing preference, and (2) keeping their current car or replacing it with a new one. The number of attributes that can be used as measures in these problems is very large, but we again simplify our presentation at this stage by considering only two attributes, namely, price and performance. We have already shown how there can be a fundamental conflict between two such attributes. What makes the ranking and replacement decisions difficult is that higher performance cars usually cost more and cheaper cars typically have lower performance.

7.4.1   Ranking Cars by Preference

Suppose you wish to rank the three cars with characteristics shown in Table* 7.1 in order of preference based on both cost and performance. Let a be the cost of a car and b its performance. Our first problem is to decide on measurement units for each of these. For simplicity assume that cost is measured by the purchase price and a is measured in thou-

TABLE 7.1
Data for the three-car ranking problem

Car i	Cost a_i ($1000)	Performance b_i (seconds)
1	17	10
2	12	12
3	10	14

sands of dollars. There are many ways to measure performance, a common one being the number of seconds it takes a given car to reach 60 miles per hour (mph) from a standing start. We assume that b has units of seconds and represents this measure. One way to view the problem is to ask, for the first two cars which combination do you prefer, (17,10) or (12,12)? Most people would clearly prefer 12 to 17 (the lower cost) and 10 to 12 (the better performance), but the problem is how to rank the two vectors when both attribute values of one vector are neither worse nor better than another. One way to do this is to suggest a trade-off between the two components. In some applications this trade-off can be determined objectively and is obvious from the formulation of the problem. But in many applications, it will not be obvious or even possible to compare directly one attribute with another. Is a one second reduction in the time a car takes to reach 60 mph worth much in terms of extra cost? For someone with considerable wealth the money attribute may be unimportant, and so they would always choose Car 1 over Car 2. For someone of very limited means the performance attribute may amount to little more than wishful thinking when compared to cost. But for many people there will be a trade-off between cost and performance that is highly dependent on the preferences of the decision maker. Our problem is to quantify this trade-off, and to do this we use a linear combination of the two attribute levels multiplied by trade-off weights in order to resolve the conflicts.

Because lower costs and better performance are both usually preferred, we measure the result using an equivalent cost function, v, and will find the ranking or replacement decision(s) that minimize the expected equivalent cost. Obviously, v must be an increasing function of both a and b. If we let w have units of thousands of dollars per second improvement in performance, and v have units of thousands of dollars, then

$$v = a + wb \tag{7.1}$$

gives the equivalent cost of the car.

The problem is how to choose w consistent with our preferences. It might help in choosing a value for w if we understood better what it represents. Suppose you compared two cars, the first with price a and the second with price $a + \delta$. Assume also that the first had performance b and the second $b - 1$. Assume that you are indifferent between these two cars. Then, you are willing to pay an additional amount δ to get a better performance of $(b - 1)$ over b. Using Equation (7.1) it is easy to show that $w = \delta$. So w is the marginal price you are willing to pay for a 1-second decrease in the time a car takes to accelerate from 0 to 60.

Assume that you prefer cheaper prices and better performance. Note that in Table* 7.1 there is no one car that dominates the others in terms of both price and performance. For each car i denote the equivalent cost by $v_i(w) = a_i + wb_i$ where the a_i's and b_i's are given in Table* 7.1. These are plotted in Figure 7-5 as functions of w. Because we wish to find the car that minimizes the equivalent cost for a given trade-off weight w, the *optimal* equivalent cost function $v^*(w)$ is given by:

$$v^*(w) = 10 + 14w \qquad \text{if } w < 1.0,$$
$$= 12 + 12w \qquad \text{if } 1.0 < w < 2.5,$$
$$= 17 + 10w \qquad \text{if } 2.5 < w.$$

This function is indicated in Figure 7-5 by the heavy line. It divides the w-axis into three intervals to determine the optimal preference:

1. If $w < 1.0$, Car #3 is preferred,
2. If $1.0 < w < 2.5$, Car #2 is preferred,
3. If $2.5 < w$, Car #1 is preferred.

By plotting the results the meaning of the parameter in the model becomes clear. There is a considerable range of w for which Car 2 is preferred. Near the boundaries of

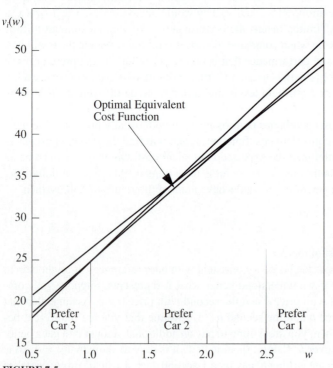

FIGURE 7-5
Optimal policies for the three-car ranking problem.

this range the values of v^* for competing cars are very close, making a decision in these areas quite insensitive to values of w. Graphical methods can quickly reveal whether or not the preferential choice is sensitive or insensitive to small changes in w.

Although Figure 7-5 helps clarify the meaning of w, *it is still up to the decision maker to choose his or her own particular value.* Whatever value is chosen, it should lead to consistent results in all reasonable decision problems. Before considering methods designed to help the decision maker in choosing a value for w we look at a car replacement problem requiring car ranking that involves uncertainty.

7.4.2 Car Replacement

Our replacement decision problem is shown graphically using a decision tree in Figure 7-6. You currently have a gas-guzzling, high-performance aging car that has become unreliable due to wear and high mileage, which with probability p will fail in the next year. It can accelerate from 0–60 mph in 8 seconds, which you like. Keeping the car will result in zero capital outlay unless the car breaks down, in which case you will have to replace it with either Car 1 or Car 2 (from Table* 7.1) at time of failure for the full price ($17,000 or $12,000). If you replace immediately, Car 1 would cost only $14,000 and Car 2 $10,000 because of current sales. Should you keep your car or replace it immediately? We keep our units of thousands of dollars and seconds.

Suppose the only attribute used to make the decision is cost. By equating expected values we see that if $p < 10/12$, the optimal immediate decision D_1 is to keep the current car, and if it fails, replace it with Car 2 (D_2). If $p > 10/12$, replace it immediately with Car 2. Now suppose the only attribute to be used is performance. For any value of p the optimal decision D_1 is to keep the current car, and if it fails, replace it with Car 1 (D_2).

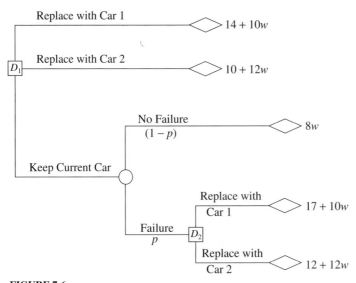

FIGURE 7-6
The car replacement decision problem.

We again learn from this simple example that the optimal decision D_1 is the same (keep the car) for any value of $p < 10/12$ using either attribute. This would indicate no conflict in making the initial decision to keep or replace under either attribute when $p < 10/12$. A conflict arises in the initial decision when $p > 10/12$. Note also that the two attributes give conflicting decisions as to *which* car to replace it with if the current one fails. We would like to resolve these conflicts.

Using Equation 7.1 for the result function let us apply the rollback algorithm to the car replacement problem in Figure 7-6. At the D_2 node we must find $Min\{17 + 10w, 12 + 12w\}$, which clearly depends on the value of w. At the D_1 node the expected minimum equivalent cost is the minimum of:

1. $14 + 10w,$
2. $10 + 12w,$
3. $8w(1 - p) + pMin\{17 + 10w, 12 + 12w\}.$

The boundaries of the region in which a given decision is optimal are determined from these expected equivalent costs. Recall that in this problem p is the probability that your current car will break down in the next year. Figure 7-7 shows a plot in the (w, p) plane indicating these boundaries and regions. The determination of these boundaries is posed as a problem at the end of the chapter. The reader should note that the solutions of the optimal policy regions make sense in that large values of p lead to the immediate replacement policy by either car 1 or 2, depending on the trade-off between price and performance.

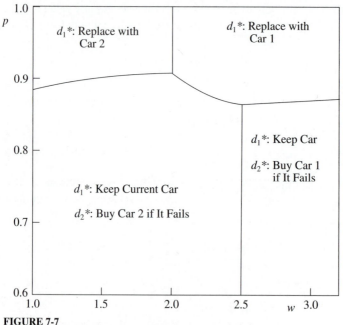

FIGURE 7-7
Optimal policy regions for the car problem.

The notion of consistency is that the chosen value of the trade-off weight is invariant to car choice. We now exploit this idea to help the decision maker determine the trade-off when he or she is unable to specify easily a particular value.

7.5   THE ADDED COST OF CONFLICT RESOLUTION

In the car replacement example we examined the trade-off between cost and performance. We now consider a more general model with n attributes and provide a method of direct assessment of the relative value in going from the least preferred level (worst) to the most preferred level (best) of each attribute.

We assume that there is a trade-off weight w_j for the jth attribute and define a vector of attribute weights $\mathbf{w} = (w_1, w_2, \ldots, w_n)$. Using our earlier notation, let $\mathcal{R}$ be a set of results that arise in a decision problem. An element of $\mathcal{R}$ is a vector $\mathbf{r}$. If r_j is the result level for attribute j, then $\mathbf{r} = (r_1, r_2, \ldots, r_n)$. There will be a range of values of r_j for every j, and we denote the "best" or "most preferred" level by $\bar{r}_j$ and the "worst" or "least preferred" level by $\underline{r}_j$. We let

$$\bar{\mathbf{r}} = (\bar{r}_1, \bar{r}_2, \ldots, \bar{r}_n) \text{ and } \underline{\mathbf{r}} = (\underline{r}_1, \underline{r}_2, \ldots, \underline{r}_n).$$

Thus $\bar{\mathbf{r}}$ represents the best levels of all attributes, and $\underline{\mathbf{r}}$ the worst levels. The reader should realize that *neither one of these may be obtainable in a given decision problem*; that is to say, they may represent hypothetical combinations of attribute levels. The reasons for considering them will become apparent as we proceed with our analysis. In the car replacement example $\bar{\mathbf{r}} = (0, 8)$ and $\underline{\mathbf{r}} = (17, 12)$. Here $\bar{\mathbf{r}}$ refers to an actual car but $\underline{\mathbf{r}}$ does not.

Assume that there is a linear equivalent value function v defined to be

$$v(\mathbf{w}, \mathbf{r}) = w_1 r_1 + w_2 r_2 + \cdots + w_n r_n = \mathbf{w} \mathbf{r}^{\mathrm{T}}, \qquad \mathbf{r} \in \mathcal{R}. \tag{7.2}$$

Note that $\mathbf{w}\mathbf{r}^{\mathrm{T}}$ indicates the inner product of the two vectors $\mathbf{w}$ and $\mathbf{r}^{\mathrm{T}}$.[1] If an attribute is expressed in some arbitrary unit, then *the trade-off weight for that attribute is expressed in units of v divided by the units of the particular attribute.*[2] In the car example already discussed, we had only one trade-off weight because one of the attributes, price, had a weight of unity. If the value is in dollars and any one of the attributes is expressed in dollars, then the corresponding multiplier is unity.

Value can be measured in many ways such as cost, reliability, performance, safety, satisfaction and so on. In the remainder of this section we measure value in dollars. The equivalent dollar value of $\underline{\mathbf{r}}$ is denoted by $\underline{v}$, where

$$\underline{v} = w_1 \underline{r}_1 + w_2 \underline{r}_2 + \cdots + w_n \underline{r}_n = \mathbf{w} \underline{\mathbf{r}}^{\mathrm{T}}.$$

---

[1] All vectors are treated as row vectors. Superscript T denotes transposition.

[2] The fact that trade-off weights have dimensions is particularly important to remember. We see later in the chapter the kind of problems one can encounter in modeling when this fact is not taken into consideration.

The equivalent dollar value of $\bar{\mathbf{r}}$ is denoted by $\bar{v}$, and

$$\bar{v} = w_1\bar{r}_1 + w_2\bar{r}_2 + \cdots + w_n\bar{r}_n = \mathbf{w}\ \bar{\mathbf{r}}^{\mathrm{T}}.$$

Let $\Delta_j = \bar{r}_j - \underline{r}_j$ and $\Delta = \bar{\mathbf{r}} - \underline{\mathbf{r}}$. Thus Δ is the n-dimensional vector of the Δ_j's. It gives the changes in all attributes required to change from the worst to the best level. The difference between the values of the most and the least preferred is therefore given by

$$\bar{v} - \underline{v} = w_1\Delta_1 + w_2\Delta_2 + \cdots + w_n\Delta_n = \mathbf{w}\Delta^{\mathrm{T}}.$$

This is the additional cost one must be prepared to spend to obtain the most preferred rather than the least preferred level in *every* attribute, if these were realizable. Let π_j be defined to be

$$\pi_j = \pi_j(\mathbf{w}, \Delta) = (w_j\Delta_j)/\mathbf{w}\Delta^{\mathrm{T}}, \qquad j = 1, 2, \ldots, n. \tag{7.3}$$

It is easy to see that π_j is the *fraction* of this additional cost that is required to bring the jth attribute from its worst to its best level. One important feature to keep in mind is that the fraction for attribute j depends not only on the trade-off weight for attribute j, w_j, but also on the weights and ranges of all other attributes. An increase in Δ_j for attribute j changes the fractional budgets for *all* attributes.

With a limited budget one is generally not able to pay the cost of obtaining the most preferred level of every attribute if in fact it were attainable. We must usually settle for something less, namely, an alternative whose attributes lie somewhere between worst and best. This means that each attribute will also lie between its least and its most preferred level. Let r_j be some realizable intermediate level between $\underline{r}_j$ and $\bar{r}_j$, and let

$$F_j(r_j) = \frac{r_j - \underline{r}_j}{\bar{r}_j - \underline{r}_j} = \frac{(r_j - \underline{r}_j)}{\Delta_j}, \qquad r_j \in [\underline{r}_j, \bar{r}_j]. \tag{7.4}$$

$F_j(r_j)$ is the fractional change of attribute j from the worst level $\underline{r}_j$ to its desired level r_j, given your budget. Note that at the most preferred level for attribute j, $F_j(\bar{r}_j)$ is 1. If $\delta v_j(r_j)$ is the equivalent dollar cost one would have to pay for this fractional change in attribute j, then

$$\delta v_j(r_j) = (r_j - \underline{r}_j) w_j = \pi_j F_j(r_j) (\mathbf{w}\Delta^{\mathrm{T}}). \tag{7.5}$$

Notice that this is the product of three terms: (1) the fraction π_j of the total budget that should be allocated to attribute j to change it from $\underline{r}_j$ to $\bar{r}_j$, (2) the desired fractional improvement in attribute j, and (3) the total budget required to change all attributes from their least to their most preferred status. The *fractional* budget allocation needed to obtain the desired level of attribute j is therefore

$$\frac{\delta v_j}{\bar{v} - \underline{v}} = \frac{\delta v_j}{\mathbf{w}\Delta^{\mathrm{T}}} = \pi_j F_j(r_j).$$

Suppose the actual budget you have available is $B \le \mathbf{w}\Delta^T$ and that you know your w_j's. Any $\mathbf{r} \in \mathcal{R}$ such that $(\mathbf{r} - \underline{\mathbf{r}})\mathbf{w}^T \le B$ is feasible. The best $\mathbf{r}$ that can be attained (not necessarily unique) will be one that minimizes the difference between B and $(\mathbf{r} - \underline{\mathbf{r}})\mathbf{w}^T$.

7.5.1   Car Ranking Revisited

To illustrate these ideas we return now to the car ranking problem of Section 7.4.1 with the cost and performance characteristics for the three cars shown in Table* 7.1. Because the least preferred cost and performance is given by the highest price and largest time in seconds to reach 60 mph, we have $\underline{\mathbf{r}} = (17, 14)$ and $\overline{\mathbf{r}} = (10, 10)$. Note that neither of these vectors represent any actual car. The vector Δ is $(-7, -4)$. With a trade-off vector given by $\mathbf{w} = (1, w_2)$ we have $\underline{v} = 17 + 14w_2$, $\overline{v} = 10 + 10w_2$, and the inner product $\mathbf{w}\Delta^T = -7 - 4w_2$.

It may be helpful at this point to comment about the signs of the multipliers and the terms in Δ. The preferred attribute levels are small values in this example. Thus, any gain in value obtained by going to a higher preference can be viewed as a reduction in equivalent dollars.

Applying Equation (7.3), the fractional increases in value (reduction in equivalent cost) if each attribute were taken to its most preferred level are

$$\pi_1 = \frac{7}{7 + 4w_2} \ge 0 ; \qquad \pi_2 = \frac{4w_2}{7 + 4w_2} \ge 0 , \qquad (7.6)$$

where we note that these two fractions add to one as they should.

Suppose a decision maker, when asked for a value of w_2, has difficulty in assigning how many thousands of dollars he or she is willing to spend in order to decrease the time to accelerate to 60 mph by 1 second. This person may find it more natural to think as follows: Of the value improvement achieved in going from a car with the highest price and poorest performance, to one with the lowest price and best performance, what fraction of that improvement should be attributed to price? If price were overwhelmingly important, the answer would be 100%, π_1 would be 1.0, and w_2 zero. If the answer is 60%, then $\pi_1 = 0.6$, which implies that $w_2 = 1.167$, or \$1,167 for every second improvement of acceleration.

Suppose that we pick a car with intermediate attribute levels (12, 12), so that $r_1 = 12$ and $r_2 = 12$. Consider the fractional reduction in "equivalent dollars" that will result from purchase of this car rather than the least preferred one. There will be two separate contributions to this reduction, one from the reduction in price from \$17,000 to \$12,000 and the other from the improved performance of 12 rather than 14. Using Equation (7.4) $F_1(12) = 5/7$ and $F_2(12) = 1/2$. Thus, the reduction in equivalent dollars from the price decrease is found using Equation (7.5) to be

$$\delta v_1 = \pi_1 F_1(12)(\mathbf{w}\Delta^T) = -5,$$

and the reduction in equivalent dollars from the improvement in performance is

$$\delta v_2 = \pi_2 F_2(12)(\mathbf{w}\Delta^T) = -2w_2.$$

TABLE 7.2
Signs of trade-off weights

Objective	Attribute j	
	$\Delta_j > 0$	$\Delta_j < 0$
Maximization	$w_j > 0$	$w_j < 0$
Minimization	$w_j < 0$	$w_j > 0$

The sum of these two terms agrees with the separate calculation that applies the trade-off weights to the (12, 12) car and compares it with the (hypothetical) worst (17, 14) car.

Suppose that we now include a new high-performance but expensive car (alternative 4) which has a price of 22 but a performance of 7.5 seconds. Notice that $\bar{\mathbf{r}}$, $\underline{\mathbf{r}}$ and Δ change and that we now obtain $\mathbf{w}\Delta^T = -12 - 6.5w$. Obviously, the fractional as well as the total equivalent dollar reductions change. The recalculation of the F_i's and the δv_i terms is left as an exercise for the reader. (See Problem 7.7 at the end of the chapter.)

Although we have described a formulation in which "lower numbers are better," it may turn out the opposite is true. One must be careful to use the proper sign convention for attribute weights as shown in Table* 7.2.

7.6   ASSESSMENT OF TRADE-OFFS THROUGH PREFERENCES

In this section we investigate in greater detail the connection between the trade-off weights, the preferences for certain alternatives, and the fractional budgets discussed in the last section. Some decision makers may find it more natural to think in terms of the π_j's rather than the w_j's. There is an important risky venture for the decision maker to consider that elicits his or her preferences as indifference probabilities; from these the fractional trade-off weights can be calculated. This technique is well-known among decision-analysts because of its simplicity and mathematical elegance; in actual implementations it is not without its difficulties. We illustrate some of the advantages as well as the difficulties that are encountered.

Recall that $\mathbf{r}$ is a vector of attribute levels, so that each element of the set $\mathcal{R}$ is an n-dimensional vector. For alternative i let $\mathbf{r}_i = (r_{i1}, r_{i2}, \ldots, r_{in})$ denote its vector of attribute levels.[3] Again we assume that we can write the value associated with this vector as a linear combination of attribute levels:

$$v(\mathbf{w}, \mathbf{r}_i) = w_1 r_{i1} + w_2 r_{i2} + \cdots + w_n r_{in} = \mathbf{w}\mathbf{r}_i^T.$$

Suppose that we now consider two alternatives, 1 and 2. We want $v(\mathbf{w}, \mathbf{r}_1) > v(\mathbf{w}, \mathbf{r}_2)$ whenever $\mathbf{r}_1$ is preferred to $\mathbf{r}_2$.

---

[3] The first index i denotes result i, and the second index j denotes attribute j.

7.6.1 Two Attributes

Consider the simple decision sapling shown in Figure 7-8. The decision to be made is whether or not to opt for a car that has a price of 10 and performance 14 or take a risky venture where, with probability π, you gain a better performing car for the same price and with probability $(1 - \pi)$ you pay much more for a car with the same performance. We emphasize that we are using attribute values listed in Table* 7.1, but in combinations not necessarily available in any given *real* car. The idea here is to use the consistency property: If cars become available with any convex combinations of the attributes values from Table* 7.1, we should make decisions concerning these cars consistent with the trade-off weight used for the combinations listed in the table.

For any particular π and w the optimal decision can be found by comparing expected values:

If $w > 7(1 - \pi)/4\pi$, take the risky venture,

If $w < 7(1 - \pi)/4\pi$, take the riskless alternative.

Denote by $\bar{\pi}$ the value of π for which the decision maker is indifferent between the risky venture and the riskless alternative, that is, when $w = 7(1 - \pi)/4\pi$. The solution to this is precisely the same as the π_1 that satisfies the first equation in (7.6). This $\bar{\pi}$ is called the indifference probability, and finding $\bar{\pi}$ is equivalent to finding w. If the reader finds it difficult to think directly in terms of w using Figure 7-5, this procedure to determine $\bar{\pi}$ using Figure 7-8 may help. The reasoning and procedure demonstrated in the following paragraph may help the reader determine their individual value for π.

You cannot lower the amount you pay for a car by taking the risky venture in Figure 7-8, but you can possibly get a better performing car for the same price. Suppose that you are much more concerned with price than with performance. Then you are unlikely to risk paying $7,000 more than you need to unless the probability of winning a better car for the same price is very high. In this case $\bar{\pi}$ would be close to 1. Now suppose that performance is very important. The only way to obtain a higher-performance car is

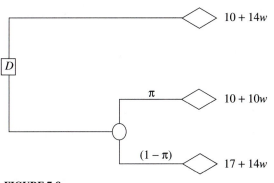

FIGURE 7-8
Decision sapling to determine *w*.

to take the risky venture and win. Even if you lose, you may pay more but still get a car with the same performance as the riskless alternative. In this case $\bar{\pi}$ will be close to 0. One way to proceed is to determine $\bar{\pi}$ by considering values of π starting at 0 and increasing in steps of 0.1 until you switch between the riskless alternative and the risky venture. This procedure can then be repeated starting at 1 and decreasing in steps of 0.1 or some other appropriately small value. In this way the indifference probability $\bar{\pi}$ can be bracketed from above and below and a final figure decided upon. Once it is determined, w can be found using the expression $7(1 - \bar{\pi})/4\bar{\pi}$.

A further consistency check can be made by comparing a certainty of a (17, 10) car with a risky venture that results in a (10, 10) car with probability ϕ and a (17, 14) car with probability $(1 - \phi)$. Let $\bar{\phi}$ be the indifference probability for this new decision problem. Then w is given by $7\bar{\phi}/4(1 - \bar{\phi})$ so that $\bar{\phi}$ is precisely the same as π_2 in the second equation in (7.6). In other words,

$$\bar{\phi} = (1 - \bar{\pi}).$$

In general if we considered any car (r_1, r_2) and applied the decision problem to this particular choice we would obtain an indifference probability $\bar{\xi}$ with

$$w = \frac{7(1 - \bar{\pi})}{4\bar{\pi}} = \frac{7\bar{\phi}}{4(1 - \bar{\phi})} = \frac{17 - r_1 - 7\bar{\xi}}{14 - r_2 - 4\bar{\xi}}$$

It is important to note that we have used a decision tree as a tool in helping determine a parameter value that is consistent in all of our car decisions. This is quite different from the original use of a decision tree to model decision problems, but it is similar to its use in eliciting utilities as we saw in Chapter 6. The simple decision sapling, together with appropriately chosen result functions, can be a useful tool to determine indifference probabilities, which in turn can be used to estimate a consistent choice of trade-off weights.

7.6.2 Many Attributes and Alternatives

For the general case there may be no immediately obvious weights that can be used. If we believe in a model of linear trade-offs and are fortunate enough to know appropriate trade-off weights to use in a given problem, then the vector of attributes can be reduced to a scalar value by using Equation (7.2). Our objective is to find the relative weight for each attribute that results in ranking the m alternatives consistently with the decision maker's preferences, which will lead to consistent results in maximizing expected value when applied to decision problems under uncertainty.

Recall that r_{ij} is the level of attribute j in alternative i and that $\mathbf{r}_i$ is the n vector of the r_{ij}'s; w_j is the relative weight assigned to attribute j, and $\mathbf{w}$ the vector of the w_j's. The value of alternative i is $v_i = \mathbf{w}\mathbf{r}_i^T$.

Using a generalization of the procedure in the preceding section we determine a set of n indifference probabilities. Let $\underline{\mathbf{r}}(k)$ be an n vector with each element set to $\underline{r}_j$ except for the kth element, which is set to $\bar{r}_k$. Then $\mathbf{w}\underline{\mathbf{r}}(k)^T$ would be the value function for a fictitious (constructed) alternative that has the least preferred level on all attributes except the kth and the most preferred level on the kth. Suppose that we consider a decision problem where this alternative is the riskless alternative against a risky venture with probability π_k of winning an alternative with the most preferred level in *all* attributes (having a value function $\mathbf{w\bar{r}}^T$) and a probability $(1 - \pi_k)$ of winning one with the least preferred level in *all* attributes (having a value function $\mathbf{w\underline{r}}^T$). If π_k is the indifference probability between the risky venture and riskless alternative, by equating expected values we see that mathematically we have

$$\pi_k = (w_k \Delta_k)/\mathbf{w}\Delta^T, \qquad k = 1, 2, \ldots, n. \tag{7.7}$$

The reader should note that the fractional budgets (π_k's) used in Section 7.5 and specified in Equation (7.3) (p. 264) are identical to the indifference probabilities in Equation (7.7). We now have two quite different interpretations of the π_k's. If a decision maker feels more comfortable thinking in terms of fractional budgets rather than indifference probabilities for risky ventures, he or she should interpret the π_k's as they were used earlier.

By summing over k we know that these indifference probabilities must add to 1. The procedure to find the π_k's uses n decision saplings such as the one shown in Figure 7-9, each with the appropriate riskless alternative. Equation (7.7) can be used to check the consistency of these indifference probabilities. In this set of n linearly dependent equations any $(n - 1)$ are linearly independent. By setting $k = 1$ and $k = k$ and dividing, we find that

$$\frac{w_k}{w_1} = \frac{\Delta_1 \pi_k}{\Delta_k \pi_1}, \qquad k = 1, 2, \ldots, n. \tag{7.8}$$

If we set $w_1 = 1$ (any arbitrarily chosen scale can be used), once the π_k's have been determined, we can determine the vector $\mathbf{w}$ using Equation (7.8).

7.6.3   Ranking Cars Using Three Attributes

To illustrate these methods, we use the data in Table* 7.3 for four cars with three attributes. The first three rows and columns contain data for the two-attribute, three-car

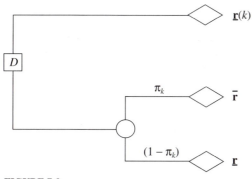

FIGURE 7-9
Decision sapling to find the π_k's.

TABLE 7.3
Data for the four-car, three-attribute problem

Car	Price	Performance	Fuel Economy
i	r_{i1}	r_{i2}	r_{i3}
	($1,000)	(seconds)	(m.p.g.)
1	17	10	18
2	12	12	22
3	10	14	28
4	22	7.5	16

example taken up earlier in this chapter. The figures for fuel economy are in miles per gallon, so that a larger value is preferred over a smaller one. For this example $\mathbf{r} = (22, 14, 16)$ and $\bar{\mathbf{r}} = (10, 7.5, 28)$; these give $\Delta = (-12, -6.5, 12)$. If the indifference probabilities are found to be $(0.7, 0.1, 0.2)$, Equation (7.8) results in a set of weights $(1.0, 0.264, -0.286)$ with the interpretation: For every second improvement in performance you are willing to pay $264, and for every mile-per-gallon decrease in fuel efficiency you are willing to pay $286.

The value function (measured in equivalent price so that smaller is better) is

$$v(\mathbf{r}) = r_1 + 0.264r_2 - 0.286r_3, \tag{7.9}$$

and the ranking problem is now solved. In decreasing order of preference the cars are 3, 2, 1, 4.

7.6.4   The Car Replacement Problem Revisited

We end this section by revisiting the car replacement problem of Section 7.4.2 but with the fuel economy attribute included. Table* 7.4 contains the data for this problem. The vectors $\mathbf{r}$ and $\bar{\mathbf{r}}$ are $(17, 12, 14)$ and $(0, 8, 22)$, respectively, so that $\Delta = (-17, -4, 8)$. Assume that using Figure 7-9 leads to indifference probabilities $\pi_1 = 0.85$, $\pi_2 = 0.05$, and $\pi_3 = 0.10$ (readers are encouraged to find their own indifference probabilities; see Problem 7.5). After setting $w_1 = 1.0$, Equation (7.8) gives $w_2 = 0.25$, and $w_3 = -0.25$, so that the value function is

$$v(\mathbf{r}) = r_1 + 0.25r_2 - 0.25r_3.$$

TABLE 7.4
Data for the three-attribute car replacement problem.

Car	Price	Performance	Fuel Economy
i	r_{i1}	r_{i2}	r_{i3}
	($1,000)	(seconds)	(m.p.g.)
1. Car #1	17	10	18
2. Car #2	12	12	22
3. Car #1 discounted	14	10	18
4. Car #2 discounted	10	12	22
5. Current car	0	8	14

The calculations are shown in Figure 7-10. The optimal decisions are:

D_1: If $p > 0.82$, replace car immediately with Car 2; if $p < 0.82$, keep your current car.
D_2: If you keep your current car and it fails, replace it with Car 2.

The weighting of the three attributes has resolved the conflicts that arise when each attribute is considered separately.

7.7   A HIERARCHICAL MULTIATTRIBUTE MODEL

Suppose that we wish to rank the three cars in Table(p. 259), but by using the five attributes shown in Table* 7.5. Following the approach of Section 7.6.3, first we would find

$$\Delta = (-7, -2, -4, -30, 60). \tag{7.10}$$

Next, we could attempt to find the indifference probabilities using a decision sapling for five separate decision problems. Once these are determined Equation (7.8) would be used to find the trade-off weights, and using these in Equation (7.2) with the data of Table* 7.5 would give us the value of each car as equivalent price in units of $1,000. The reader can check that if

$$\pi = (0.30, 0.20, 0.15, 0.25, 0.10), \tag{7.11}$$

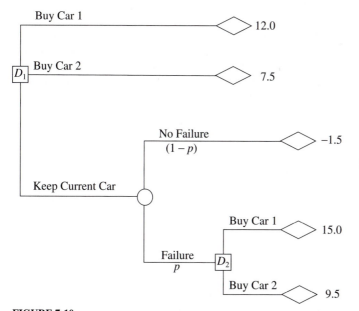

FIGURE 7-10
The car replacement decision resolved.

TABLE 7.5
Five-attribute data for the three-car ranking problem

	Cost		Performance (benefit)		
Car i	Purchase price ($1,000)	Operating cost ($1,000/yr)	Acceleration (secs, 0–60)	Braking (feet, 60–0)	Horsepower
1	17	3.5	10	120	180
2	12	2.2	12	135	140
3	10	1.5	14	150	120

then

$$\mathbf{w} = (1, 2.333, 0.875, 0.194, -0.039),$$

and the values are

$$v(17, 3.5, 10, 120, 180) = 50.2,$$

$$v(12, 2.2, 12, 135, 140) = 48.4, \tag{7.12}$$

$$v(10, 1.5, 14, 150, 120) = 50.2.$$

When there are numerous attributes to be considered, it becomes increasingly difficult to consider all of them simultaneously. It may be easier for the decision maker to group attributes that have some common basis, make attribute trade-offs within groups, and then make trade-offs among these groups. This results in a hierarchical approach to finding the overall value. Note that in Table* 7.5 the attributes are shown in two groups, Cost, consisting of purchase price and annual operating cost, and Performance (or more generally Benefit), consisting of acceleration, braking, and engine horsepower. The idea is illustrated in Figure 7-11 for this car problem. If we can find v_C and v_B, the equivalent cost and benefit values, respectively, we then need to combine them in a way that gives an overall value consistent with our earlier approach. Because of the special attention paid to cost–benefit analysis in the literature, and because the mathematical notation required for the general case is more complex, we first consider a cost–benefit model.

7.7.1 A Hierarchical Cost–Benefit Model

Suppose there are $n(C)$ attributes that measure the cost of a given alternative, and $n(B)$ that measure the benefit. Again we let an alternative result in a vector $\mathbf{r}$ that has a total of $n = n(C) + n(B)$ attributes. If we list all the cost attributes first, followed by the benefit attributes, we can partition both $\mathbf{r}$ and a corresponding trade-off weight vector $\mathbf{w}$ into two vectors,

$$\mathbf{r} = (\mathbf{r}_C, \mathbf{r}_B), \quad \text{and} \quad \mathbf{w} = (\mathbf{w}_C, \mathbf{w}_B). \tag{7.13}$$

Let v_C and v_B be the equivalent cost and equivalent benefit of some alternative, where

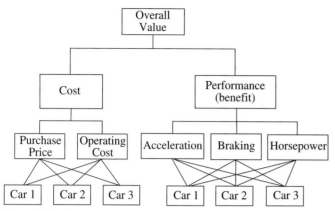

FIGURE 7-11
Hierarchical structure of car ranking problem.

$$v_C(\mathbf{w_C}, \mathbf{r_C}) = w_{C,1}r_{C,1} + w_{C,2}r_{C,2} + \cdots + w_{C,n(C)}r_{C,n(C)} = \mathbf{w_C}\mathbf{r_C}^T,$$

and (7.14)

$$v_B(\mathbf{w_B}, \mathbf{r_B}) = w_{B,1}r_{B,1} + w_{B,2}r_{B,2} + \cdots + w_{B,n(B)}r_{B,n(B)} = \mathbf{w_B}\mathbf{r_B}^T.$$

The methods and results of Section 7.5 and 7.6 can be applied to each group separately. First, we find the Δ vector and partition it into two parts, Δ_C for cost and Δ_B for benefit. Once we have decided on a particular attribute by which to measure each group (assumed to be the first attribute in each group), we then determine the fractional weights. Let $\pi_{C,k}$ be the fractional weight assigned to the kth cost attribute, and $\pi_{B,k}$ that assigned to the kth benefit attribute, where the sum over k for the cost and benefit groups must each add to 1. These could be found as indifference probabilities using a decision sapling as was demonstrated in Figure 7-9 (p. 269). Equation 7.8 (p. 269) is then used to determine the trade-off weights within each group; that is, after setting $w_{C,1} = w_{B,1} = 1$, we find

$$w_{C,k} = \frac{\Delta_{C,1}\pi_{C,k}}{\Delta_{C,k}\pi_{C,1}}, \qquad \text{and} \qquad w_{B,k} = \frac{\Delta_{B,1}\pi_{B,k}}{\Delta_{B,k}\pi_{B,1}}. \tag{7.15}$$

The equivalent payoffs are then found using Equations (7.14). We illustrate using the car example.

Using the groupings in Table* 7.5 and Figure 7-11, suppose the fractional weights shown in column 3 of Table* 7.6 are found as indifference probabilities using a decision sapling as described. We must first determine the weights $\mathbf{w_C}$ and $\mathbf{w_B}$. Equation (7.10) gives the deltas, and using these in Equations (7.15) with the fractional weights gives the trade-off weights for each attribute in each group shown in column 4. Note that the unit of equivalent cost is $1,000, and that for benefit is seconds (to reach 60 mph). Using these in Equation (7.14) gives the values for each car for each group,

TABLE 7.6
Trade-off weights for the two-hierarchy car example

Group	Attribute	Fractional weights	Trade-off weights
Cost (C)	Purchase price (C,1)	0.6	1.000
	Operating cost (C,2)	0.4	2.333
Benefit (B)	Acceleration (B,1)	0.3	1.000
	Braking (B,2)	0.5	0.222
	Horsepower (B,3)	0.2	−0.044

Cost Value	Benefit Value
Car 1: $v_C(17, 3.5) = 25.16$,	$v_B(10, 120, 180) = 28.72$,
Car 2: $v_C(12, 2.2) = 17.13$,	$v_B(12, 135, 140) = 35.81$,
Car 3: $v_C(10, 1.5) = 13.50$,	$v_B(14, 150, 120) = 42.02$.

Although it is often easier to think of trade-offs among attributes that essentially measure a common characteristic, this decomposition and simplification of the problem has a price; to find the overall payoff from an alternative, we now have to find a way to trade off the equivalent cost and the equivalent benefit.

Using the notation and definitions in Equations (7.13) and (7.14), we can define an overall equivalent cost function,

$$v(\mathbf{w},\mathbf{r}) = v_C(\mathbf{w}_C,\mathbf{r}_C) + w_{C,B}v_B(\mathbf{w}_B,\mathbf{r}_B). \tag{7.16}$$

Note that this has a new trade-off weight $w_{C,B}$, the number of units of equivalent cost required to obtain one unit of increased benefit. The weight $w_{C,B}$ is often referred to as the *cost–benefit ratio*. To find it, we use the same ideas and procedures used to find the $w_{C,k}$'s and $w_{B,k}$'s.

Having found the equivalent cost and equivalent benefit of every alternative, we can find the deltas for each of these. Let $\bar{v}_C$ and $\underline{v}_C$ be the best and worst equivalent cost obtainable, respectively, and let $\bar{v}_B$ and $\underline{v}_B$ be the best and worst equivalent benefits. Let $\pi_{C,B}$ be the indifference probability for the risky venture shown in Figure 7-12, where the riskless alternative has the best payoff for cost and worst for benefits. The risky venture will result in the best of both with probability $\pi_{C,B}$ and in the worst of both with probability $(1 - \pi_{C,B})$. By equating expected payoffs, the reader can show that

$$w_{C,B} = \frac{(\bar{v}_C - \underline{v}_C)\,(1 - \pi_{C,B})}{(\bar{v}_B - \underline{v}_B)\,\pi_{C,B}}. \tag{7.17}$$

We illustrate again with the car example. From the above equivalent cost and value functions, $\bar{v}_C - \underline{v}_C = -11.66$, and $\bar{v}_B - \underline{v}_B = -13.30$. Assume that our value for the in-

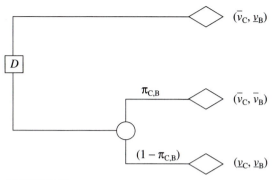

FIGURE 7-12
Decision sapling to find $\pi_{\mathrm{C,B}}$.

difference probability $\pi_{\mathrm{C,B}}$ is 0.5. Using Equation (7.17) we find that $w_{\mathrm{C,B}} = 0.875$. The reader should check that Equation (7.16) when used with these results gives the same overall values for the three cars as we obtained with the single hierarchy model in Equation (7.12).

Some readers may be surprised that the one- and two-hierarchy models give the same value for each car, as $\pi_{\mathrm{C,B}}$ was determined using Figure 7-12 independently of the $\pi_{\mathrm{C},k}$'s and $\pi_{\mathrm{B},k}$'s. In fact, we chose the value of 0.5 for $\pi_{\mathrm{C,B}}$ in order to obtain consistent results! The values of vector π in Equation (7.11), the $\pi_{\mathrm{C},k}$'s and $\pi_{\mathrm{B},k}$'s in column 3 of Table* 7.6, and $\pi_{\mathrm{C,B}}$ are related by the laws of probability and cannot be chosen arbitrarily. This is shown following the next subsection where a more general multigroup model is considered.

7.7.2* The Two-Hierarchy Multigroup Model

Let $\mathcal{A}$ be the set of n attributes. This is partitioned into m groups or subsets $\mathcal{A}_j$, where $\mathcal{A}_i \cap \mathcal{A}_j = \emptyset$ when $i \neq j$, $\mathcal{A} = \cup \mathcal{A}_j$, and $|\mathcal{A}_j| = n(j)$, the number of attributes in group j. Note that an attribute is assigned uniquely to a group. Let the $n(j)$ vector $\mathbf{r}_j$ contain the results obtained from the attributes in group j for a particular alternative. If $r_{j,k}$ is the result for attribute k in group j, then

$$\mathbf{r}_j = (r_{j,1}, r_{j,2}, \ldots, r_{j,n(j)}), \qquad j = 1, 2, \ldots, m.$$

For each group j, let the value of an alternative be $v_j(\mathbf{r}_j)$, measured in units appropriate to the particular group. Let $w_{j,k}$ be the weight associated to attribute k in group j, with

$$\mathbf{w}_j = (w_{j,1}, w_{j,2}, \ldots, w_{j,n(j)}), \qquad j = 1, 2, \ldots, m.$$

Note that $w_{j,k}$ has units of group j value per unit of attribute k in group j. The group j value of the alternative is given by

$$v_j(\mathbf{r}_j) = \sum_{k \in \mathcal{A}_j} w_{j,k} r_{j,k} = \mathbf{w}_j \mathbf{r}_j^T . \tag{7.18}$$

Let $v(\mathbf{r})$ be the overall value assigned to the alternative. Using our earlier model, recall that the overall value is found by

$$v(\mathbf{r}) = \mathbf{wr}^{\mathrm{T}} = \sum_{i=1}^{n} w_i r_i. \tag{7.19}$$

Here w_i has units of overall value per unit of attribute i in the single-hierarchy model. If g_j is the weight assigned to group j (in units of overall value per unit of group j), from Equations (7.18) and (7.19) we see that

$$v(\mathbf{r}) = \sum_{j=1}^{m} g_j v_j(\mathbf{r}_j) = \sum_{j=1}^{m} g_j \sum_{k \in \mathcal{A}_j} w_{j,k} r_{j,k}. \tag{7.20}$$

By comparing the coefficients of the same attribute in $\mathbf{r}$ and the appropriate $\mathbf{r}_j$ in Equations (7.19) and (7.20) for all possible values of the attribute, we find that

$$w_i = g_j w_{j,k}. \tag{7.21}$$

On the left side of this equation i is the index of a particular attribute in the single-hierarchy model, and on the right side (j,k) is the index of the *same* attribute in the two-hierarchy model. The careful attention we pay to ensuring consistent measurement units is crucial in ensuring consistency both between models and within models.

7.7.3* Tradeoff Weights through Indifference Probabilities

Suppose that we determine a set of indifference probabilities $\{\pi_i, i = 1, 2, \ldots, n\}$ for the single-hierarchy model of Section 7.6. Recall that we can use these with Equation (7.8) (p. 269) to find the w_i's. Let us now follow this same procedure for each of the groups in the two-hierarchy model. Let $\{\pi_{j,k}, k = 1, 2, \ldots, n(j)\}$ be the set of indifference probabilities for group j; $\pi_{j,k}$ is that success probability for which you are indifferent between:

1. A certainty of obtaining an alternative that has every attribute in set j at its worst level except the kth that is at its best level.
2. A risky alternative for which a success will give you the best level of all attributes and a failure the worst of all.

If we do this for every group, we have m sets of $\pi_{j,k}$'s that sum to 1 over k in $\mathcal{A}_j$ for each j. The $\pi_{j,k}$ has a fractional budget interpretation similar to that of π_i for the same attribute. If we denote by $\Delta_{j,k}$ the difference $\bar{r}_{j,k} - \underline{r}_{j,k}$ and form the vector

$$\Delta(j) = (\Delta_{j,1}, \Delta_{j,2}, \ldots, \Delta_{j,n(j)}),$$

then

$$\pi_{j,k} = (w_{j,k} \Delta_{j,k}) / \mathbf{w}_j \Delta(j)^T. \tag{7.22}$$

If an attribute has labels i and j,k in the single- and two-hierarchy models, respectively, using Equations (7.3) and (7.21) in (7.22) gives

$$\pi_{j,k} = \frac{\pi_i}{\displaystyle\sum_{h \in \mathcal{A}_j} \pi_h} \,. \tag{7.23}$$

Equation (7.23) shows that the fractional weights (the indifference probabilities) in the single- and two-hierarchy models are related by the laws of conditional probability.

If we let the denominator of Equation (7.23) be ϕ_j so that

$$\phi_j = \sum_{i \in \mathcal{A}_j} \pi_i, \qquad j = 1, 2, \ldots, m, \tag{7.24}$$

then these add to 1 over j. One can show that ϕ_j is the fractional increase in overall payoff associated with group j. Let $v_j(\bar{\mathbf{r}}_j)$ and $v_j(\underline{\mathbf{r}}_j)$ be the group j values obtained from the best level and the worst level of every attribute in group j, respectively, and let

$$\Delta v_j = v_j(\bar{\mathbf{r}}_j) - v_j(\underline{\mathbf{r}}_j).$$

Then

$$\phi_j = \frac{g_j \Delta v_j}{\displaystyle\sum_{h=1}^{m} g_h \Delta v_h}, \qquad j = 1, 2, \ldots, m. \tag{7.25}$$

Consider the decision sapling in Figure 7-13, which the reader should compare to that of Figure 7-9 (p. 269) for the single-hierarchy model. The risky venture results in the best level of all attributes with probability p and the worst level of all attributes with probability $(1-p)$. The riskless alternative results in the worst level of all attributes except those in group j that are at their best level. The value of p for which the

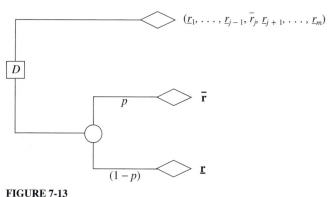

FIGURE 7-13
Decision sapling to find the group ϕ_j's.

expected values of the two decisions are equal is precisely ϕ_j. Again we see that the fractions can be found as indifference probabilities on an appropriate decision sapling.

For our car example, let g_C and g_B be the weights for the cost and performance groups, respectively. From our earlier results these are 1.000 and 0.875. The reader should check that Δv_C and Δv_B are -11.67 and -13.30, respectively, and that Equation (7.25) gives $\phi_C = \phi_B = 0.5$. These agree with their values obtained using Equations (7.11) and (7.24).

We end this section with a suggested method for finding the trade-off weights in a single-hierarchy model by using a two-hierarchy model:

1. Determine the ϕ_j's using the decision sapling in Figure 7-13 for $j = 1, 2, \ldots, m$.
2. For each group determine the $\pi_{j,k}$'s using the decision sapling in Figure 7-9 for $k = 1, 2, \ldots, n(j)$.
3. Determine $\pi_i = \phi_j \pi_{j,k}$ where attribute i in the single-hierarchy model corresponds to attribute (j,k) in the two-hierarchy model.
4. Set $w_1 = 1$ and determine $w_i = \Delta_1 \pi_i / \Delta_i \pi_1$, $i = 2, 3, \ldots, n$.
5. Set $v(\mathbf{r}) = \sum_{i=1}^{n} w_i r_i$.

7.8 THE ANALYTIC HIERARCHY PROCESS

The analytic hierarchy process (AHP) is a method developed to produce a ranking of decision alternatives with a mathematical structure that appears to be similar to that used in the car ranking example in Section 7.4.1. The essential idea is to find attribute trade-off weights through pairwise comparisons of attributes and, in addition to find the value of each decision alternative for each given attribute through pairwise comparisons of decision alternatives on that attribute. AHP has a large and devoted following and in some organizations has become the institutionalized procedure of choice.

One might ask why we should introduce material on AHP in a text that deals with sequential decision making under uncertainty and the forecasts associated with the uncertain quantities. We have discussed the formulation and use of trade-offs between different attributes in Section 7.5. The linear value function given in Equation (7.2) (p. 263) has the same mathematical structure as is assumed in AHP. Because the ranking of alternatives is required, we therefore include an examination of AHP to compare and contrast it with the ideas contained in this chapter. We demonstrate what we view as its inconsistencies with the concepts promulgated in this book. For simplicity our comparison is limited to the single-hierarchy case. The problems we describe can also occur with multiple hierarchies.

Using the notation from earlier sections, suppose we have m decision alternatives, each with n attributes. Let w_i be the weight associated with attribute i and $\mathbf{w}$ the n vector of these weights. Let $r_{i,j}$ be the result (level of attribute i) obtained from decision alternative j and $\mathbf{r}_j$ the n vector of these results. The overall value obtained from decision alternative j is given by (Equation (7.2) (p. 263)):

$$v(\mathbf{w},\mathbf{r}_j) = w_1 r_{1,j} + w_2 r_{2,j} + \cdots + w_n r_{n,j}, \qquad j = 1, 2, \ldots, m. \tag{7.26}$$

The problem is to find both the weights and the result levels in order to rank the alternatives.

Suppose for a moment that the vector of attribute weights $\mathbf{w}$ is known, and define the ratio of the weights w_i/w_j to be $\rho_{i,j}$. Let $\mathbf{R}$ be the $n \times n$ matrix

$$\mathbf{R} = \begin{bmatrix} \rho_{1,1} \cdots \rho_{1,n} \\ \vdots \qquad \vdots \\ \rho_{n,1} \cdots \rho_{n,n} \end{bmatrix}.$$

It follows by using matrix multiplication and the definitions that

$$\mathbf{Rw} = n\mathbf{w}. \tag{7.27}$$

The simple identity in Equation (7.27) is central to understanding what follows and forms the basis for the AHP procedure. If the matrix $\mathbf{R}$ is constructed from a vector $\mathbf{w}$, it has three important properties:

1. The diagonal terms are all 1 as each is the ratio of a weight with itself.
2. Corresponding terms above and below the diagonal of $\mathbf{R}$ are reciprocals, as $\rho_{i,j} = w_i/w_j = 1/(w_j/w_i) = 1/\rho_{j,i}$.
3. For every i, j, and k, $\rho_{i,k}$, $\rho_{k,j}$ and $\rho_{i,j}$, must satisfy $\rho_{i,j} = \rho_{i,k}\rho_{k,j}$.

We must digress here for a moment to introduce a mathematical technique often used in AHP computations. Any square matrix such as $\mathbf{R}$ has what are known in the field of linear algebra as eigenvalues. The reader does not need to know the theory behind their existence or how to find them, but if $\mathbf{A}$ is a square matrix and it has an eigenvalue λ, then one can find a vector $\mathbf{x}$ that satisfies $\mathbf{Ax} = \lambda\mathbf{x}$; $\mathbf{x}$ is called the corresponding eigenvector.[4] If $\mathbf{A}$ were a matrix for which the above three properties held, its largest eigenvalue would be the dimension of the matrix. This result is used in the following.

Returning to our problem, the vector $\mathbf{w}$ is *not* known, and we must develop a procedure to find it. The approach used in AHP is to have the decision maker compare every pair of attributes, independent of the set of possible decision alternatives, and determine the relative preference of one over the other using some standard numeric scale. If every distinct pair of attributes is compared, a total of $n(n-1)/2$ comparisons have to be made

---

[4] Strictly speaking λ is a right eigenvalue and $\mathbf{x}$ a corresponding right eigenvector. Similar equations can be found by multiplying $\mathbf{A}$ on the left by $\mathbf{x}$. None of this is important for understanding the use of eigenvalues in AHP.

by the decision maker to estimate the $\rho_{i,j}$'s.[5] We denote these estimates by $\hat{\rho}_{i,j}$ to differentiate them from the $\rho_{i,j}$'s, and denote the corresponding matrix $\hat{\mathbf{R}}$. This matrix will always satisfy the first two of the three properties. If the decision maker is consistent in making judgments concerning every pair of attributes, the third property will also hold. Equation (7.27) can then be used to find the vector of weights $\mathbf{w}$. Note that if $\mathbf{w}$ is any weight vector that satisfies Equation (7.27), then so does $a\mathbf{w}$ for any scalar a. The reader should keep this in mind, as we shall see later that the way a is chosen in AHP leads to difficulties in obtaining consistent results.

In practice it is very difficult to estimate a complete set of $\hat{\rho}_{i,j}$'s that are consistent in the sense that they satisfy property 3. Recognizing this, AHP uses the following procedure:

1. Start with the matrix $\hat{\mathbf{R}}$, and find its largest eigenvalue.
2. If this value is "close enough" to n (the number of attributes), calculate the corresponding eigenvector.[6]
3. The elements of the eigenvector are normalized to add to 1, and the elements used as weights $\mathbf{w}$.

To find the vectors $\mathbf{r}_j$, $j = 1, 2, \ldots, m$, in Equation (7.26) every pair of decision alternatives are compared for each attribute separately, and the procedure repeated. Finally the $\mathbf{w}$ and $\mathbf{r}_j$ vectors are combined using Equation (7.26) for each decision alternative, and the m values obtained are used to rank them. The claim is made that these values fall on a ratio scale, meaning that if, for example, values for two alternatives are found to be 0.1 and 0.3, then the alternative corresponding to 0.3 is three times as preferred (or is three times as important) as the one corresponding to 0.1. The following example demonstrates the procedure and illustrates some of the problems one encounters in using AHP.

7.8.1   Ranking Alternatives with AHP

In Section 7.4.1 we ranked cars using attributes that are measurable, namely, cost and performance. Here we choose two attributes ($n = 2$), color and prestige, that have no obvious measurement scale. Our first task is to compare every pair of attributes, and as there are only two, there is only one comparison to make. Suppose that in comparing color (attribute 1) with prestige (attribute 2) we decide

Prestige is 1.25 times as important as color.[7]

---

[5] The estimate for $\rho_{j,i}$ is taken to be $1/\rho_{i,j}$.

[6] What this means is defined in Saaty (1990). Roughly speaking the largest eigenvalue should be within 10% of the number of attributes. The reader should consult the references for a precise definition.

[7] There is a standard scale given in the AHP literature using the integers from 1 through 9, but in commercial software implementations any nonnegative numeric values are allowed.

Then using the first row and column for color and the second for prestige,

$$\hat{\mathbf{R}} = \begin{bmatrix} 1 & 0.8 \\ 1.25 & 1 \end{bmatrix}.$$

Note that for this matrix all three properties hold (when n is 2, the matrix must always be consistent) and that the solution $\mathbf{w}$ to Equation (7.27) with $n = 2$ and elements that add to 1 is

$$\mathbf{w} = (4/9, 5/9).$$

Clearly, the second weight is 1.25 times the first weight.

Next, we turn to comparing decision alternatives on each attribute. Suppose that we are considering two cars, where car 1 is a Blue Buick and car 2 a White Chevy; the make of car will be used to assess the prestige attribute. First, we compare colors and decide that

White is preferred 1.5 over Blue.

This leads to the consistent matrix

$$\hat{\mathbf{R}} = \begin{bmatrix} 1 & 1.50 \\ 0.67 & 1 \end{bmatrix},$$

where the first row and column are for white and the second for blue. The normalized weights for white and blue, respectively, are,

$$(0.6, 0.4).$$

Next, we compare makes of cars and decide that

Buick is preferred 1.5 over Chevy.

This leads to the consistent matrix

$$\hat{\mathbf{R}} = \begin{bmatrix} 1 & 1.50 \\ 0.67 & 1 \end{bmatrix},$$

and the normalized weights for Buick and Chevy, respectively, are

$$(0.6, 0.4).$$

Thus a Blue Buick has a vector $\mathbf{r}_1 = (0.4, 0.6)$ and a White Chevy has a vector $\mathbf{r}_2 = (0.6, 0.4)$. Finally, using Equation (7.26) we obtain the values

For the Blue Buick $\quad v(\mathbf{w}, \mathbf{r}_1) = 0.511,$
For the White Chevy $\quad v(\mathbf{w}, \mathbf{r}_2) = 0.489.$

Thus we conclude that:

The Blue Buick is preferred to the White Chevy.

Recall the claim that the results fall on a ratio scale. If this is so, the Buick is preferred 1.045 times over the Chevy.

Suppose that in addition to the two cars we consider a third alternative, a Green Lexus. Moreover, let us assume that when comparing colors, White is still preferred 1.5 over Blue, but Blue is preferred 2 over Green. When comparing makes of cars, a Buick is still preferred 1.5 over a Chevy, but a Lexus is preferred 1.4 over a Buick. Adding a third row and column for Green let us assume that the comparison of White with Green is consistent with the previous comparisons so that

$$\hat{\mathbf{R}} = \begin{bmatrix} 1 & 1.50 & 3.0 \\ 0.67 & 1 & 2.0 \\ 0.33 & 0.5 & 1 \end{bmatrix}.$$

The normalized weights for White, Blue, and Green, respectively, are

$$(0.500, 0.333, 0.167).$$

Adding a third row and column for Lexus and again assuming consistency, for makes of cars

$$\hat{\mathbf{R}} = \begin{bmatrix} 1 & 1.40 & 2.1 \\ 0.714 & 1 & 1.5 \\ 0.476 & 0.667 & 1 \end{bmatrix}.$$

The normalized weights for Lexus, Buick, and Chevy, respectively, are

$$(0.457, 0.326, 0.217).$$

Thus a Blue Buick has a vector $\mathbf{r}_1 = (0.333, 0.326)$, a White Chevy has a vector $\mathbf{r}_2 = (0.500, 0.217)$, and a Green Lexus has a vector $\mathbf{r}_3 = (0.167, 0.457)$. Finally, using Equation (7.26) we obtain the values

For the Blue Buick $v(\mathbf{w}, \mathbf{r}_1) = 0.329,$
For the White Chevy $v(\mathbf{w}, \mathbf{r}_2) = 0.343,$
For the Green Lexus $v(\mathbf{w}, \mathbf{r}_3) = 0.328.$

Thus we conclude that:

The White Chevy is preferred to the Blue Buick, which in turn is preferred to the Green Lexus.

Not only is a ratio scale not maintained, the consideration of the Lexus causes a *reversal in ranking* of the Buick and the Chevy.

To illustrate this rank reversal characteristic of AHP further, suppose we consider a fourth alternative, a Cream Ford, and decide that we are indifferent between a Chevy and a Ford and indifferent between White and Cream. It is left as an exercise to the reader to show that the values for the four cars are now

For the Blue Buick	0.248,
For the White Chevy	0.247,
For the Green Lexus	0.258,
For the Cream Ford	0.247.

We now must conclude that by considering a Cream Ford in addition to the other three cars, the ranking of the Lexus changes from being least preferred to most preferred!

This reversal of rank by the addition (or deletion) of alternatives in AHP has been known for some considerable time and has resulted in considerable literature. It is the most disturbing feature of AHP and one of the easiest to fix. But AHP proponents argue that it is to be expected that, for example, adding the Cream Ford in our example is the same as adding a duplicate White Chevy and that such "duplications" cause "dilution." There have been attempts to show that as long as a decision alternative is added that is not within 10% of one currently under consideration, rank reversal will not occur. Depending on how one interprets "not within 10%" we believe our example of adding the Green Lexus shows that this is not true.

One example given to argue that rank reversal is to be expected is the case of a lady entering a store to buy a new hat. She is offered two hats, A and B, and after considering them both, states that she prefers hat A to hat B. However, when she finds there is a duplicate of hat A she changes her mind and prefers hat B; thus rank reversal occurs sometimes in real problems, and duplicates of a given alternative can have a diluting effect. This appears to be presented as a serious example and not simply a way to explain away the rank reversal feature. It is certainly true that in many situations uniqueness increases value and is a prized *attribute*; works of art and handmade artifacts are examples. But a little thought should convince the reader that the use of the uniqueness argument does not explain rank reversal. One can easily find examples where the availability of large numbers of an item is a distinct advantage. Suppose your problem is to decide on a supplier of a given commodity required to run your operation. The fact that one supplier has the item available in thousands whereas another only in hundreds could certainly affect the value assigned to each in a ranking scheme. In our opinion, uniqueness can and should be included as an *attribute* when it is important. The literature is rife with confusing and poorly defined assumptions and representations that the reader must be very careful to discern. Of course, if the addition of an alternative causes changes in the decision maker's relative rankings of attributes or of alternatives on each attribute, then one can expect changes in rank. But it is very important for the reader to distinguish between such changes and the fact that in AHP rank reversal can and does occur *even when the decision maker does not change relative rankings*. This is clearly demonstrated by the car example. Even though the decision maker made no changes in pairwise comparisons of colors or makes of car, significant rank reversals occurred. Even if the lady were not concerned by the fact that two A-type hats existed, AHP could still reverse the ordering independent of her feelings about the duplication.

Notwithstanding the considerable effort that has gone into explaining rank reversal as evidenced by the quantity of literature on this topic, it is our position that no model for rational decision making should have this characteristic. If one pays careful attention to measurement units, rank reversal can be easily avoided, as we now demonstrate.

7.8.2   Avoiding Rank Reversal in AHP

The reversal in preference order of the first two cars by the addition of a third in our above example is caused by the disregard for measurement units in AHP. Even though there are no natural units for measuring color or prestige, one should not assume that the weights are dimensionless quantities. One must define an underlying scale based on an arbitrary standard. Both the scale intervals and standard can be arbitrary, but once chosen, measurements must be made consistent with this scale. The reader is referred to the principle stated on page 253 as a guide to understanding our approach.

Before demonstrating how a simple modification to AHP can avoid rank reversal using the car example, consider the car ranking problem in Section 7.4.1. There we had two attributes that are both measurable, cost and performance. Our units chosen for these attributes were thousands of dollars and seconds. In Equation (7.1) (p. 259) we denoted the levels of these in a given alternative as a and b. As was stated in part c of the procedure on page 280, what AHP does is to make the result measurements add to 1 in every application. This would mean that if AHP were used, a and b would add to 1, so

1. The units of a would be \$1,000/(\$1,000 + Seconds).
2. The units of b would be Seconds/(\$1,000 + Seconds).

Clearly, these make no sense, and one should never normalize in this way. Applying the principle stated on page 253 we offer the following procedure when the attributes have no obvious measurement units:

1. Rank the possible outcomes of each attribute by preference.
2. Assign a value 1 to the least preferred level of each attribute, and assign it a unit of measure (at least conceptually). For our car example we define a Col for a unit of color and Pre for a unit of prestige.
3. Using a system of pairwise comparisons, for each alternative determine the level of each attribute relative to the least preferred level. If we denote attribute i level by x_i, this value is denoted $v_i(x_i)$.
4. In order to assign an overall goal value to an alternative, combine attribute units in a logical way. Suppose that we call our unit of overall value the Val and let $V(\mathbf{x})$ be the Vals obtained from a vector $\mathbf{x}$ of attribute outcomes. Then,

$$V(\mathbf{x}) = \sum_{i=1}^{n} a_i v_i(x_i),$$

where a_i is measured in Vals/unit of attribute i. For our example $n = 2$, a_1 is measured in Vals/Col and a_2 in Vals/Pre.

5. To determine the a_i's it simplifies the procedure if we let $a_1 = 1$. This is equivalent to measuring the overall goal value in units of the first attribute.[8] One must compare a (possibly fictitious) alternative that has the least preferred levels of every attribute. *The interattribute weights cannot be consistently determined in the abstract, unrelated to attribute ranking and levels.* For our example we must consider a green Chevy or Ford, even though neither of these is an alternative being considered.

Rank reversal can never occur using this approach.

For our example:

$v_1(\text{Green}) = 1$ Col, $v_1(\text{Blue}) = 2$ Cols, $v_1(\text{White}) = 3$ Cols, $v_1(\text{Cream}) = 3$ Cols.

$v_2(\text{Ford}) = 1$ Pre, $v_2(\text{Chevy}) = 1$ Pre, $v_2(\text{Buick}) = 1.5$ Pres, $v_2(\text{Lexus}) = 2.1$ Pres.

As Vals are arbitrary units, let $a_1 = 1$ and $a_2 = 1.25$ so that 1 Val is equal to 1 Col and prestige contributes 1.25 as much as does color. These are consistent with our original relative weights of prestige to color. The results are shown in Table* 7.7. The details are left to the reader who should also check that if any two of the cars are considered without the third, the same values are obtained in column 4 so that rank reversal does not occur when any of the cars is removed from consideration.

Before leaving this example we demonstrate in another way the importance of maintaining units. Suppose that a blue Dodge is also being considered. We have a measure of Blue, namely, 2 Cols, but what measure do we put on prestige for Dodge. If a Dodge is ranked at least as high as a Chevy, there is no problem including this alternative. For example, if Dodge and Chevy are ranked equally in a pairwise comparison, the Dodge would have a value 1. But what if you rank a Dodge below a Chevy by saying that you prefer a

TABLE 7.7
Car ranking using artificial units

| Car | Attributes | | Value | Relative |
	Color ($a_1 = 1$)	Prestige ($a_2 = 1.25$)	(Vals)	value
Blue Buick	2	1.5	3.875	0.33
White Chevy	3	1.0	4.250	0.36
Green Lexus	1	2.1	3.625	0.31

---

[8] Using this procedure one should choose as attribute 1 that which is most easily understood and measurable.

Chevy by a factor of 2 over a Dodge. To be consistent you must assign a Dodge 0.5 Pres and determine that a Blue Dodge has a value of 2.625. It would be a mistake to assign a Dodge a prestige value of 1 because it is the lowest ranking make of car, and then adjust the prestige values of the other makes to this new standard, *without changing the value of a_2*. If a Dodge is given a prestige value of 1, a_2 must be raised to 2.5 to compensate for the fact that prestige is now being measured in units half the size of those used when a Chevy was assigned a 1. It is left as a problem for the reader to show that failure to change a_2 if a Dodge is assigned a prestige value of 1 will lead to rank reversal caused by ignoring the units.

The reader will recall that the relative values assigned to color and prestige were determined *with no reference to any of the decision alternatives*. The above example with the Dodge shows that this cannot be done in a consistent way. *When making judgments concerning the relative preference of one attribute over another, one must consider the range of possible decision alternatives*. From Equation (7.8) (p. 269) it follows that the ratio of weights must satisfy

$$\rho_{i,j} = \frac{w_i}{w_j} = \frac{\pi_i \Delta_j}{\pi_j \Delta_i}. \tag{7.28}$$

Because the determination of the delta terms requires consideration of the least and most preferred level of each attribute, Equation (7.28) demonstrates the italicized statement mathematically. The fact that the ratio weights in AHP are determined without consideration of the decision alternatives is a problem. The reader will find in Section 7.10 that multiattribute utility models specifically include these important interactions.

7.8.3 Finding the Weights in AHP

There are numerous criticisms to be found of AHP in addition to rank reversal. One that is of lesser importance, but nevertheless significant, is the way in which the weights in the vectors $\mathbf{w}$ and the $\mathbf{r}_j$'s, are determined from the pairwise ranking matrix $\hat{\mathbf{R}}$. For simplicity we consider only the vector $\mathbf{w}$; the same comments apply to the $\mathbf{r}_j$'s.

If one chooses a particular attribute, sets its weight to 1, and performs pairwise comparisons of the remaining attributes only with this particular one, we would obtain a set of data that uniquely determine the remaining weights. Remember we are soliciting strengths of preferences from a decision maker and this is not an easy task. By comparing every possible pair of attributes we obtain a much larger data set that presumably contains more information than would be obtained by comparing each attribute to a particular one. What we have is an overdetermined system of linear equations to which we need to find the best fit of weights, rather like fitting the best line to a set of points using regression.

Suppose that we use some system to produce the weights $\mathbf{w}$ from the pairwise comparison data in $\hat{\mathbf{R}}$. What should then be done is to compare w_i/w_j with $\hat{\rho}_{i,j}$ for every i and j to see how close they are. A common and widely accepted way to find $\mathbf{w}$ such that w_i/w_j and $\hat{\rho}_{i,j}$ are "as close as possible" is to minimize the squared error of the difference between them; in other words, find the $\mathbf{w}$ such that

$$\sum_{i=1}^{n}\sum_{j>i}\left(\hat{\rho}_{i,j}-\frac{w_i}{w_j}\right)^2 \qquad (7.29)$$

is a minimum. If this is done instead of taking the more difficult eigenvalue approach central to AHP, one can use a well-understood norm and also find a way to determine where the major discrepancies in consistency of rankings occur (from the relative size of terms in this equation). There seems to be little justification for using the eigenvalue approach other than the observation from Equation (7.27) that n is the largest eigenvalue of $\hat{\mathbf{R}}$ and $\mathbf{w}$ is a corresponding eigenvector when $\hat{\mathbf{R}}$ is consistent. There are many ways to find the "best fit" of $\mathbf{w}$ to the pairwise comparison data. We are not claiming that using the least-squares fit (LSF) is superior to all others; one must first define what one means by best fit.

We demonstrate these comments by using an example taken from Saaty (1990). In that article AHP is applied to ranking three houses using eight attributes. One of these attributes is "size of house".[9] Comparing the three houses on this attribute the pairwise comparison matrix is given as

$$\hat{\mathbf{R}} = \begin{bmatrix} 1 & 6 & 8 \\ 0.167 & 1 & 4 \\ 0.125 & 0.25 & 1 \end{bmatrix}. \qquad (7.30)$$

The reader can check that this is not a consistent matrix ($6 \times 4 \neq 8$). Its largest eigenvalue is $3.136 > 3$. The corresponding vector of weights (right eigenvector) for the three houses on this attribute is

$$\mathbf{r}(AHP) = (0.754, 0.181, 0.065),$$

which leads to a ratio matrix

$$\mathbf{R}\,(AHP) = \begin{bmatrix} 1 & 4.17 & 11.60 \\ 0.24 & 1 & 2.78 \\ 0.09 & 0.36 & 1 \end{bmatrix}. \qquad (7.31)$$

Using Equation (7.29) with (7.30) we obtain an LSF set of weights

$$\mathbf{r}(LSF) = (0.778, 0.137, 0.085),$$

which leads to a ratio matrix

$$\mathbf{R}\,(LSF) = \begin{bmatrix} 1 & 5.68 & 9.17 \\ 0.18 & 1 & 1.61 \\ 0.11 & 0.62 & 1 \end{bmatrix}. \qquad (7.32)$$

---

[9] No attempt is made in the article to use a physical measure of size such as square feet, number of bedrooms, or number of bathrooms, even though these are generally accepted measures.

The reader can compare the matrices in Equations (7.31) and (7.32) with the decision maker's preference matrix in (7.30). Of course, it is not obvious how to measure whether $\mathbf{R}(AHP)$ or $\mathbf{R}(LSF)$ is "closer" to $\hat{\mathbf{R}}$. One way is to count the number of entries in each that are closest; of the three terms above the diagonal two in $\mathbf{R}(LSF)$ and one in $\mathbf{R}(AHP)$ are closer. Another widely accepted way is to compare the sums of squares of the differences of all elements. For $\mathbf{R}(LSF)$ it is 7.295 and for $\mathbf{R}(AHP)$ it is 17.820. It is not surprising that $\mathbf{R}(LSF)$ gives the smaller number because the weights were found to minimize this number. What may be surprising to the reader is the large difference between the two. In the final analysis it is left to the decision maker to decide which set of weights is preferable.

In summary our analysis of AHP leads us to conclude that:

1. The inherent assumption of dimensionless measures for attributes, together with the method of normalizing the attribute weights, violates our principle of consistent modeling stated on page 253.
2. The normalization method used for the weights is the cause of rank reversal.
3. The use of the largest eigenvalue to measure consistency is not essential, nor is the determination of the corresponding eigenvalue. There are simpler norms to use for finding the weights that, in addition to being more widely understood, (a) can lead to a better fit to the decision maker's preferences, and (b) can be used to rank those pairwise comparisons that have led to inconsistencies.

The interested reader can find more material on AHP in the references at the end of the chapter.

7.9 A BUDGET PLANNING EXAMPLE WITH THREE ATTRIBUTES

Our next example in this chapter is based on the budget planning problem described in Section 4.8 (p. 182). In this section we consider an objective function with three attributes.

In Section 4.8 it was assumed that the objective was to minimize the expected overexpenditure in the given fiscal year. Recall that the notation used was

D_1 = Initial expenditure rate (decision 1)
X = Midyear budget figure
D_2 = Midyear corrected expenditure rate (decision 2)
Y = End-of-year (EOY) funds.

Let Z_1 be the amount overspent at the end of the year. This is a random variable given by

$$Z_1 = Max\{0, D_2 - (X + Y)\}.$$

Let Z_2 be the amount left unspent at the end of the year. This is a random variable given by

$$Z_2 = Max\{0, (X + Y) - D_2\}.$$

Let Z_3 be the amount of the originally planned program not completed in the year. We assume that any program not completed in the first half of the year because of a too low initial expenditure rate is lost and cannot be completed in the second half, no matter how much money is available. The reason for this is that the maximum rate at which program can be carried out is the originally planned rate of $54 (millions). At midyear the amount of program not completed is $27 - D_1/2$. Because D_2 is the adjusted *annual* rate, the amount not completed in the second half of the year is $Max\{0, 0.5D_1+27-D_2\}$. Thus,

$$Z_3 = (27 - D_1/2) + Max\{0, 27 - D_2 + D_1/2\}.$$

Note that Z_1 and Z_2 are both independent of the initial allocation D_1, so the influence diagram in Figure 4-39 (p. 185) is the appropriate one to use when any function or combination of Z_1 and Z_2 are used in a payoff function. The decision maker can make any of the alternate decisions at the start of the year, make a corrective action at midyear, and come out with the same over- or underexpenditure. In this case the decision tree with twenty-two terminal nodes in Figure 4-40 (p. 186) is the appropriate one to use. But program completion Z_3 depends on the initial expenditure rate D_1, so when this is used in the payoff function the appropriate influence diagram is the one in Figure 4-38 (p. 184) that contains a directed arc from D_1 to R. The corresponding decision tree has sixty-six terminal nodes consisting of three subtrees, each with the same structure as the one in Figure 4-40 (p. 186). We now develop an appropriate loss function based on these three random variables Z_1, Z_2, and Z_3.

It is often the case that small deviations from a goal are not as serious as large ones. This is assumed to be the case in this budget planning problem, and so we choose to measure our losses proportional to the square of the Z_i's. Because all three measures are important, we consider all three as possible result functions,

$$R = E[Z_1^2] \text{ or } E[Z_2^2] \text{ or } E[Z_3^2].$$

The reader is encouraged to solve the problem for each of these payoff functions. It will be found that an initial allocation rate of $54 is an optimal solution for the initial decision *for all three cases*. However, there are conflicts in deciding the optimal midyear decision; it depends on which of the three measures we use.

The advisory group meets to discuss the situation. The comptroller still insists on minimizing overspending, but some in the group worry that underspending will jeopardize next year's budget. Not surprisingly, the line managers worry that the reason the activity exists, namely, to perform a planned program, should be the important factor in deciding the budget expenditure rate. Although the group can agree on an initial expenditure rate, they are concerned about the midyear conflicts. At this point they call in their analyst for help.

The analyst points out that they need to compromise and suggests looking at a linear combination of the three measures. This leads to an overall result function

$$R = w_1 E[Z_1^2] + w_2 E[Z_2^2] + w_3 E[Z_3^2],$$

where, as all three random variables are measured in monetary units, we set $w_1 + w_2 + w_3 = 1$. Here w_i is the *relative* weight attached to measure i. To the comptroller

w_1 should be 1. To the line managers w_3 should be close to 1. How is the advisory group to decide on appropriate values for the w_i's? At this point the analyst produces Figure 7-14.

Because the w_i's add to 1, we can write $w_3 = 1 - w_1 - w_2$. The triangle that forms each of the three graphs in Figure 7-14 is the area of feasible w_i's, with the corner points $(1,0)$, $(0,1)$, $(0,0)$ representing $w_1 = 1$, $w_2 = 1$, and $w_3 = 1$, respectively. The diagonal boundary represents all sets of weights where $w_3 = 0$. Each of the shaded areas repre-

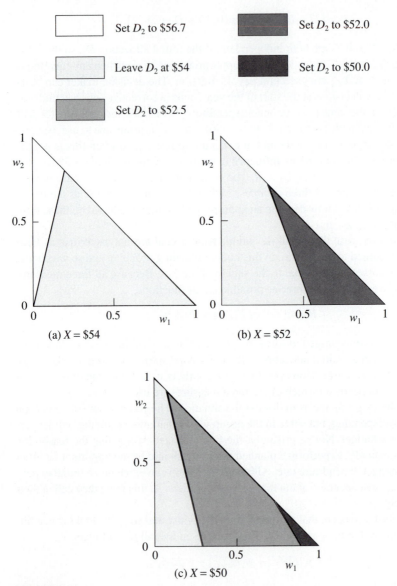

FIGURE 7-14
Midyear decisions in the (w_1, w_2) space.

sents a combination of weights for which the midyear policy given in the legend is optimal. The boundaries between regions are found by finding the minimum expected payoff at the D_2 nodes. For example, suppose that at midyear it is known that $X = 54$ (recall that an initial allocation, $D_1 = 54$ is made). At the D_2 node on this path the expected payoff is $Min\{1.458w_2, 5.832w_1\}$. If the first term is the smaller of the two, choose D_2: Leave the annual spending rate at $54. If the second term is the smaller of the two, choose D_2: Set the annual spending rate to $1.05X = \$56.7$. We are indifferent between these decisions when equality holds, that is, when $w_2 = 4w_1$. This gives the equation for the line through the origin in Figure 7-14(a). The boundaries in the other two graphs are found in a similar way (see Problem 7.11).

The advisors can now see whether their individual choices of w_i's do in fact lead to conflicting policies. If $X = 52$, the weights $(0.55, 0, 0.45)$ and $(0.3, 0.1, 0.6)$, for example, give the same policy. Thus, the advisors can present to their superior not only a single "optimal" solution but also the ranges of the w_i's (Figure 7-14) for which their recommendations are optimal.

7.10* MULTIATTRIBUTE UTILITY

As described in the earlier discussion of AHP, it is not always possible to quantify every attribute of interest in a decision problem. Often, the higher the level of the decision, the less likely an attribute is susceptible to direct measurement. In such cases we can adapt and extend to multiattribute problems the ideas of utility presented in Chapter 6 for measuring a single-attribute result.

It is relatively simple for most people to put a preference ordering on a single attribute. But most people have great difficulty comparing results with multiple attributes. For example, most would prefer spending less for a car rather than more; most would prefer a higher-performance car to a lower-one; most would prefer better fuel economy. But how many can rank the four cars in Table(p. 270) directly using all three attributes? Which car (or three-dimensional vector) do you prefer, $(17, 10, 18)$, $(12, 12, 22)$, $(10, 14, 28)$, or $(22, 7.5, 16)$? Can you pick out the most preferred and least preferred vectors? Clearly, this is not an easy problem. For this reason we consider using a simple linear model where each attribute can be considered independently of the others.

We present a model where we assign a separate (nonlinear) utility function to each attribute. Let R_j be the set of possible results of attribute $j = 1, 2, \ldots, n$. We assume that a preference ordering is defined on each of these sets. Let u_j be a utility function defined on R_j, and let k_j be a relative weight assigned to attribute j. Let $r_j \in R_j$ and $\mathbf{r} = (r_1, r_2, \ldots, r_n)$. The (scalar) utility resulting from this vector of attributes is

$$U(\mathbf{r}) = \sum_{j=1}^{n} k_j u_j(r_j). \tag{7.33}$$

It is not obvious that we can decompose $U(\mathbf{r})$ in this way and still be consistent with the preference ordering placed on each of the sets R_j. We now present an overview of the additional assumptions that need to be made on each ordering so that the decomposition shown in Equation (7.33) is valid.

We assume that a transitive preference ordering (i.e., the first two assumptions of utility theory stated on page 236 hold) is defined on each set $\mathcal{R}_j$. Thus if $r_j^{(1)}$ and $r_j^{(2)}$ are any two results in $\mathcal{R}_j$, we assume that one of the following holds:

$$r_j^{(1)} > r_j^{(2)}, \qquad r_j^{(1)} \sim r_j^{(2)}, \qquad \text{or } r_j^{(1)} < r_j^{(2)}.$$

For each attribute j let $\bar{r}_j$ be the "Best Result" and $\underline{r}_j$ be the "Worst Result" of j. Now let $\bar{\mathbf{r}} = (\bar{r}_1, \bar{r}_2, \ldots, \bar{r}_n)$ and $\underline{\mathbf{r}} = (\underline{r}_1, \underline{r}_2, \ldots, \underline{r}_n)$. Four important concepts follow that are used in determining the restrictions on the preference ordering and utility functions necessary for our decompositions to hold.

1. *Preferential independence.* Let $r_j^{(1)} \in \mathcal{R}_j$, $r_j^{(2)} \in \mathcal{R}_j$, and $r_1 \in \mathcal{R}_1$. If $r_j^{(1)} > r_j^{(2)}$ for *every* r_1, attribute j is said to be preferentially independent (PI) of attribute 1.[10]
2. *Mutual preferential independence.* If j is PI of 1 and 1 is PI of j, j and 1 are said to be mutually preferentially independent (MPI).

Both concepts extend to subsets of attributes. Let $\mathcal{M}$ be a subset of $\mathcal{N}$, the set of all attributes, and $\mathcal{M}^c$ its complement. $\mathcal{M}$ is PI of $\mathcal{M}^c$ if for

$$\mathbf{r}^{(1)} \in \prod_{i \in \mathcal{M}} \mathcal{R}_j, \qquad \mathbf{r}^{(2)} \in \prod_{i \in \mathcal{M}} \mathcal{R}_j, \qquad \text{and } \mathbf{s} \in \prod_{i \in \mathcal{M}^c} \mathcal{R}_j,$$

then, $\mathbf{r}^{(1)} > \mathbf{r}^{(2)}$ for every $\mathbf{s} \in \prod_{i \in \mathcal{M}^c} \mathcal{R}_i$. If $\mathcal{M}$ is PI of $\mathcal{M}^c$ and $\mathcal{M}^c$ is PI of $\mathcal{M}$, $\mathcal{M}$ and $\mathcal{M}^c$ are MPI. Finally, we say that the set of all attributes, $\mathcal{N}$, has the MPI property if every proper subset and its complement are MPI.

Consider the three-attribute car ranking problem in Section 7.6.3 (p. 269). Consider a car with a fixed price and performance. If you continue to prefer higher miles per gallon to a lower value no matter what the price is in the range of possible prices, then fuel economy is *preferentially independent* of price.

Because there are 2^n subsets of n attributes, one needs to check all $2^n - 2$ nonempty proper subsets to ensure that the MPI property holds. We return to this problem later in the section.

In most practical applications MPI holds, but not always. For many people food preferences commonly change when combinations are considered. Think of items as common as bread and meat being combined into a sandwich. A person may prefer white bread to rye bread for a roast beef sandwich, but rye to white for a corned beef sandwich. For this person the attribute Bread is not preferentially independent of the attribute Meat. For a possibly more serious example, suppose that on leaving graduate school you are

---

[10] Here we use PI as an acronym for *preferential independence*. The reader should not confuse this with the use of PI in the remainder of the book and the literature at large for *perfect information*.

considering two alternate careers, one in investment banking and the other in the federal government in the treasury department. Suppose that another area of concern is where you live and that you are considering both New York and Washington, DC. If you fix your career choice as investment banking you would probably prefer New York to Washington, whereas if you fix your career as one in government you would probably prefer Washington to New York. In this case PI of career and living area would not hold.

3. *Utility independence.* This concept is most easily understood using a decision sapling in Figure 7-15. Choose any attribute from $\mathcal{N}$; we choose $j = 1$ for convenience to simplify the notation. From $\mathcal{R}_1$ choose three values such that $r_1^{(1)} > r_1 > r_1^{(2)}$, and choose any vector $\mathbf{s} \in \mathcal{R}_2 \times \mathcal{R}_3, \ldots, \times \mathcal{R}_n$. If the indifference probability p stays unchanged for every possible $\mathbf{s}$, then attribute 1 is said to be utility independent of its complement.

We can extend this concept to any subset of attributes and its complement; when comparing three vector results of a given subset of attributes, if the indifference probability remains unchanged for all possible results of the complement attributes, this subset is said to be utility independent of its complement.

4. *Mutual utility independence.* If a subset of attributes is utility independent of its complement and the reverse is also true, then the subset and its complement are said to be mutually utility independent (MUI).

Finally, we say that $\mathcal{N}$ has the MUI property if every proper subset and its complement are MUI.

Utility independence is a much stronger assumption than preferential independence. It can be shown that $\mathcal{N}$ must have the MPI property for the decomposition of the utility function into the additive form to be valid, but it is not sufficient. In order to determine further conditions that are sufficient, we consider the more complex form of $U(\mathbf{r})$ called the *multiplicative form*:

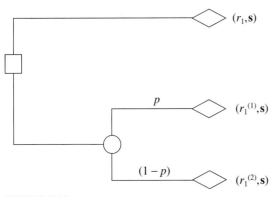

FIGURE 7-15
Illustration of utility independence.

$$U(\mathbf{r}) = \frac{\displaystyle\prod_{j=1}^{n} [1 + Kk_j u_j(r_j)] - 1}{K}. \tag{7.34}$$

Note that $U(\underline{\mathbf{r}}) = 0$. To scale $U(\mathbf{r})$ on $(0, 1)$ we require $U(\overline{\mathbf{r}}) = 1$. Thus, K must satisfy

$$1 + K = \prod_{j=1}^{n} (1 + Kk_j). \tag{7.35}$$

An important result in multiattribute utility theory is that the multiplicative decomposition shown in Equation (7.34) holds if and only if the set of attributes $\mathcal{N}$ has the MUI property. Thus, single-attribute utilities can be combined in a multiplicative way to calculate the multiattribute utility when MUI holds. For the additive form of Equation (7.33) to be valid, certain conditions must hold on the weights k_j as we now demonstrate.

We now turn our attention to the determination of the k_j's. For an arbitrary attribute j if we let $\underline{\mathbf{r}}(j)$ be the vector with all attributes set to their worst result *except* the jth which is set to its *best*, Equation (7.34) gives

$$U(\underline{\mathbf{r}}(j)) = k_j.$$

By comparing a certainty of getting $\underline{\mathbf{r}}(j)$ to a gamble between $\overline{\mathbf{r}}$ and $\underline{\mathbf{r}}$ on a decision sapling, the indifference probability is k_j. An important result of applying one's judgment in determining the k_j's is to discover whether or not these weights add to 1. It can be shown that if they do, the solution to Equation (7.35) gives $K = 0$, and in this case Equation (7.34) simplifies to (7.33), namely,

$$U(\mathbf{r}) = \sum_{j=1}^{n} k_j u_j(r_j),$$

and additive utility holds.

If the k_j's do not add to 1, but MUI holds, U can be written in terms of the individual u_j's, but in the multiplicative form of Equation (7.34). It can be shown that

1. If $\displaystyle\sum_{j=1}^{n} k_j > 1$, K is the unique root of (7.35) in $(-1, 0)$.

2. If $\displaystyle\sum_{j=1}^{n} k_j < 1$, K is the unique root of (7.35) in $(0, \infty)$.

As pointed out, to show that $\mathcal{N}$ has the MUI property using the definition of MUI requires checking $2^n - 2$ subsets. The following result, which we state without proof, reduces the amount of checking considerably: Find an attribute that is UI of its complement; number this 1. Then check to see if each of the $(n-1)$ pairs $\{1, j\}$ is preferentially independent of its complement. This requires making a total of n checks of which $(n-1)$ are for the simpler verification of preferential independence rather than utility independence.

7.10.1   An Example with Two Attributes

For the two-attribute case Equation (7.34) simplifies to

$$U(\mathbf{r}) = k_1 u_1(r_1) + k_2 u_2(r_2) + K k_1 k_2 u_1(r_1) u_2(r_2).$$

But from Equation (7.35) $K k_1 k_2 = 1 - (k_1 + k_2)$, so

$$U(\mathbf{r}) = k_1 u_1(r_1) + k_2 u_2(r_2) + [1 - (k_1 + k_2)] u_1(r_1) u_2(r_2),$$

or

$$U(\mathbf{r}) = k_1 [u_1(r_1) - u_1(r_1) u_2(r_2)] \tag{7.36}$$

$$+ k_2 [u_2(r_2) - u_1(r_1) u_2(r_2)] + u_1(r_1) u_2(r_2).$$

This shows that the utility is linear in the two weights k_1 and k_2 and that if they add to 1, we have additive utility. But k_1 and k_2 must be found as indifference probabilities on a decision sapling, a procedure that is typically not easy for a decision maker. Equation (7.36) allows us to plot the utility as a function of k_1 and k_2 and to demonstrate why precise determination of these weights may not be necessary.

   Suppose that a farmer can plant any of three crops. The success of the planting will be measured by two attributes, (1) the net income obtained per acre (Income), and (2) the length of time it takes before the crop can be harvested (Harvest time). Both these will depend on the weather in the growing season, which of course cannot be observed before the crop is planted. Let the weather conditions be modeled simply by categories poor, normal, and good. Formally we can let X be the unobserved state of the weather and

$$\mathbf{r}_i(x) = [r_{i1}(x), r_{i2}(x)],$$

the income and harvest time obtained when crop i is planted and weather conditions x occur. Table* 7.8 contains the data for this example. The farmer's decision problem, as

TABLE 7.8
Data for crop planting decision example

Weather	Probability	Crop type		
x	$p(x)$	1	2	3
poor	0.25	(−400, 16)	(50, 20)	(−250, 18)
normal	0.50	(20, 14)	(100, 17)	(0, 11)
good	0.25	(200, 12)	(150, 16)	(160, 8)

illustrated in the decision tree in Figure 7-16, is to choose the best crop to plant using his estimate of what the uncertain future will bring.

In order to analyze this problem we need to decide on an appropriate measure for the results. It is agreed that nonlinear utility functions should be determined and used on both income and harvest time and further agreed that the exponential form of Equation (6.3) (p. 242) is applicable. For the income measure, $\underline{r}_1 = -400$ and $\bar{r}_1 = 200$, and using $\beta = 5$,

$$u_1(r_1) = 1.25\left(1 - 0.2^{\left(\frac{r_1 + 400}{600}\right)}\right), \qquad -400 \le r_1 \le 200. \qquad (7.37)$$

For the harvest time measure, $\underline{r}_2 = -400$ and $\bar{r}_2 = 200$, and using $\beta = 9$,

$$u_2(r_2) = 1.125\left(1 - 0.111^{\left(\frac{20 - r_2}{12}\right)}\right), \qquad 8 \le r_2 \le 20. \qquad (7.38)$$

These are shown plotted in Figure 7-17.

The next step is to determine if MUI holds. Because there are only two attributes, we need to show that they are MUI using the method described with a decision sapling. Assuming MUI holds, the multiplicative form of a two-attribute utility can be used. We use the form shown in Equation (7.36).

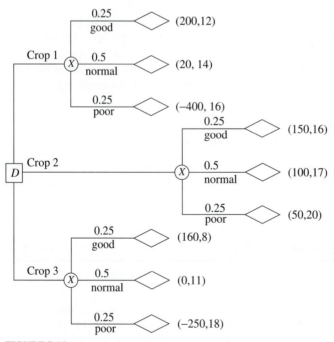

FIGURE 7-16
Decision tree for crop planting example.

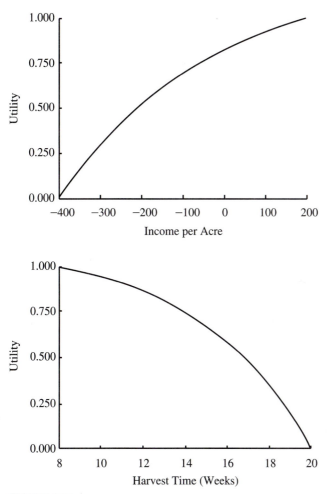

FIGURE 7-17
Utility functions for crop planting example.

The next task is to determine the weights k_1 and k_2. This could be done using methods described with a decision sapling, but we take an approach similar to that taken in earlier chapters. Using Equations (7.37) and (7.38) with (7.36), and taking expected values, the expected utilities at each X node in Figure 7-16 in order from top to bottom are

1. Crop 1: $0.533 + 0.139k_1 + 0.204k_2$,
2. Crop 2: $0.360 + 0.561k_1 + 0.024k_2$,
3. Crop 3: $0.652 + 0.105k_1 + 0.138k_2$.

The optimal policies for any values of the k_i's in $(0,1)$ are shown in Figure 7-18. The reader can show that it is never optimal to plant Crop 1. The boundary where we are in-

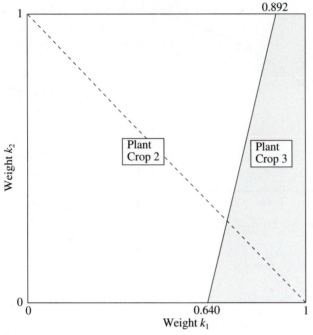

FIGURE 7-18
Optimal crop planting policy.

different between planting Crops 2 and 3 is found by equating the last two of these three expressions (see Problem 7.13). We note again that there are large ranges of values for both k_1 and k_2 for which it is optimal to plant the same crop. Our final comment is that additive utility holds only when k_1 and k_2 are such that they lie on the diagonal dashed line shown in Figure 7-18.

7.11   SUMMARY AND INSIGHTS

When results are measured using more than one attribute, conflicts in choosing the "best" decisions usually occur. When faced with such conflicts it pays to analyze the different attributes independently to determine if and where conflicts occur in the decision choices. Most of the literature on multiattribute problems assumes that trade-offs are to be made without first checking to see whether or not decision conflicts exist. In many cases there is no conflict in the optimal policies and there is no need to become involved in assessing trade-offs between different attributes. In other problems there may only be conflicts among a small subset of the attributes of interest; moreover, the conflicts may arise only under certain conditions. In such cases it is helpful to focus on those decisions and outcomes where policies conflict to better understand where trade-offs or improved forecasts can influence later decision making.

Policy conflicts are always improved or worsened by improved discrimination in forecasts. In the crop protection problem in Figure 7-3 increases are preferred to decreases in both yield and quality; thus, as we move to the northwest corner in the (p_0, p_1) plane, we enter a region of complete policy agreement. Had the result been measured with two attributes where an increase was preferred in one and a decrease preferred in the other we would move toward an area where conflict was certain.

In cases where conflicts cannot be avoided, the decision maker will need to think carefully about the interpretation of trade-offs among the attributes. Does preferential independence hold so that linear trade-offs are valid? Can the trade-offs be directly assessed? Can the attributes be readily measured on a common scale such as money? When they cannot, the simple decision sapling may be used to quantify trade-offs through indifference probabilities. By equating expected values and assuming that a decision maker is consistent in his or her preferences, these indifference probabilities can be used to determine the trade-off weights.

In many practical problems the attributes fall into two major groups: costs and benefits. After selecting any one of the attributes within a group by which to measure that group, trade-off weights can be found among the remaining attributes within the group. Having done this for both cost and benefit groups, one can find a single trade-off weight between the two often called the cost–benefit or benefit–cost ratio. By modeling a multiattribute problem in this way it is often easier to understand and interpret the meaning of the weights.

When large amounts of resources are at risk, where the uncertainty in outcomes places a premium for accepting risky ventures, and when it is difficult to quantify result outcomes, the decision maker may have to consider the possibility of utilities that are both nonlinear for each attribute and nonlinear in the trade-off weights. To do this in a consistent way one must be more concerned with interactions among attributes than is the case for the linear model. One is usually required to make assumptions about interactions of preferences, including the preferential independence and utility independence of attributes.

More detailed discussion and proofs of results in multiattribute utility theory can be found in Keeney and Raiffa (1976), and Chankong and Haimes (1983). The crop planting decision example in Section 7.10.1 is based on one in Chankong and Haimes (1983). A number of aspects on risk taking and the role of multiattribute preference theory are contained in Kirkwood (1992a,b).

An early recognition of the occurrence of rank reversal in AHP was published in Belton and Gear (1983) for the case where duplicates of existing alternatives were added. Hijn and Johnson (1988) compare sixteen different methods of fitting weights to an overdetermined ratio matrix and conclude that there is no basis for believing that the method used in AHP is superior or even comparable to other commonly used methods. The equivalence between the ratio weights of AHP and the indifference probabilities for finding trade-off weights in a multi-attribute value function is shown in Equation (7.28) (p. 286); the important observation is that meaningful ratio weights are scale dependent and directly related to the range of alternatives being considered.

There is a considerable literature on the subject of AHP. For a succinct description and example of its use see Saaty (1990). The entire issue of the journal *Mathematical and Computer Modelling*, 17 (4/5), is devoted to articles on AHP, Vargas and Zahedi (1993). A comparison of AHP with multiattribute utility can be found in Winkler (1990) and an extensive discussion can be found in French (1993).

PROBLEMS

7.1 For the two-attribute crop protection problem, let the return per acre be 10 bushels if the crop is protected, two if it is not and freezing weather occurs, and 12 if it is not protected and the weather is good. Let the quality of the crop be measured on a scale from 1 to 5, where 5 is top quality. Let the quality be 4 if the crop is protected, 1 if it is not and freezing weather occurs, and 5 if it is not protected and the weather is good.

   a. Find s_1 and s_2.

   b. Using each attribute separately, find the optimal policy and any conflicts if (1) $p_1 = 0.8$, $p_0 = 0.1$, (2) $p_1 = 0.7$, $p_0 = 0.3$, (3) $p_1 = 0.6$, $p_0 = 0.22$.

   c. When $p_X = 0.3$, plot the appropriate graph in the (f_0, f_1) plane, and find the optimal policy and any conflicts if (i) $f_1 = 0.7$, and $f_0 = 0.2$, (ii) $f_1 = 0.6$, and $f_0 = 0.4$, and (iii) $f_1 = 0.5$, and $f_0 = 0.45$.

7.2 Find the equations of the boundaries and the coordinates of the intersection points of all optimal policy regions in Figure 7-7.

7.3 For the car replacement example in Figure 7-6 (p. 261), in addition to the uncertainty in the failure of the current car, include an uncertain reliability of the high performance Car 1. With probability q, Car 1 has low reliability, and the estimated future cost of repairs effectively increases the price by $4,000. With probability $(1 - q)$ it has high reliability, and the cost and performance data are as before.

   a. Draw the influence diagram for the decision problem in the original problem as illustrated in Figure 7-6 (p. 261).

   b. Draw the influence diagram and decision tree for the problem when Car 1 can fail with probability q.

   c. Find and plot the optimal policy regions in the (w, p) plane, and indicate how these regions compare with Figure 7-7 (p. 262). (*Hint:* Try solving the case $q = 1$ first.)

7.4 You are to make a decision on whether to keep your existing car or buy a new one. You realize that both old and new cars are unreliable. The old car breaks down with probability p, and if this occurs you *must* (i.e., you have no choice) purchase the new car at $3,000 more than it now costs. The new car breaks down with probability $q \leq p$ and is instantly repaired free of charge. The cost (in thousands of dollars) and performance (seconds to reach 60 mph) are:

Car type	Cost ($000)	Performance
Old	0	10 sec.
New	10	12 sec.

   a. Draw the simplest influence diagram and decision tree that describe this decision problem.

   b. Assume a linear trade-off w between performance and cost. When are you indifferent to keeping your existing car or immediately buying the new car? Derive any equations and/or inequalities that must be satisfied.

   c. Plot the indifference curves obtained in (b) in the (w, p) plane.

   d. How much money can you save if you are given perfect information on the failure or non-failure of the new car just before you make your decision to buy or keep?

   e. What are your expected savings if you are given perfect information about the failure or nonfailure of the old car just before you make your decision to buy or keep?

   f. Plot the lines of constant expected savings (isocontours) in the (w, p) plane.

7.5   For the car replacement example with three attributes in Section 7.6.4, use a decision sapling to find your own indifference probabilities, and from these find your own ranking of the cars in Table(p. 270).

7.6   Using two attributes in the one-stage crop protection problem in Section 2.2 (p. 46) where p_X is the probability of freezing weather, denote the indifference probability by $\bar{p}_1$ for attribute 1 and $\bar{p}_2$ for attribute 2, where $\bar{p}_1 > \bar{p}_2$. The decision maker's assessment for the probability of freezing weather lies between these two critical probabilities so that there is a conflict and she is unable to decide whether to protect the crop or not. She decides to use a trade-off weight w for attribute 2 in terms of attribute 1. Plot the optimal policy regions in the (w, p_X) plane.

7.7   Using the price and performance data for the four cars listed in Table(p. 270), calculate the reduction in equivalent dollars if the (12, 12) car is selected. Compare your answers with the earlier example where the (12, 12) car is purchased but we use only the data in Table(p. 259). Explain the differences.

7.8   Verify Equation (7.27) (p. 279) that $\mathbf{Rw} = n\mathbf{w}$.

7.9   Calculate the values of the four car shown on p. 283.

7.10   For the car ranking example in Table(p. 285), show that if a Blue Dodge is added as an alternative, a Dodge rates 1/2 a Chevy, and since Dodge is now the least preferred make, it is assigned a weight 1. For the values in column 4 to remain unchanged a_2 must now be 2.5. Show that keeping it at 1.25 leads to rank reversal.

7.11   Find the equations of all the boundary lines shown in the three graphs for the budget allocation problem in Figure 7-14 (p. 290).

7.12   (Covaliu) You have to choose among three topics A, B, or C, for a project in an important graduate course. You are concerned about the grade (on a scale of 0 to 4), the time it will take to complete the project (2 to 6 weeks), and knowledge acquired (on a scale of 1 to 10). Topics A and B are sure to give you (3.5, 3, 6) and (3, 4, 6), respectively in the attributes (grade, time, knowledge). With topic C, however, you believe that with probability p you would get (3.5, 2, 2), and with $(1 - p)$ (3, 6, 8). You want to maximize the expected grade and knowledge acquired but minimize the expected time to completion.

   a.   For what values of p is there no conflict among attributes in selecting the topic? In this case what is the optimal choice and the optimal expected attribute vector?

   b.   Suppose you assess p to be equal to 0.25. What is your optimal decision if you do not care about acquired knowledge?

   c.   With $p = 0.25$ let w_t and w_k denote your constant linear trade-offs between time and grade and between knowledge and grade respectively, with dimensions grade per week and grade per unit of knowledge. Draw the optimal decision regions in the (w_t, w_k) space. Is the answer in (b) consistent with your graph? Explain carefully.

7.13   Determine the boundary between the optimal policy regions for the crop planting example in Figure 7-18.

7.14   (Clemen) A hospital administrator is making a decision regarding the policy of treating uninsured patients. There are two conflicting objectives: to maximize the revenue for the hospital and to provide as much care as possible for the uninsured poor. The decision to admit or not is based on an examination of a prospective patient's financial resources. There are two attributes to be measured; let x_R = revenue for the hospital and x_P = number of poor who are admitted. The utilities of these, $u_R(x_R)$ and $u_P(x_P)$, are assumed to have already been assessed. The attributes are believed to be mutually utility independent. Two additional assessments are then made. First, Lottery A and its certain alternative B,

A: Best on revenue, worst on treating poor, probability 0.65; worst on revenue, best on treating poor, probability 0.35.

B: Levels of revenue and treatment of poor that give $u_R = u_P = 0.5$.

In **B**, it is understood that the certainty equivalent is $\mathbf{x}^* = (x_R{}^*, \; x_P{}^*)$ such that $u_R(x_R{}^*) = u_P(x_P{}^*) = 0.5$.

Second Lottery **C** and its certain alternative **D**,

C: Best on revenue, best on treating poor, probability π; worst on revenue, worst on treating poor, probability $1 - \pi$.

D: Worst on revenue and best on treatment of poor.

a. Write out the mathematical formula for the two-attribute utility (1) as a function of k_R and K, and (2) as a function of π.

b. Plot values of k_R and K as a function of π in the second lottery.

c. Defend those cases where the administrator would be justified in using additive utility.

d. Under what conditions do the attributes complement or substitute for one another. How are these reflected in the answer to part (a)?

CHAPTER
8

FORECAST
PERFORMANCE

8.1   INTRODUCTION

Up to this point we have concentrated on showing how forecasts play a role in decision making and how one can sometimes get substantial improvement in results by improving baseline forecasts. We have shown that forecasts may result from subjective estimates, expert opinions, probabilities estimated from historical data, or likelihood estimates and the use of Bayes' rule to calculate posterior probabilities. We have assumed that estimates of conditional probability distributions or odds are available for the random quantities and events of interest. In many problems we are unable to make perfect forecasts or even to have enough data to obtain good estimates. For this reason we now turn to the issues of forecast quality and performance and to the important concepts of calibration, discrimination, and correlation. Although we do not discuss the design of forecasting systems or the analysis and use of specific forecasting formulas, one can address many issues of forecast performance and quality independently of the detail that goes into a particular forecast or a full knowledge of the decision problem where the forecasts are used. We attempt to clarify further the important relationships between probability and categorical forecasts.

No matter what type of forecast is used, assessment and validation of forecast performance and quality are important. This is particularly true when the decisions and the forecasts are made by different people or organizations. We have already considered the economic value of perfect information and how forecasts affect decisions. But how does one measure quality or the economic value of forecasts when the forecaster may not know the structure or objectives of the decision problems in which the forecasts are being used? One might suspect that for a forecast to be useful it should be highly correlated with the random event being predicted, should not be too biased in any particular direction and should be reliable in the sense that the user can trust the forecast. Moreover, when given an event outcome, one should be able to discriminate easily between forecasts that predicted the event. It is not always obvious how one should measure the quality of a forecast as there are many possible measures. In Section 8.2 we introduce the concept of forecast calibration in which we compare the frequency an event actually occurs relative to the probability forecast we make for the occurrence of that event; we then discuss the related concept of calibration in expectation. The definition of calibration is different for categorical and probability forecasts. In Section 8.3 forecast discrimination is introduced to help us measure the ability of a forecast to identify known event outcomes. In Section 8.4 we compare and contrast these two performance measures and demonstrate that they often conflict. Forecast correlation is introduced in Section 8.5 and the composition of Brier score in terms of quadratic discrimination and calibration terms is shown in Section 8.6. In the final four sections we emphasize the effect that calibration and discrimination have upon optimal decisions. Section 8.7 demonstrates the effect of calibration on optimal decisions in problems where probability forecasts are used. In Section 8.8 we show how an equivalent and coherent two-category forecast can be obtained directly from a detailed probability forecast. In Section 8.9 we show how probability forecasts can be aggregated to obtain coherent categorical forecasts and conclude by illustrating the effects of this aggregation on optimal decisions.

8.2   FORECAST CALIBRATION

In evaluating the performance and quality of a forecast what is important is the relationship between the forecast and the random event being predicted. The basic source of information to measure performance is therefore the joint distribution of the forecast and the random event. From this, summary measures can be derived. In this section our concentration is on calibration, a concept that applies to both categorical and probability forecasts but in quite different ways.

8.2.1   Calibration of Categorical Forecasts

We begin by considering categorical forecasts with square matrices $\mathbf{P}$ and $\mathbf{F}$. Recall that $\mathbf{P}$ has (i,j)th element equal to $Pr\{X=j|F=i\}$, and $\mathbf{F}$ has (i,j)th element equal to $Pr\{F=i|X=j\}$. A categorical forecast is said to be *perfect* when $\mathbf{P}$ and $\mathbf{F}$ are identity matrices, that is, when

$$Pr\{X=i|F=i\} = Pr\{F=i|X=i\} = 1,$$

and

$$Pr\{X=j|F=i\} = Pr\{F=i|X=j\} = 0 \text{ when } i \neq j.$$

These equations hold only if the forecast is correct every time. For real-world forecasts it is unrealistic to expect this to occur. Every forecaster knows that forecasts correctly predict outcomes only a certain fraction of the time. But even though it is unrealistic, the concept of a perfect forecast helps point the way to defining a highly desirable property for a categorical forecast to have. The law of total probability tells us that $\mathbf{p}_X = \mathbf{p}_F\mathbf{P}$ and $\mathbf{p}_F = \mathbf{p}_X\mathbf{F}^T$ so that in the case of perfect categorical forecasts where both $\mathbf{P}$ and $\mathbf{F}$ are identity matrices, $\mathbf{p}_X$ must always equal $\mathbf{p}_F$. But the marginal distributions of F and X can be equal, even though $\mathbf{P}$ and $\mathbf{F}$ are not identity matrices. This leads us to the following definition:

> A *well-calibrated* categorical forecast is one where the fraction of time that the outcome is forecast to be in a certain category will equal the fraction of time the outcome actually falls into that category. Mathematically, a categorical forecast is said to be well calibrated if $\mathbf{p}_F = \mathbf{p}_X$.

Let us look again at the minimum temperature example introduced in Chapter 3 with the data reproduced in Table 8.1. Note that $\mathbf{p}_F$ and $\mathbf{p}_X$ are (0.100, 0.300. 0.400, 0.200) and (0.090, 0.282, 0.403, 0.225), respectively. Clearly, the forecast is not quite well calibrated. The forecasts predict "−5° or below" and "−4°–0°" more frequently and the higher temperature ranges less frequently than they actually occur.

8.2.2 Calibration of Probability Forecasts

With a probability forecast the concept of a perfect forecast makes little sense because its output is itself a probability statement. The $\mathbf{P}$ matrix in this case has n rows and 2 columns. Only in the first and last rows can we have a value of 1 or 0 because only in these can we predict an outcome with probability 1 or 0. All other rows have entries between 0 and 1. The decision maker is presented with probability statements that he or she uses in a decision tree to help choose among alternative decisions. If the frequency of actual outcomes were close to the probability statements used by the decision maker, he or she would feel confident about using the forecasts; if the frequency of actual outcomes were very different from the probability statements, the decision maker would not want to rely on the forecasts. What we want to measure is the reliability of the forecast to the decision maker. We therefore make the following definition:

TABLE 8.1
Data for minimum temperatures

Forecasted minimum temperature		Actual minimum temperature category j				
Category i	Interval	1	2	3	4	Total days
1	−5° or below	75	20	5	0	100
2	−4°–0°	15	240	30	15	300
3	1°–5°	0	20	360	20	400
4	6° or above	0	2	8	190	200
Total days		90	282	403	225	1,000

A probability forecast is *well calibrated* when conditional expected outcomes, given the forecast, are equal to the probability statement in the forecast. Mathematically, a probability forecast is well calibrated when

$$Pr\{X = 1|F = f(i)\} = f(i), \quad i = 1, 2, \dots, n,$$

where $f(i)$ is a probability statement of event outcome.

The forecast is well calibrated when the first column of the decision probability matrix **P** and the forecast probability statements are identical. For example, a forecaster using the states defined in Table 3.1 (p. 91) would be well calibrated if and only if it never rained when "No Rain" was predicted, it rained on 25% of the occasions when "Rain Unlikely" was predicted, it rained on 50% of the occasions when "Moderate Chance" was predicted, it rained on 75% of the occasions when "Rain Likely" was predicted, and it always rained when "Rain Certain" was predicted. To state it another way, a forecast is well calibrated if the relative frequency of the event of interest is equal to the probability value that defines the forecast. The observed outcomes in Table 3.1 show that those forecasts are not well calibrated; columns 2 and 3 are not equal.

A simple and powerful check on calibration of a probability forecast is to plot the $p_{X|F}(1|f(i))$'s against the $f(i)$'s and compare the result with a straight line having a 45° slope; this is called a calibration plot. The difference between the relative frequency of occurrence and the forecast probability is an indication of the degree of miscalibration. The power and usefulness of this graphical check cannot be overemphasized. Figure 8-1 shows the plot for the rain example from Table 3.1.

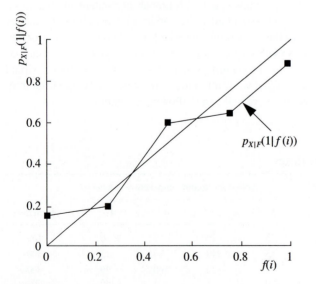

FIGURE 8-1
Calibration plot for the rain probability forecast.

When the probability forecasts are well calibrated, the first column of $\mathbf{P}$ agrees with the probability statements themselves, and we denote this special probability matrix by $\mathbf{P}^{(w)}$. With $\mathbf{P}^{(w)}$ and $\mathbf{p}_X$ known, the results of Section 3.4 tell us that any $\mathbf{p}_F$ that satisfies

$$\mathbf{p}_F \mathbf{P}^{(w)} = \mathbf{p}_X$$

$$\mathbf{p}_F \mathbf{1} = 1,$$

$$\mathbf{p}_F \geq \mathbf{0}.$$

is a feasible distribution of the probability statements. [1]

It is common practice to use linear regression to obtain probability forecasts from data. One can show that when one uses probability forecasts derived from linear regression, if the forecasts are well calibrated, the regression line passes through the origin with unit slope, in which case the standard deviation of the forecast is equal to the product of the standard deviation of the event outcome times the correlation coefficient (between forecast and event). Because the correlation is less than or equal to 1, the calibrated forecast has a variance that is less than or equal to the variance of event observations (see Problem 8.1).

8.2.3 Calibration in Expectation

A less restrictive form of calibration is calibration in expectation. This requirement is simply that "on average" the forecaster be well calibrated or unbiased. Mathematically, we say that a forecast is calibrated in expectation if $E[F] = E[X]$. For categorical forecasts $E[F] = E[X]$ only when

$$\sum_{i=1}^{n} i p_F(i) = \sum_{j=1}^{n} j p_X(j) \tag{8.1}$$

or equivalently

$$\sum_{i=1}^{n} i \sum_{j=1}^{n} p_{F|X}(i|j) p_X(j) = \sum_{j=1}^{n} j \sum_{i=1}^{n} p_{X|F}(j|i) p_F(i). \tag{8.2}$$

For the minimum temperature example of Table 8.1, we find that $E[F]$ is 2.70 and $E[X]$ is 2.76; on average the probability forecast is a good one, even though it is not perfectly calibrated.

In the two-category case where both F and X take on only values 1 and 0, Equations (8.1) and (8.2) simplify to

---

[1] The vector $\mathbf{1}$ is a column vector all of whose entries are 1.

$$p_F = p_{X,}$$

$$f_1 = p_1, \qquad\qquad (8.3)$$

$$f_0 = p_0.$$

This shows that in two-state categorical forecasts calibration in expectation is equivalent to being well calibrated. We stress that this equivalency holds *only* in the two-state case.

In probability forecasts X takes on the value 1 if the event of interest occurs and 0 if it does not. So a probability forecast that is calibrated in expectation must have

$$E[F] = \sum_{i=1}^{n} f(i)\, p_F(f(i)) = p_X(1) = E[X]. \qquad (8.4)$$

Because a well-calibrated probability forecast has $f(i) = p_{X|F}(1|f(i))$, we see that it is also calibrated in expectation, but the reverse is not necessarily true. For the rain example in Table 3.1 (p. 91) for which $\mathbf{p}_F = (0.1,\ 0.2,\ 0.4,\ 0.2,\ 0.1)$ and $\mathbf{p}_X = (0.515,\ 0.485)$, Equation (8.4) shows that $E[F]$ and $E[X]$ are 0.500 and 0.515, respectively, so that it is neither well calibrated nor calibrated in expectation.

A well-calibrated forecast does not by itself ensure a useful forecast or one that provides economic value, as we demonstrate with the following simple example. Suppose that you live in an area of the country where rain occurs on 25% of the days of the year; defining $X = 1$ if it rains and 0 if it does not, $\mathbf{p}_X = (0.25, 0.75)$. Suppose that a forecast is made each day simply by choosing a random number from 1 to 100; on any day when a number is chosen between 1 and 25 the forecast calls for rain, whereas if the number is between 26 and 100 it calls for fine weather. The vector $\mathbf{p}_F$ will equal $\mathbf{p}_X$, and the forecast is therefore well-calibrated even though there is clearly no correlation between the actual weather and the forecast on a given day.

This example suggests that although calibration is an important property of a forecast, we must consider additional properties that are useful and desirable. The concepts of forecast discrimination and correlation are covered in the next three sections.

8.3   FORECAST DISCRIMINATION

Calibration measures the degree to which outcomes actually occur given that we forecast their occurrence. Another major concern for a decision maker is the degree to which a forecast can discriminate well between different predictions; for example, do scores developed from credit records of individuals known to be good differ significantly from those known to be bad? Do probability forecasts of "Below freezing weather" associated with actual "Below freezing" records differ significantly from the forecasts associated with "Above freezing" records? As we shall see, there is often a trade-off between probability forecasts that discriminate well and forecasts that are well calibrated. As with calibration, the definition of discrimination is different for probability and categorical forecasts. We begin by considering probability forecasts.

8.3.1   Discrimination in Probability Forecasts

Let us assume in our probability forecast of rain that we have access to historical data records of forecasts and actual outcomes of Rain and No Rain such as are shown in Table 3.4 (p. 98) for which the **F** matrix is

$$\mathbf{F} = \begin{bmatrix} 0.175 & 0.021 \\ 0.252 & 0.144 \\ 0.466 & 0.330 \\ 0.078 & 0.330 \\ 0.029 & 0.175 \end{bmatrix}. \tag{8.5}$$

The first and second column gives us the conditional probability that a probability statement $f(i)$ was made, given rain occurred and rain did not occur, respectively. The column frequencies add to 1. In general, these columns differ, and if we plotted the conditional probabilities $p_{F|X}(f(i)|x)$ as a function of the probability statements $F = f(i)$ separately for $x = 1$ and $x = 0$, we would obtain two pmf's which have different shapes, and that in some sense are separate from each other. With good discrimination the high probability statements are more likely to have been forecast given that the event occurs than they would be if the event had not occurred. These observations lead us to the following definition:

> A probability forecast is said to be *perfectly discriminating* if an $f(k)$ can be found that partitions the rows of **F** into two nonoverlapping subsets such that $Pr\{F \le f(k)|X = 1\} = 0$ and $Pr\{F > f(k)|X = 0\} = 0$.

A perfectly discriminating probability forecast has the property that the product of terms in each row of the **F** matrix equals 0, that is, $p_{F|X}(f(i)|1)p_{F|X}(f(i)|0) = 0$ for all i. This means that the occurrence of the event could never lead to the same probability statement as the nonoccurrence of the event and there would never be any confusion as to which probability statement was made. Obviously, our probability forecast of rain with the **F** matrix in Equation (8.5) is not perfectly discriminating. An example of a perfectly discriminating forecast using the same probability statements is

$$\mathbf{F} = \begin{bmatrix} 0.4 & 0.0 \\ 0.3 & 0.0 \\ 0.3 & 0.0 \\ 0.0 & 0.5 \\ 0.0 & 0.5 \end{bmatrix}. \tag{8.6}$$

It is common practice in many engineering and business settings to illustrate discrimination by plotting the conditional cdf $Pr\{F \le f(k)|X = 0\}$ on the vertical axis against the conditional cdf $Pr\{F \le f(k)|X = 1\}$ on the horizontal axis. The ensuing curve is known as a Lorenz or receiver operating characteristic (ROC) curve; it is often used in signal detection theory to show what fraction of real signals a particular receiving de-

TABLE 8.2
Conditional cdf's for Equations (8.5) and (8.6)

Probability statement $f(i)$	Rain probability forecast		A perfect discriminating forecast	
	$Pr\{F \leq f(i)|X = 0\}$	$Pr\{F \leq f(i)|X = 1\}$	$Pr\{F \leq f(i)|X = 0\}$	$Pr\{F \leq f(i)|X = 1\}$
1.00	1.000	1.000	1.0	1.0
0.75	0.979	0.825	1.0	0.6
0.50	0.835	0.573	1.0	0.3
0.25	0.505	0.107	1.0	0.0
0.00	0.175	0.029	0.5	0.0

vice identifies correctly as a function of the fraction of spurious signals identified as real signals. Starting with the **F** matrices in Equations (8.5) and (8.6) we find the partial column sums, which are shown together with the probability statements in Table 8.2. The results are shown plotted in Figure 8-2, where the probabilities in columns 3 and 2 are plotted on the horizontal and vertical axes, respectively, for each $f(i)$, as are those in columns 5 and 4.

No discrimination occurs when both columns in the **F** matrix are identical, that is, when records of actual data indicate that a given forecast probability statement was forecast the same fraction of time no matter whether the event did or did not occur. In this case the ROC curve is a 45° line as shown in Figure 8-2 and the marginal distribution $\mathbf{p}_F$ would be equal to either column vector in the **F** matrix. Using Bayes' rule one can show that when both columns of **F** are identical, all rows of the **P** matrix are identical and each is equal to $\mathbf{p}_X = [p_X(1), p_X(0)]$. Thus, every probability statement yields the same conditional probability of observing the event. This means that $p_{X|F}(1|f(i)) = p_X(1)$ for every $f(i)$ and the conditional forecasts cannot improve on baseline forecasts. Recall that a plot of $p_{X|F}(1|f(i))$ against $f(i)$ is called a calibration plot (see Figure 8-1). For a

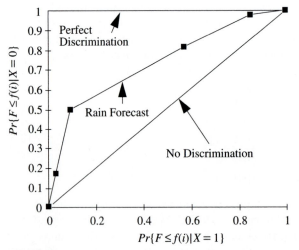

FIGURE 8-2
ROC curve for three probability forecasts.

well-calibrated forecast this is a straight line with a slope of 45°. For a probability forecast that does not discriminate the ROC curve is a 45° line and the calibration plot is a horizontal line with constant value $p_X(1)$. So not only do we have no discrimination, we also have calibration.

With no discrimination the ROC curve yields a straight line from (0,0) to (1,1), whereas if there is perfect discrimination, the plot jumps vertically from (0,0) to (0,1), then follows a horizontal line from (0,1) to (1,1). One way of comparing the discrimination of two forecasts of the same event is to measure the area between each forecast's ROC curve and the diagonal line and compare it with 1/2, which is the total area in the triangle between the diagonal line and the ROC curve of a perfectly discriminating forecast. This ratio is called the Gini coefficient and is often used in the economics and statistical literature to provide a single scalar measure for comparison. However, there is no substitute for looking at the joint distribution of F and X as it contains all distributional information.

The ROC curve in Figure 8-3 is a plot of the cumulative fraction of bads versus the cumulative fraction of goods for the credit data discussed in Section 3.7. Note that we have not plotted the cumulative distributions from their origin at (0,0) but only shown the interesting part of the curve that tells us that most of the bads are found before too large a fraction of goods is detected. This discriminatory power is important when we adopt policies that attempt to exclude or reject bads in the decision problem. The graph is obtained by plotting the partial sums (i.e., the cumulative distributions) from columns 6 and 7 in Table 3.11 (p. 123).

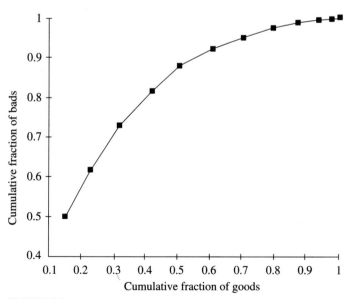

FIGURE 8-3
ROC curve for credit portfolio.

8.3.2 Discriminating Categorical Forecasts

For a categorical forecast we would prefer that, over repeated uses of the forecast, when a certain category occurred it was predicted to occur more often than when it did not occur. This leads us to the definitions:

> A categorical forecast is said to be *discriminating* if, for every category i, $Pr\{F = i|X = i\} > Pr\{F = i|X = j\}$ for every $j \neq i$. A categorical forecast is said to be perfectly discriminating if for every category i, $Pr\{F = i|X = i\} = 1$ and $Pr\{F = i|X = j\} = 0$ for every $j \neq i$.

The last part of the statement means that for a perfectly discriminating forecast **F** must be an identity matrix.

For the general case of many categories there is no two-dimensional plot such as an ROC curve to graphically demonstrate discrimination in a categorical forecast. One should look at the **F** matrix of forecast likelihoods for a given forecast; the closer this is to an identity matrix, the more discriminating the forecast. For example, the matrix in Equation (3.21) (p. 100) shows that the four-category minimum temperature forecast has a high degree of discrimination. We return to measuring discrimination in the general case at the end of this section, but first consider the two-category case where discrimination can quite easily be measured.

Suppose that in a population of objects (people, institutions, electronic signals) there is a subset of particular interest. For example, in order to determine how people in a given geographical area with specific characteristics will respond to a marketing campaign, we must attempt to discriminate between different populations. The insurance business may wish to identify a group of clients that offer low casualty risk and who therefore can be offered discounts and lower premiums in their insurance policies. The military may wish to identify hostile aircraft in a mixture of military and civilian aircraft that appear on a radar screen.

A bank may wish to identify individuals that are good credit risks and therefore qualify for special authorization limits or payment terms on their credit line; others may have a history of not making timely payments on their loans and may therefore be more likely to be in default of payments and therefore a bad risk. Although the definitions may be difficult to make one can "identify" good and bad credit risks from a population of individuals and test a particular classification scheme. In medical applications it is common practice to test the effectiveness of new drugs on people by considering individuals with and without a given disease who have or have not been treated with the drug. When one can actually determine object type in the population or the historical data, a typical method of evaluating a forecast is to test its performance against known object types. This is typically done in a laboratory, on a test range, in a test market, or by examining historical data records on performance. The system (possibly a combination of one or more forecasting procedures) will identify some targets correctly and some incorrectly, as is also the case with nontargets.

Suppose that the following forecast likelihoods are estimated from data:

$$Pr\{F = 1|X = 1\} = f_1 \qquad \text{Correct identification of event,}$$

$Pr\{F = 1 | X = 0\} = f_0$ False Positive,[2]

$Pr\{F = 0 | X = 1\} = (1-f_1)$ False Negative,[3]

$Pr\{F = 0 | X = 0\} = (1-f_0)$ Correct identification of no event.

The term *discriminating* is used in the same manner as it was earlier to describe a sensor or forecast that identifies a target correctly more often than it identifies a nontarget as a target. Mathematically, the two-category forecast is discriminating when $f_0 < f_1$, so we can use $f_1 - f_0$ to quantify the amount of discrimination. Thus, a two-category forecast is perfectly discriminating only when $f_1 = 1$ and $f_0 = 0$. The reader should check that in this case the (2×2) **F** matrix is an identity matrix.

One can show that the inequality $f_0 < f_1$ implies an ordering of the corresponding decision probabilities: $p_0 < p_1$ (see Problem 8.2). Because

$$p_X = p_0(1 - p_F) + p_1 p_F,$$

p_X is a convex combination of p_1 and p_0. Thus, for a discriminating two-category forecast the following ordering holds:

$$0 \le p_0 \le p_X \le p_1 \le 1. \tag{8.7}$$

We find that discrimination in forecasts is closely tied to the structure of the optimal decisions that are a direct result of the decision probabilities derived from forecasts and used by the decision maker.

In Section 3.4.3 we showed the relation between the forecast likelihoods, the decision probabilities, and the marginals p_X and p_F (see Equation (3.17) (p. 99)). It is shown that there is no reason to assume that the sensor predicts the same fraction of targets as are present in the population. However, the sensor may be a type where it predicts an object as a target whenever it receives a signal (e.g., electromagnetic or acoustic) above a certain threshold, in which case it might be possible to "calibrate" the sensor and thus change the relative values of f_1 and f_0. Recall that categorical forecasts are calibrated in expectation when $p_X = p_F$. This states that, when used against a certain population, the forecasts on average predict the same fraction of targets as there are targets in the population.

Recall that for the two-category case p_X, p_F and the forecast likelihoods are related by

$$p_F = f_1 p_X + f_0(1 - p_X).$$

Using this expression it is straightforward to show that the forecast is calibrated in expectation if and only if

---

[2] In medical and military applications this is often referred to as a false alarm.

[3] In military applications this is referred to as a leaker because a real target escapes detection.

$$\frac{f_0}{(1 - f_1)} = \frac{p_X}{(1 - p_X)} .$$
(8.8)

In words, Equation (8.8) states that the sensor is calibrated only when the ratio of the probability of a false positive to the probability of a false negative is equal to the odds of an object in the population being a target, or because $p_X = p_F$, the odds of the sensor identifying an arbitrary object as a target. Equation (8.8) shows that with categorical forecasts an imperfect sensor (one with $f_1 < 1$ and/or $f_0 > 0$) can be calibrated only against a single-population mix. If it is calibrated against a mix that contains 30% target objects, it will not be calibrated against any other mix. Although calibration is an important characteristic, we shall see that a forecast that on the average overpredicts or underpredicts the presence of a target may in fact have substantial advantages in a given decision problem over one that is calibrated.

We end this section by returning to the problem of how to measure discrimination in n-category forecasts. Results for the two-category case suggest finding a way to combine all categories other than a particular one of interest. Suppose that category i is chosen. Instead of making $n-1$ comparisons of $Pr\{F = i | X = i\}$ with $Pr\{F = i | X = j\}$ for every $j \neq i$, we combine these latter probabilities using the laws of probability into

$$P\{F = i | X \neq i\} = \sum_{j \neq i} \frac{P\{F = i | X = j\} Pr\{X = j\}}{\sum_{j \neq i} Pr\{X = j\}} .$$
(8.9)

Now we can make the single comparison of $Pr\{F = i | X = i\}$ with $Pr\{F = i | X \neq i\}$. This is analogous to comparing f_1 with f_0 in the two-category case.

Table 8.3 gives the results of using Equation (8.9) for the minimum temperature forecast shown in Equations (3.19) and (3.21) (p. 100). The reader is encouraged to check these results (see Problem 8.3). Although not a perfectly discriminating forecast, the probabilities in Table 8.3 indicate considerable discrimination. By again using the laws of probability, one can demonstrate consistency by combining columns 2 and 3 using the weights $Pr\{X = i\}$ and $Pr\{X \neq i\}$ to obtain the distribution of the forecasts in Equation (3.18) (p. 99).

TABLE 8.3
Calibration of minimum temperature forecast

Category i	$Pr\{F = i | X = i\}$	$Pr\{F = i | X \neq i\}$
1. (−5° or below)	0.833	0.027
2. (−4° − 0°)	0.851	0.084
3. (1° − 5°)	0.893	0.067
4. (6° and above)	0.844	0.013

8.4   COMPARING DISCRIMINATION AND CALIBRATION

We use two numeric examples to illustrate that a forecast may not exhibit both good discrimination and good calibration. In the first forecast we have perfect discrimination and large miscalibration, and in the second we have less than perfect discrimination but well-calibrated forecasts. Both forecasts attempt to predict the occurrence ($X = 1$) or non-occurrence ($X = 0$) of an event; we assume that without obtaining a forecast our prior probability is $p_X = Pr\{X = 1\}$, which for simplicity in this section we simplify to p. We shall see the role of p in our examples.

Our first example is summarized in Figure 8-4, where the top row shows a set of forecast likelihoods **F** that are associated with the probability statements $\mathcal{F} = \{0.8, 0.6, 0.4, 0.2\}$. These summarize the performance of the forecaster; when the event of interest occurred, 70% of the time the probability statement 0.8 was forecast, and 30% of the time 0.6 was forecast. When the event did not occur, 20% of the time the probability statement 0.4 was forecast and 80% of the time 0.2 was forecast. This is an example of perfect discrimination. There are no two data points, one where the event occurred and one where it did not, in which the same probability statement is recorded for both. Using the equation $\mathbf{p}_F = \mathbf{F}\mathbf{p}_X^T$ the vector $\mathbf{p}_F$ is calculated. Equation (3.12) (p. 97)

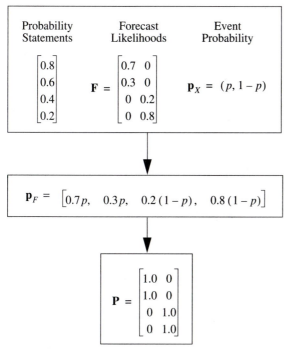

FIGURE 8-4
Perfect discrimination and poor calibration.

is then used to find the corresponding decision probabilities. Alternatively, we can argue the result in another way; whenever there is a zero in one of the entries in **F**, Bayes' rule tells us that the entries in the same row and column location of the **P** matrix will also be zero. Therefore, the other entry in the two-column **P** matrix *must* be equal to 1. If the baseline forecasts of the event are $(p, 1 - p)$ we get the same **P** matrix for every value of p. In other words, *the decision probabilities are independent of the baseline forecast.* This occurs only for perfectly discriminating forecasts. A well-calibrated forecast would have $p_{X|F}(1|f(i)) = f(i)$, but in this forecast $p_{X|F}(1|f(i)) = 1$ or 0 depending on whether the low or high probability statements were forecast. Thus, the decision probabilities are very badly calibrated while the likelihoods in the **F** matrix offer perfect discrimination. Although this example shows that perfect discrimination leads to very poor calibration, it does not follow that a well-calibrated forecast always has poor discrimination.

Our second example, shown in Figure 8-5, has the same probability statements and $\mathbf{p}_F$ vector as the first example, but when used with a prior probability p equal to 0.5, illustrates a well-calibrated **P** matrix. The likelihoods are shown in the **F** matrix; if we plot $p_{F|X}(f|0)$ as a function of the cumulative distribution $p_{F|X}(f|1)$, it is clear that we no longer have perfect discrimination. The ROC curve lies between the straight line with slope of unity and the right-angled curve of a perfect discriminator. Thus, we see that we have obtained a well-calibrated forecast at the expense of perfect discrimination.

If i and j are two different outcomes of an event in a categorical forecast, the forecast likelihoods $p_{F|X}(j|j)$ and $p_{F|X}(j|i)$ should be very different, as should the pair $p_{F|X}(i|i)$ and $p_{F|X}(i|j)$ if we are to have good discrimination. For each pair the first probability

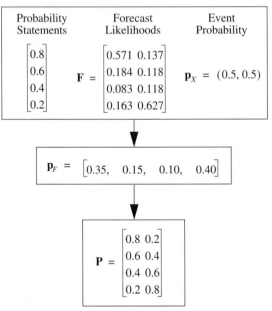

FIGURE 8-5
Discrimination for a well-calibrated forecast.

should be close to 1 and the second close to 0. For the two-category Bernoulli case this is easy to check. Using the notation in Equations (3.22) and (3.23), the larger the difference between f_1 and f_0, the more discriminating the forecast. Of course, if we had perfect discrimination in the probability forecasts used to determine the categorical forecasts, then f_1 should be equal to 1 and f_0 to 0 for the $f(k)$ that partitioned the probability statements.

8.5   FORECAST CORRELATION

A commonly used measure of dependence between two random variables is their correlation coefficient; when the two random variables of interest are the forecast F and the event outcome X, we denote the covariance by

$$Cov[F,X] = E[FX] - E[F]E[X]. \qquad (8.10)$$

Denote the variances of F and X by σ_F^2, and σ_X^2, respectively, and their correlation coefficient by ρ_{FX}. Then,

$$\rho_{FX} = Cov[F,X]/\sigma_F\sigma_X. \qquad (8.11)$$

It is well-known that the correlation coefficient always lies between -1 and $+1$, with 0 indicating no correlation between the random variables. The more frequently the forecast and event outcome agree or disagree, the closer the correlation coefficient will be to 1 or -1, respectively. It is also well-known that ρ_{FX} is invariant to changes in scale of the random variables F and X (see Problem 8.6).

Some insight can be gained by studying ρ_{FX} for the categorical forecast with two categories, that is, the Bernoulli (0,1) random variable case. Using the notation on page 312 for two-state categorical forecasts, Equation (8.11) can be written as

$$\rho_{FX} = (p_1 - p_0)\sqrt{\frac{p_F(1 - p_F)}{p_X(1 - p_X)}} \qquad (8.12)$$

or

$$\rho_{FX} = (f_1 - f_0)\sqrt{\frac{p_X(1 - p_X)}{p_F(1 - p_F)}}. \qquad (8.13)$$

When the forecast is calibrated in expectation, $p_F = p_X$ so that

$$\rho_{FX} = p_1 - p_0 = f_1 - f_0.$$

We illustrate the use of these equations with the four forecast examples in Figure 3-3 (p. 88). For case (a) ρ_{FX} is $+1.0$, and for (c) -1.0. For case (d) it must be that $p_X = 0.8$ and that ρ_{FX} is 0. For case (b), if $p_X = 0.20$ (freezing weather occurs on average 20% of the time), $p_F = 0.143$, and both equations give $\rho_{FX} = 0.612$. Note that if $p_X = 0.45$ (freezing weather occurs on the average 45% of the time), $p_F = 0.357$ and $\rho_{FX} = 0.674$.

The reader should recall that in the preceding sections on forecast discrimination, for the two-category forecast this was measured by the distance between f_1 and f_0, that is, $f_1 - f_0$. We see in Equation (8.13) that this coincides with large correlations of F and X. This equation shows that for the two-category case one can think of correlation as incorporating both discrimination (in the $(f_1 - f_0)$ term) and calibration (in the term under the square root).

8.6 MEASURING FORECAST PERFORMANCE WITH BRIER SCORES

To gain some additional insight into forecast quality and performance, we consider measures of mean squared error. Even though a single measure cannot reveal the complete structure of forecast performance and quality, mean squared error (MSE) has a long history of common usage and affords us an opportunity to see how discrimination, calibration, and correlation are interrelated. To accurately measure forecast performance we should always return to the joint distribution of F and X.

If F is a probability forecast of an uncertain event X, the MSE, referred to in the forecast literature as the Brier score, is

$$E[(X - F)^2] = (E[X] - E[F])^2 + Var[X - F]. \qquad (8.14)$$

Clearly, this is zero only if the forecast and event outcome always agree. The larger it is, the larger the discrepancy between forecast and outcome. The first term on the right-hand-side of Equation (8.14) measures what is commonly known as the unconditional bias or miscalibration in expectation, whereas the rightmost term measures the unconditional variance between the forecasts and the outcomes. Although the bias term may increase at the expense of the variance (or vice versa), the trade-off between these two terms may not be obvious.

Mean squared error can be decomposed in several different ways, each one providing us different insight on how squared error depends on miscalibration or poor discrimination. It is possible to decompose the MSE by expressing the result in terms of expectations that are conditional on F. By adding and subtracting $E[X]^2$ on the right side of Equation (8.14) and noting that we can write $E[X] = E_F E[[X|F]]$ and $E[FX] = E_F[FE[X|F]]$, we find that

$$E[(X - F)^2] = Var[X] + E_F[(E[X|F] - F)^2] - E_F[(E[X|F] - E[X])^2]. \qquad (8.15)$$

Consider the three right-hand terms separately. The first one is the unconditional variance of the random event, is independent of the forecast and is therefore the same for every forecast. The second term measures the expected squared error of miscalibration. This term would be identically zero if F were a well-calibrated probability forecast; in general the term is positive and adds to the variance. The final term on the right is a measure of the resolution capability of the probability forecasts and tells us how much the conditional expectation differs from the unconditional expectation. The higher its value, the better the conditional forecast can discriminate from the baseline. Note also that it is the only term that can *reduce* the MSE. Even though there is significant miscalibration, the third term may be a very large positive term and may thereby offset the effects of miscalibration. In Section 8.3 it was shown that the worst discrimination occurs when $E[X|F] = E[X]$, in which case the third term would be zero.

Since squared error is symmetric in F and X, by interchanging them the decomposition can also be expressed in terms of the variance of forecast probability statements plus two terms, one of which is equal to $E_X[(E[F|X] - X)^2]$. This term measures the discrimination of the forecast conditional on known outcomes. Note that when $X = 1$, the smallest contribution arises when the probability statements are also close to 1, whereas if $X = 0$, one would like to see a high probability of a small forecast probability statement.

A third decomposition allows us to write the expected MSE in terms of both discrimination and calibration terms,

$$E[(X - F)^2] = E_F[(E[X|F] - F)^2] - E_F[(E[X|F] - E[X])^2]$$

$$+ 2E[FX] - (E_F[E[X|F]^2] + E_X[E[F|X]^2]). \tag{8.16}$$

What is interesting about this expression is that if we focus attention on well-calibrated probability forecasts, where $E[X|F] - F = 0$, $E[X] = E[F] = p_X(1)$ and $Var[X|F] = p_X(1)(1 - p_X(1))$, the Brier score simplifies to

$$E[(X - F)^2] = p_X(1) - E[F^2] = E[F] - E[F^2]. \tag{8.17}$$

Thus, Brier score for well-calibrated probability forecasts can never be zero unless there is no uncertainty in the event being forecast!

We emphasize that Brier score is only one measure of forecast quality and may not be appropriate for different users of a given forecast. If we had the luxury of designing a different forecast model for each decision maker, we would focus on the value of the forecast in achieving the objectives of each decision maker. But we do not usually have this luxury and seldom know the specific use to which our forecasts will be put or even the structure of the forecast-decision problem faced by an individual or institution. Thus, our continuing interest in monitoring forecast performance and quality and making such measures available to decision makers.

8.6.1 An Example

With the perfect discrimination probability forecasts in Figure 8-4, let $p_X(1) = E[X] = 0.5$, so that $Var[X] = 0.25$, and

$$E[(X - F)^2] = 0.35(1 - 0.8)^2 + 0.15(1 - 0.6)^2 + 0.10(-0.4)^2 + 0.4(-0.2)^2 = 0.07.$$

Thus, the Brier score is seen to be significantly smaller than the variance of X. The expectation and variance of the probability forecasts are $E[F] = 0.49$, $Var[F] = 0.07$. By using Equation (8.15) we see that the contribution of squared error from this very badly calibrated forecast is about one-fourth the variance of X even though the data records show that the forecaster provided perfect discrimination and the ability to have $E[X|F]$ differ from $E[X]$.

On the other hand if we look at the data in Figure 8-5 we see that the corresponding Brier score is now given by

$$E[(X - F)^2] = 0.50 - 0.31 = 0.18,$$

which means that the MSE has more than doubled even though the contribution due to calibration is reduced to zero. How has this increase come about? The answer lies in seeing that the discrimination terms involving $E_F[(E[X|F] - E[X])^2]$ have decreased. These terms, which are always subtracted from the right-hand side of Equation (8.15), are now reduced from 0.25 to 0.07 and offset the large decrease of MSE resulting from better calibration. The net effect is that the MSE more than doubles.

8.7   CALIBRATION EFFECTS IN DECISION MODELS

In this section we show how an uncalibrated probability forecast affects the optimal decisions in a decision model. To do this we first return to the crop protection model described in Section 3.6 (p. 116). Following this analysis we revisit the credit problem of Section 3.7 (p. 121).

8.7.1   Effect of Calibration on Crop Protection Policies

Assume that an expert forecaster provides the decision maker with one of the four freezing predictions from column 1 in Table 8.4 together with the corresponding probability statement from column 2; for this example $\mathcal{F} = \{0.8, 0.6, 0.4, 0.2\}$. The decision tree for this problem is identical to the one we used in Figure 3-16 (p. 117) and is reproduced in Figure 8-6. As before, the probability statements $f(i)$ and $1 - f(i)$ are assigned to the appropriate branches of the i th decision sapling in the decision tree and the decision maker solves a set of optimal decision problems each similar to the one used when only the baseline or climatological forecast was available.

When we studied this problem in Chapter 3 we assumed that data from historical forecasts were those shown in Table 3.9 (p. 119). Without knowledge of this information the probability statements were assigned to the branches of the decision tree and used by the decision maker to find the optimal policy. This resulted in the discontinuous optimal expected value curve shown in Figure 3-17 (p. 120). The discontinuities were caused by the fact that the forecast was not well calibrated.

Given historical data for a forecast, there are two approaches to correcting for miscalibration. The first of these is for the decision maker to take the forecast historical data as given but to use the actual distribution of forecast probability statements as the correct statements in place of the forecaster's statements themselves. In Table 3.9 the probability statements would become (0.85, 0.57, 0.45, 0.16). The effect of using these is to redefine the ranges used the optimal expected loss function, which becomes

$$l_P^* = r_2 \qquad \text{when } 0 < r_2/r_1 < 0.16,$$

$$= 0.07r_1 + 0.57r_2 \qquad \text{when } 0.16 < r_2/r_1 < 0.45,$$

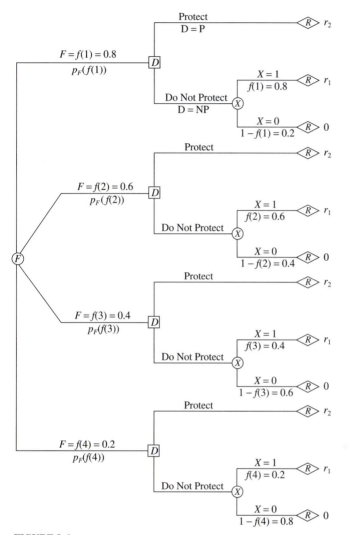

FIGURE 8-6
Crop protection decision tree using probability forecasts.

$$= 0.23r_1 + 0.21r_2 \quad \text{when } 0.45 < r_2/r_1 < 0.57,$$

$$= 0.30r_1 + 0.08r_2 \quad \text{when } 0.57 < r_2/r_1 < 0.85,$$

$$= 0.37r_1 \quad \text{when } 0.85 < r_2/r_1 < 1.0.$$

This function and the one from Figure 3-17 (p. 120) are shown plotted in Figure 8-7 with r_1 set to 1. The effect is to extend the lines for the uncalibrated forecast until a continuous function is obtained. The shaded triangles in this figure illustrate the reduction in the minimum expected loss that could be obtained by the decision maker if access to

TABLE 8.4
Historical data for the calibrated probability forecast

	Forecast outcomes		Temperature outcome			
Description of outcome i	$f(i)$	Below freezing $(X = 1)$	Above freezing $(X = 0)$	Total days	$E[X	F = f(i)]$
1: High chance	0.80	64	16	80	0.80
2. Good chance	0.60	78	52	130	0.60
3. Moderate chance	0.40	144	216	360	0.40
4. Low chance	0.20	86	344	430	0.20
Total days		372	628	1,000	1.00

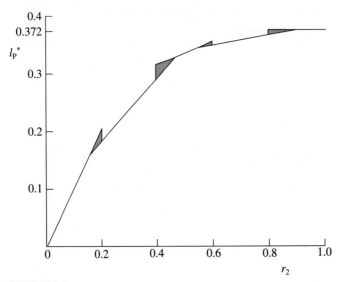

FIGURE 8-7
Probability forecast calibration by decision maker.

the historical records were available. The reduction is positive only in the range changes due to miscalibration.

In real applications, the $f(i)$ statements for the forecasts are usually determined prior to model building and data collection and are not changed when the data show that the forecasts are uncalibrated. It is more usual for the forecaster to adjust the forecast model to bring it into calibration even though the forecaster could have made the same recalibration as the decision maker. He or she is unlikely to change the probability statements to those in column 6 of Table 3.9 (p. 119). This is the second approach to correcting for miscalibration. Assume that this is done so that the data in Table 3.9 become that shown in Table 8.4. Note that the number of days on which freezing occurred is unchanged at 372, but the fractions in columns 2 and 6 now agree so that the forecast is calibrated. The optimal expected loss function,

$$l_p^* = r_2 \qquad \text{when } 0 < r_2/r_1 < 0.20,$$
$$= 0.09r_1 + 0.57r_2 \quad \text{when } 0.20 < r_2/r_1 < 0.40,$$
$$= 0.23r_1 + 0.21r_2 \quad \text{when } 0.40 < r_2/r_1 < 0.60,$$
$$= 0.31r_1 + 0.08r_2 \quad \text{when } 0.60 < r_2/r_1 < 0.80,$$
$$= 0.37r_1 \qquad \text{when } 0.80 < r_2/r_1 < 1.0.$$

is shown plotted in Figure 8-8. Again the result is a continuous function. This is achieved by vertically moving the various segments of the curve in Figure 3-17 (p. 120). The reader should check the above equations and confirm that they describe a continuous function on (0, 1). Note that if calibration is achieved by this second approach, there is a gain in the minimum expected loss in the regions when $0.20 < r_2/r_1 < 0.40$ and $0.60 < r_2/r_1 < 0.80$. One explanation for this is that the correlation between event and forecast is 0.38 for the data in Table 8.4, whereas it is 0.44 for the data in Table 3.9 (p. 119) when the fractions in column 6 are used as the probability statements (see Problem 8.8).

8.7.2 Uncalibrated Forecasts in a Credit Portfolio

In this section we revisit the credit scoring problem to better understand how forecast calibration and discrimination affect portfolio profits. To do so, we introduce a second population of credit applicants some of whom will be accepted and constitute a new portfolio that we refer to as the N portfolio. The D (development sample) portfolio is the original one described in Section 3.7 (p. 121) and is usually obtained from historical

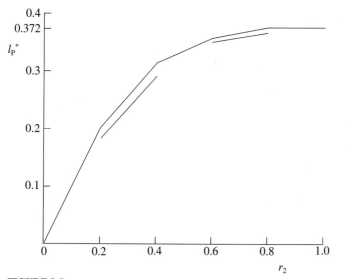

FIGURE 8-8
The expected cost of several probability forecasts.

data provided to the forecaster. He or she is responsible for developing a forecasting model but not for its later implementation in constructing and monitoring a new portfolio. Typically, the manager (decision maker) of the N portfolio uses decisions and cutoff rules based on what he or she believes to be well-calibrated probability forecasts developed by the forecasting expert from the historical development sample or D portfolio.[4] The data in Table 3.11 (p. 123) consisted of approximately 10,000 historical credit applicant records. The analysis led to a set of probability forecasts and a cutoff score $\bar{s}$ associated with the very simple decision rule: accept applicants with scores higher than or equal to $\bar{s}$ and reject all others. This cutoff rule is then applied to an incoming stream of applicants that make up the new N portfolio.

The sole purpose of the development sample is to develop, test, and evaluate a scoring procedure from historical records that typify the population of interest; applicants in the development sample may not represent an actual portfolio even though the data can be used to evaluate on paper what the structure and likely profits for the development sample would have been had a particular cutoff score and acceptance policy been adopted. The advantage in using the development sample as a test bed is that it can provide a simple and inexpensive way to study different policies and perform sensitivity analyses on expected returns. The disadvantage is that even if the forecasts are well calibrated for the development sample, they cannot reveal to the decision maker how or why the forecasts may become uncalibrated or may discriminate poorly when applied to a new portfolio.

In credit portfolios the likelihood that a good or bad will yield a particular score may remain stable over time even though there may be a shift in the composition of the population, that is, the baseline forecast of a good. The reader should realize that the distribution of applicant scores in the new portfolio (denoted N) can be assessed long before the new performance data for goods and bads is known. Unfortunately, good/bad performance may not be known to the portfolio manager for as long as six months to a year. It may take this time for the credit institution to establish whether an individual is or is not meeting his or her credit obligations. Therefore, it should be common practice to monitor the distribution of scored and accepted applicants in the new portfolio and compare it with the distribution of scores obtained from the original development sample to see whether there is some suggestion of change in the composition of goods and bads. Although the problems of early inference of changing behavior of applicants is an important and difficult subject, we concern ourselves primarily with the miscalibration problem that arises when there is a shift in the prior probability (baseline forecast) of a randomly selected individual being good.

It is tempting to believe that the new incoming distribution of applicant scores, because it is observable, has a "causal" effect on the distribution of goods and bads and the likelihoods that affect decision probabilities and eventually portfolio performance; in fact, the reverse may be true. The decision rules, the likelihoods that reflect new behav-

---

[4] Throughout this and the following section probabilities, odds and expectations are conditioned on either the D or N portfolio.

ioral patterns and shifts in composition of the population will determine the new score distribution and eventually, through Bayes' rule, the decision probabilities. A portfolio manager and forecaster must distinguish between these influences and effects and not confuse observability with "causality."

8.7.3 Stable Likelihoods

When likelihoods are stable over time it is a straightforward matter to relate cutoff scores and expected profits of the N portfolio in terms of the expected profit for the D portfolio. Notice that Equation (3.41) (p. 123) together with Equation (3.28) (p. 107) show us how a change in the population odds affects the distribution of applicant scores, the decision probabilities, and the cutoff scores. Also, recall that the scores and probability forecasts of the development sample were designed so that they are well calibrated in the D portfolio with $p_{X|S}(G|s) = f(s)$ but that this is no guarantee they will remain so for the N portfolio.

In the N portfolio, we assume that the fraction of goods in the population decreases by slightly more than two percentage points from 0.944 to 0.920, which means that the population odds of a good has dropped from 16.6 to 1 to 11.5 to 1. In all other respects, the N portfolio is identical to the hypothetical D portfolio constructed from the development sample. Comparisons between the D and N portfolios are shown in the six columns of Table 8.5. The first two columns contain the raw and log-odds scores obtained in the original development sample of Tables 3.11 (p. 123) and 3.12 (p. 127), the only difference being that we have replaced the log-odds score by the probability $f(s)$ derived from Equation (3.40) (p. 122). The third and fourth column contain the marginal distribution of scores and the conditional probability of being a good for each score in-

TABLE 8.5
New credit portfolio with uncalibrated probabilities

Applicant score (s')	Probability forecast	Development sample (D) (from Table 3.12)		New portfolio (N) (new population of goods and bads)	
	$f(s)$	$p_S(s; D)$	$p_{X|S}(G|s; D)$	$p_S(s;N)$	$p_{X|S}(G|s;N)$
Below 170	0.831	0.166	0.831	0.174	0.774
170–179	0.916	0.083	0.916	0.084	0.883
180–189	0.931	0.091	0.931	0.092	0.903
190–199	0.951	0.099	0.951	0.098	0.931
200–209	0.959	0.094	0.959	0.094	0.941
210–219	0.974	0.094	0.974	0.093	0.962
220–229	0.982	0.095	0.982	0.093	0.974
230–239	0.986	0.087	0.986	0.085	0.980
240–249	0.991	0.078	0.991	0.076	0.987
250–259	0.992	0.059	0.992	0.057	0.988
260–289	0.998	0.043	0.998	0.042	0.997
Over 290	0.991	0.011	0.991	0.011	0.987

terval, obtained from columns 3 and 4 of Table 3.12. Even though the likelihoods for the N portfolio are the same as those shown in Table 3.11, we see that the small change in population odds has had a significant effect on the distribution of scores and the conditional probability of being a good (columns 5 and 6). We assume that the portfolio manager uses the decision probabilities and cutoff scores obtained from the development sample when accepting or rejecting applicants from the new population. If the population odds for N were known to the portfolio manager, a cutoff score that maximizes expected profits, although still a solution of Equation (3.42) (p. 124), would require the as yet unavailable decision probabilities of column 6, not the original ones obtained from the development sample in column 4.

To see precisely how the new population of applicants leads to uncalibrated decision probabilities, we express the posterior odds of a good, given a score s, in terms of the prior population odds times a likelihood ratio. The common $p_S(s)$ terms in numerator and denominator cancel so that we have

$$O(G|s;N) = \frac{p_{X|S}(G|s;N)}{p_{X|S}(B|s;N)} = \frac{p_X(G;N)}{p_X(B;N)} \times \frac{p_{S|X}(s|G;D)}{p_{S|X}(s|B;D)} . \tag{8.18}$$

The posterior odds for the N portfolio is the new population odds times the old likelihood ratio; by substituting the expression for the posterior odds in the development sample into Equation (8.18) we find that the posterior odds for the new portfolio is a constant less than or greater than 1 multiplied by the posterior odds from the development sample. The effect is shown in Figure 8-9, which compares the calibration curves for the N portfolio (bottom line) with the original development sample (top line). Although the original probability forecasts continue to be well calibrated at values close to 1, they become less well calibrated at smaller values. The N calibration line is obtained by rotating the D line about the point $(1, 1)$. Because the population odds has changed, it is clear from Bayes' rule that the score distribution and the decision probabilities must also change; thus using Equation (3.42) there will be a new optimal cutoff score that moves up or down depending on whether the fraction of goods in the population increases or decreases.

On taking logarithms of both sides of Equation (8.18), the score consists of the log of the population odds plus the log of the likelihood ratio,

$$s = log\,O(G|s;N) = log\,O(G|N) + log\,\frac{p_{S|X}(s|G;D)}{p_{S|X}(s|B;D)} . \tag{8.19}$$

If the likelihood ratio remains unchanged but only the population odds changes, then a new score for N would differ from the old by a constant that is a positive number if the population odds of a good increases and is negative if the population odds decreases. It is well-known in the credit industry that a plot of log-odds in the new portfolio versus score s obtained from the development sample is often approximated by a line parallel to the original shifted upward or downward by a constant amount depending on whether the odds of a good are increased or decreased. This effect is explained by Equation (8.19).

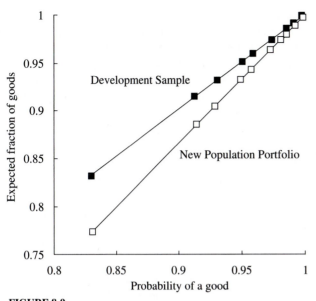

FIGURE 8-9
Calibration curves for D and N portfolios.

Denote the optimal expected profit of the development sample portfolio by $r_D{}^*$ and the expected profit of the new investment portfolio using the D portfolio cutoff score by r_N. Using Equation (3.46) (p. 126) the expected portfolio profit for the new portfolio can be written as the linear function

$$r_N = (l+g) \sum_{s > \bar{s}} [p_{X|S}(G|s;N) - \bar{f}] \, p_S(s;N), \qquad (8.20)$$

where the cutoff score $\bar{s}$ and indifference probability $\bar{f}$ are calculated from the development sample but are applied to the applicants for the new portfolio. Note that, for the reasons just described, r_N does not represent an optimal portfolio. Of course, the decision probabilities and the score distribution refer to the N portfolio. By adding and subtracting $r_D{}^*$ from the right-hand side of (8.20) we obtain an equivalent equation that can be written as

$$r_N = r_D{}^* + (l+g) \sum_{s > \bar{s}} [p_{X|S}(G|s;N) - f(s)] \, p_S(s;N)$$

$$+ (l+g) \sum_{s > \bar{s}} (f(s) - \bar{f}) \, [p_S(s;N) - p_S(s;D)]. \qquad (8.21)$$

Score distributions refer to both the development sample, $[p_S(s; D)]$, and the applicants in the new portfolio $p_S(s; N)$.

On the right-hand side of Equation (8.21) the first term is the optimal expected profit for a portfolio constructed from the development sample. The second term is the contribution from the miscalibration of the original probability forecasts with the decision probabilities in the new portfolio. The third term is the contribution due to the change in score distributions resulting from the shift in population of goods and bads. We should never assume that because a discriminating and well-calibrated forecast or score has been obtained from a large development sample, it follows that optimal policies derived from the D portfolio will yield optimal decisions and optimal expected profits in a new N portfolio.

8.7.4 Numeric Example

If we use the numbers from Table 3.11 (p. 123) and Table 8.5, we can get some idea of the importance of the contribution of miscalibration and other terms in Equation (8.21). Let us assume that the loss for each bad is $l = 1.00$ and the gain for each good is $g = 0.05$. These numbers mean that for every dollar lost on a bad we make 5 cents for each good. Obviously, the numbers can be scaled to represent hundreds or thousands of dollars, but in any case we would only want to select credit applicants with scores leading to very high probabilities of being a good.

Either of the last two terms in Equation (8.21) may be positive or negative, but, in practical applications, the miscalibration term is typically much larger than the latter, possibly because the changes in applicant score distributions are fairly small in the third term.

8.8 COHERENT CATEGORICAL AND PROBABILITY FORECASTS

There are situations where a decision maker may prefer to use a categorical rather than a probability forecast. For example, a decision maker may find that probability forecasts with values in certain ranges always lead to the same decision. As a result the decision maker becomes primarily interested in whether the probability forecast lies in a given interval rather than in knowing the detailed prediction or how it was derived. From the decision maker's point of view he or she is much more interested in what different decision rules the forecast may lead to rather than the structure of the forecasts themselves. We illustrate using a crop harvesting example with a decision sapling.

Suppose a farmer has a crop that appears ready for harvesting. Harvesting today will carry no risk but will result in an average crop return, r. Two more weeks of growth will lead to a 50% increase in harvest if the current weather conditions hold, but if rain comes in that time, it could result in a 30% loss. If we denote the (unknown) probability of rain by p, the farmer's decision problem is shown in Figure 8-10.

The optimal decision is to harvest immediately if and only if $p > 5/8$. Suppose that the farmer uses a forecast with the probability statements in Table 3.4 (p. 98), namely, $\mathcal{F} = \{1, 0.75, 0.50, 0.25, 0\}$, and that these are only information available to the farmer. The particular element of $\mathcal{F}$ forecast is what the farmer would have to use for p in Figure 8-10, and so would

Harvest Immediately   if $F = 1.00, 0.75,$
Wait Two Weeks   if $F = 0.50, 0.25, 0.$

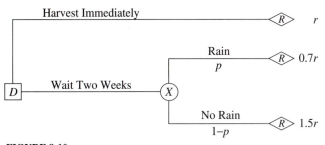

FIGURE 8-10
The farmer's harvesting decision.

The farmer would gain the same return from a simpler categorical forecast of "rain" or "no rain" where a "rain" forecast was equivalent to obtaining one of the probability statements 1 or 0.75. If the farmer had access to the forecaster's historical performance shown in Table 3.4 (p. 98) the value to use for p would be the number in column 3 divided by the number in column 5 for the row corresponding to the probability statement obtained in the forecast. Note that in this example the farmer's optimal policy would not change even if these decision probabilities were available, even though they do not coincide with the probability statements (why?).

We have shown that it is easy to obtain a categorical forecast from a probability forecast once the decision alternatives are known. The particular return values and risky venture structure will result in the decision maker setting a threshold, denoted $\bar{f}$, that will be an indifference probability obtained from a decision tree. For all $f(i) > \bar{f}$, we define one category and for all $f(i) \le \bar{f}$ the other.

If we have access to the more detailed data in Table 3.4, we know that Rain actually occurs 51.5% of the time, and the forecast likelihoods are given in matrix form in Equation (3.16) (p. 98). The element at the intersection of row i and first column of that matrix is $p_{F|X}(f(i)|1)$. If we use a categorical forecast "rain" to replace $f(i)$ whenever it is above $\bar{f} = 5/8$, then

$$Pr\{\text{"rain"}\} = p_F(1.00) + p_F(0.75) = 0.300,$$

$$Pr\{\text{"rain"}|X = 1\} = p_{F|X}(1.00|1) + p_{F|X}(0.75|1) = 0.427.$$

These show that we can simply add the appropriate elements of $\mathbf{p}_F$ and rows of $\mathbf{F}$ to obtain a new vector and new 2×2 matrix designated $\mathbf{p}_F^{(c)}$ and $\mathbf{F}^{(c)}$, respectively, for the equivalent categorical forecast.[5] We get

$$\mathbf{p}_F^{(c)} = (0.3, 0.7),$$

---

[5] We have used the notation $\mathbf{P}$ and $\mathbf{F}$ for both categorical and probability forecasts; in this section we use superscript (c) to emphasize that we have derived a categorical forecast from a probability forecast.

$$\mathbf{F}^{(c)} = \begin{bmatrix} 0.427 & 0.165 \\ 0.573 & 0.835 \end{bmatrix}.$$

Once we have calculated the appropriate likelihoods for this decision maker, we can then calculate the correct categorical decision probabilities that we designate by $\mathbf{P}^{(c)}$. Bayes' rule yields the result

$$\mathbf{P}^{(c)} = \begin{bmatrix} 0.733 & 0.267 \\ 0.422 & 0.578 \end{bmatrix}.$$

We point out that $\mathbf{P}^{(c)}$ could not have been obtained directly from the original $\mathbf{P}$ matrix associated with either the probability statements or the probability forecasts (Equation (3.15) (p. 98)). Clearly, the probabilities $p_{X|F}(j|f(i))$ for the probability forecasts cannot be added to obtain the $p_{X|F}(j|i)$ for the categorical forecasts. Figure 8-11 shows schematically the order in which the calculations involving Bayes' rule and aggregation must be carried out.

8.8.1   The Protect Decision with a Categorical Forecast

It is important for the reader to understand the different methods by which a threshold value can be obtained. We distinguish between those situations where there is good communication between the decision maker and the forecaster in providing categorical forecasts and those where the forecaster is left to decide how to modify a probability forecast independent of the decision maker.

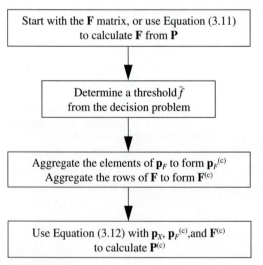

FIGURE 8-11
Calculating categorical from probability forecasts.

In the example we have just considered the farmer is aware of the operational effects of the forecast and so chooses for $\bar{f}$ the optimal cutoff value of 0.625. The farmer could then either continue to obtain the probability forecast and make a decision as to the best harvesting procedure based on this cutoff. Another situation that arises in practice is that the decision maker realizes that a forecast of "rain" or "no rain" is all that is required from a forecast and asks the forecaster to replace the probability forecast with this categorical forecast. In this case the forecaster may choose a threshold that is the "best" in the eyes of the forecaster but that is not optimal for the particular decision problem. For example, the forecaster may choose the cutoff threshold that leads to the two-category forecast that is closest to being calibrated in expectation; that is, the forecaster chooses $\bar{f}$ so that p_F is as close as possible to p_X.

From Equation (3.14) (p. 98) we see that in the region of interest it rains 51.5% of the time. Using the data in Table 3.4 the forecaster can make p_F equal 0.6 by using a cutoff threshold between 0.25 and 0.5. This will result in the two-category forecast best calibrated in expectation, but one that is not optimal for the farmer. The reader can show that the forecast probabilities become (see Problem 8.10)

$$\mathbf{P}^{(c)} = \begin{bmatrix} 0.657 & 0.343 \\ 0.183 & 0.817 \end{bmatrix} \tag{8.22}$$

and that this forecast results in a lower expected return to the farmer. The forecaster is doing a splendid job at forecasting, but it is hurting the client!

What is extremely important to realize about the use of categorical forecasts is that there are as many different $\mathbf{F}^{(c)}$ and $\mathbf{P}^{(c)}$ matrices as there are different thresholds or cutoffs where the optimal decision changes from protect to not protect. To convince yourself of this important feature, repeat the analysis of the decision problem described for the remaining threshold possibilities using the data in Table 3.4 (p. 98) (See Problem 8.11).

8.9   COHERENT AGGREGATION OF CATEGORICAL FORECASTS

In a decision problem where a categorical forecast is already available it may be that the categories being forecast are not directly useful for the decision problem. For example, suppose that the four-category forecast of minimum daily temperature in Table 3.3 (p. 94) is available to a decision maker but that the only critical factor about temperature in the particular problem is whether or not it exceeds freezing. All that is needed is the probability that the actual temperature will or will not be above or below 0° C. Thus, we need consider only a Bernoulli random variable that takes on value 1 if the event of interest occurs and 0 if it does not.

We illustrate coherent aggregation with an example. A general formulation would require more complex notation, but the basic ideas and use of Bayes' rule would be the same. The problem is how to aggregate correctly the probabilities in Equations (3.18), (3.19), (3.20), and (3.21) (p. 100).

Let Y be the new event random variable so that

$$Y = 1 \text{ if Minimum Temperature} \leq 0°,$$

$$= 0 \text{ if Minimum Temperature} > 0°.$$

In terms of the original outcome random variable X,

$$Y = 1 \text{ if } X = 1 \text{ or } 2,$$

$$= 0 \text{ if } X = 3 \text{ or } 4.$$

It is easy to see that the new marginal distribution of outcomes is [6]

$$p_Y(1) = p_X(1) + p_X(2),$$

$$p_Y(0) = p_X(3) + p_X(4),$$

and from Equation (3.19)

$$\mathbf{p}_Y = (0.372, 0.628).$$

To find the new decision probabilities from Equation (3.20), note first that $p_{Y|F}(1|i) = p_{X|F}(1|i) + p_{X|F}(2|i)$, $i \in \mathcal{F}$. If the forecast categories remain unaggregated, the new decision probabilities are

$$\mathbf{P}^{(T)} = \begin{bmatrix} 0.95 & 0.05 \\ 0.85 & 0.15 \\ 0.05 & 0.95 \\ 0.01 & 0.99 \end{bmatrix}. \tag{8.23}$$

At this stage the $\mathbf{P}^{(T)}$ matrix has four forecast categories and two outcomes so that it is not square.[7] Because forecasts in each 5° interval are not required, we should aggregate over the forecast categories as well.

Let G be the new forecast category (random variable) with

$$G = 1 \text{ if } F = 1 \text{ or } 2,$$

$$= 0 \text{ if } F = 3 \text{ or } 4.$$

Using arguments similar to those above, from Equation (3.18) the new marginal distribution of the forecast is

$$\mathbf{p}_G = (0.4, 0.6).$$

---

[6] The events $\{X = i\}$ for $i = 1,2,3,4$ are mutually exclusive, so that $Pr\{X = i \text{ or } X = j\} = Pr\{X = i\} + Pr\{X = j\}$, $i \neq j$.

[7] Superscript (T) is used to denote an intermediate stage in the calculation. Do not confuse this with a superscript T used to denote matrix transposition.

To calculate $Pr\{G=1|Y=1\} = Pr\{F=1|Y=1\} + Pr\{F=2|Y=1\}$, note that in the above $\mathbf{P}^{(T)}$ matrix we have $Pr\{Y=1|F=i\}$. So using Bayes' rule we obtain the forecast likelihood

$$p_{G|Y}(1|1) = \frac{p_{Y|F}(1|1)\,p_F(1)}{p_Y(1)} + \frac{p_{Y|F}(1|2)\,p_F(2)}{p_Y(1)}.$$

The reader can derive similar expressions for the other three probabilities in a new 2×2 aggregated $\mathbf{F}^{(A)}$ matrix of forecast likelihoods [8]

$$\mathbf{F}^{(A)} = \begin{bmatrix} 0.941 & 0.080 \\ 0.059 & 0.920 \end{bmatrix}.$$

Again we use Bayes' rule to obtain the aggregated decision probabilities

$$\mathbf{P}^{(A)} = \begin{bmatrix} 0.875 & 0.125 \\ 0.037 & 0.963 \end{bmatrix}. \tag{8.24}$$

Thus, if the forecast calls for the minimum temperature to be freezing or below, it will be freezing or below 87.5% of the time and above freezing 12.5% of the time. Figure 8-12 shows schematically the order in which the calculations must be carried out.

8.10 FORECAST AGGREGATION AND OPTIMAL DECISIONS

We return to the crop protection problem described in Sections 3.6 (p. 116) and in 8.7.1 (p. 320). In this section we compare the two-category forecast obtained in Section 8.9 with the original four-category forecast of temperature. The forecast outcomes are denoted $G=1$ for freezing weather, and $G=0$ otherwise. Recall that the decision probabilities for the two-category case are denoted $p_1 = Pr\{X=1|G=1\}$ and $p_0 = Pr\{X=1|G=0\}$. The decision tree for the problem using a two-category forecast is shown in Figure 8-13. Notice that from Equation (8.24) $p_1 = 0.875$, $p_0 = 0.037$. The optimal decisions are:

1. If freezing weather is forecast,
    $d^* = 0$ (protect)　　　　when $r_2/r_1 \leq p_1 = 0.875$,
    $d^* = 1$ (do not protect)　when $0.875 < r_2/r_1$.
2. If non-freezing weather is forecast,
    $d^* = 0$ (protect)　　　　when $r_2/r_1 \leq p_0 = 0.037$,
    $d^* = 1$ (do not protect)　when $0.037 < r_2/r_1$.

---

[8] In this section we use the superscript (A) to denote aggregated categories.

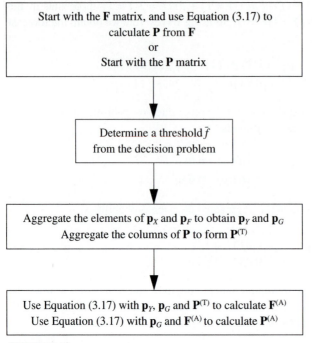

FIGURE 8-12
Aggregating categorical forecasts.

If we denote the minimum expected loss incurred by $l_{C2}{}^*$ (the subscript C2 indicating a two-category forecast), then

$$l_{C2}{}^* = r_2 \qquad \text{when } 0 < r_2/r_1 < 0.037,$$
$$= 0.022r_1 + 0.4r_2 \quad \text{when } 0.037 < r_2/r_1 < 0.875,$$
$$= 0.372r_1 \qquad \text{when } 0.875 < r_2/r_1 < 1.$$

For the case where $r_1 = 1$ this function is shown plotted in Figure 8-14 with a heavy line; the upper light line is the minimum expected cost using only the climatological or baseline forecast. Note that there is no value to the forecast when r_2/r_1 is larger than p_1 or smaller than p_0. When the cost of protection is very high compared to the potential crop loss ($r_2 > p_1r_1$), it never pays to protect. When the cost of protection is very low compared to the potential crop loss ($r_2 < p_0r_1$), it always pays to protect. The forecast has maximum value when r_2/r_1 is equal to the climatological forecast p_X. The more discriminating the forecast the longer is the interval in which there is economic value to the forecast.

Using Equation (8.12) (p. 317) the correlation coefficient ρ_{GX} is seen to be 0.85, indicating a highly correlated forecast. Because the forecast is not a probability forecast, the concept of being well calibrated does not generally apply. In fact, when there are only two categories with G and X both Bernoulli random variables, one can think of a categorical forecast as an uncalibrated probability forecast with $f(1) = 1$ and $f(0) = 1$. It

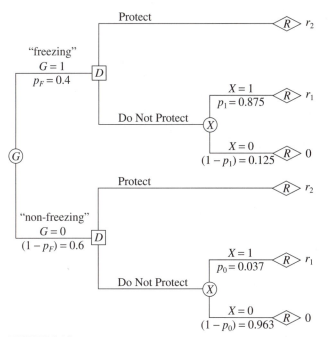

FIGURE 8-13
Crop protection with a two-category forecast.

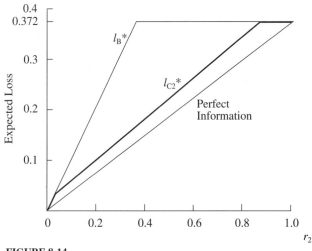

FIGURE 8-14
The value of a two-category forecast.

is stressed that this can only be done in the two-category case. From the definition in Section 8.2 the forecast is well calibrated when $E[X|G = f(i)] = f(i)$. Therefore, the two-category categorical forecast is well calibrated only when the forecast is perfect.

Because both X and G are Bernoulli random variables, we have

$$E[G] = p_G = 0.4 > 0.372 = p_X = E[X],$$

and it follows that the forecast is not calibrated in expectation. Although the forecaster tends to overpredict inclement (crop-damaging) weather, this does not invalidate the forecast. When comparing forecasts one might expect that the best-calibrated one would be preferred. This may not necessarily be the case depending on their discrimination.

We now apply the four-category forecast based on data in Table 3.5 and partially aggregated in Equation (8.23) (p. 332). The decision tree is shown in Figure 8-15 with four branches emanating from the forecast node, each terminating with a decision sapling. The reader may wonder whether there is any advantage to having a temperature forecast with four categories when crop damage depends only on whether or not the temperature drops to or below freezing. The details of the solution to this problem are left for Problem 8.14 at the end of this chapter; the minimal expected loss function using this forecast, denoted l_{C4}^*, is given by:

$$
\begin{aligned}
l_{C4}^* = r_2 && \text{when} && r_2/r_1 < 0.01, \\
= 0.002r_1 + 0.800r_2 && \text{when } 0.01 < r_2/r_1 < 0.05, \\
= 0.022r_1 + 0.400r_2 && \text{when } 0.05 < r_2/r_1 < 0.85, \\
= 0.277r_1 + 0.100r_2 && \text{when } 0.85 < r_2/r_1 < 0.95, \\
= 0.372r_1, && \text{when } 0.95 < r_2/r_1 < 1.
\end{aligned}
$$

The reader is encouraged to plot this function with $r_1 = 1$ in Figure 8-14 and compare it with the two-category forecast. You will see that for values of r_2/r_1 between 0.010 and 0.050 and between 0.850 and 0.950 the four-category forecast reduces the expected minimum loss. Only in these ranges is the added information in the four-category forecast of value. For all other values of r_2/r_1, the added information has no value.

Consider the discriminating power of the four-category forecast. Using the basic data in Table 3.5 (p. 99) one can find the joint distribution of F and X and hence the covariance (see Problem 8.7, part (b)). The variances of F and X are found from the probability vectors in Equations (3.18) and (3.19), respectively. Equation (8.11) is then used to show that the correlation coefficient $\rho_{FX} = 0.878$ is slightly higher than in the two-category case. Finally, the forecast is not quite calibrated in expectation because

$$E[F] = 2.700 < 2.763 = E[X].$$

8.11 SUMMARY AND INSIGHTS

In Chapter 3 various types of forecasts were discussed, with emphasis on probability, categorical, and odds forecasts. In this chapter emphasis is on how to measure the "quality" of a forecast. The three concepts introduced are calibration (including calibration in expectation), discrimination, and correlation. All three are important, and in two-state categorical forecasts one can see how correlation encompasses both calibration and

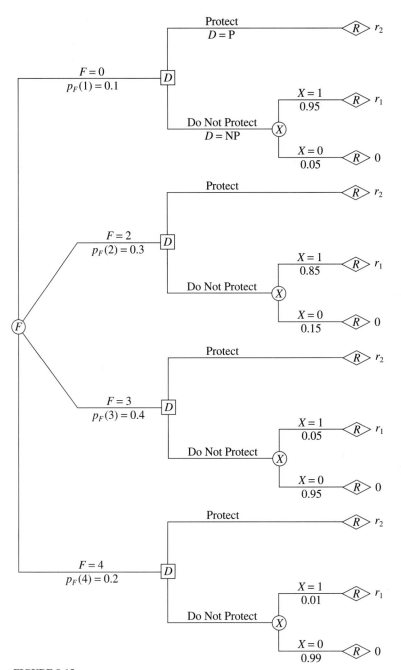

FIGURE 8-15
Crop protection for four-category forecast.

discrimination terms. It should be remembered that correlation is a quadratic measure whereas discrimination compares the distributional performance of a probability forecast.

A well-calibrated categorical forecast is one in which the probability distribution of forecast outcomes is the same as the distribution of the event being forecast. A well-calibrated probability forecast is one in which, for every forecast probability statement, the conditional probability that the event of interest occurs, given the forecast probability statement, is equal to the forecast probability statement. A less restrictive requirement is that a forecast be calibrated in expectation, or $E[F] = E[X]$.

Although calibration is a desirable quality for a forecast to have, we have shown that this property by itself is not sufficient to ensure that the forecast is useful or has economic value. To see this, assume there is some event of interest that can occur daily such as whether or not it rains. Suppose that in your part of the world in an average year it rains on 20% of the days. Now create a forecast by picking a random number each day that is 1 with probability 0.2 and 0 with probability 0.8, and use this to forecast "rain" the next day if it is 1 and "no rain" if it is 0. This two-category forecast is well calibrated, but on any given day any relation between the event and the forecast is purely coincidental. This simple example helps motivate the concept of discrimination.

Roughly speaking we call a forecast discriminating when the prediction agrees more often than not with the actual outcome. To define this rigorously we must use different definitions for probability and categorical forecasts. For a probability forecast a plot of the ROC curve shows how well it discriminates as the curve approaches 1 with a sharp knee. For a categorical forecast it is harder to demonstrate discrimination. The closer the $\mathbf{F}$ matrix resembles an identity matrix, the more the forecast discriminates. When comparing two forecasts it is often difficult to compare two $\mathbf{F}$ matrices. When there are only two categories, one need only look at the difference in the two forecast likelihoods. By using the laws of probability the $n \times n$ matrix of probabilities can be reduced to an $n \times 2$ matrix to allow for comparison of forecasts more easily.

These two measures of forecast performance, calibration, and discrimination, often conflict in that a better calibrated forecast often has poor discrimination whereas a forecast that discriminates well may be poorly calibrated. Another well-known measure of forecast performance, correlation, can be shown to include both concepts. For the two-category forecast, the correlation coefficient between forecast and event is seen to be the product of two terms, the first being discrimination, and the second a function of the calibration. More generally, it has the desirable features of being a single number for any forecast, making comparisons of forecasts easy, and one which remains unchanged by any change in the scale of how the forecast and/or the random event is measured. On the other hand it has the great disadvantage that it does not look at the distributional structure of the relationship between forecasts and outcomes but only at a quadratic norm that penalizes outliers in proportion to their distance from expected value. We emphasize that it is the joint distribution of F and X that carries all the information about forecast and outcome performance.

A common measure of performance of probability forecasts is the mean squared error between the forecast and the uncertain event, known as the Brier score. Analysis shows that this comprises three components, the first of which is the variance of the uncertain event, a factor not controllable or changeable by any change in a forecast. The second term adds to the Brier score increases as forecast calibration decreases and is zero for a perfectly calibrated forecast. The third term subtracts from the score and increases with increasing discrimination. Thus, the Brier score also incorporates both concepts and can be decomposed into separate discrimination and calibration terms.

In some decision problems the forecaster and decision maker may be the same individual or organization. In such cases the forecaster can focus attention on those aspects of the problem where measures of forecast performance are closely linked with the objectives of the decision maker. But in most cases with which we are familiar the forecaster and the decision maker represent different individuals or different institutions with different objectives. The forecaster may have to provide good-quality forecasts for a large and diverse set of decision makers. For example, there are many potential users of weather forecasts. One user might be a farmer, with decision problems and payoffs obviously related to weather outcomes but whose objectives are economic in nature and are not closely linked with measures of forecast quality. Another user of a weather forecast might be a commercial airline company whose choice of routes might lead to greater or less fuel economy. Still another user might be an organization planning a large outdoor function in which a last-minute decision has to be made to rent space that can be used as an alternative venue. Few of these different decision makers would probably agree on the same measures of forecast quality.

It is important to recognize that quality measures for a forecaster seldom coincide with the value of forecasts to decision makers. What may be statistically or mathematically of interest to a forecaster may not be relevant to the policies, decisions, or economic implications of a decision problem. We have considered three important quality and performance measures: discrimination, calibration, and correlation; we have shown how, with mean squared error measures, there are trade-offs between the first two in most forecasting problems.

We have seen how probability and categorical forecasts can be used by a decision maker and how optimal decisions are affected by calibration and discrimination. Further, we have shown how to derive coherent categorical forecasts from probability forecasts in those situations where no information is lost by using the categorical forecast in place of the probability forecast.

The ROC curve is often used in signal detection theory (the name ROC stands for receiver operating characteristic) and can be found in the following references: Urkowitz (1983), Wilkie (1992), Nelson (1986), Swets (1986), Rao and Tam (1987), and Kmietowicz and Ding (1993).

A small amount of literature on discrimination appears in a very large variety of literature sources. Although these are by no means exhaustive treatments of the subject, the reader might want to refer to the publications by Hand (1981), McLachlan (1992), Nelson (1986), Swets (1979, 1986), Wilkie (1992), and Murphy and Winkler (1992). For further reading on the Brier score, see Brier (1950).

A detailed examination of diagnostic aids to test for calibration and discrimination goes beyond the scope of this book. However, see Brier (1950), McLachlan (1992), Murphy, Brown, and Chen (1989), Murphy and Winkler (1987, 1992), Rao and Tam (1987), Swets (1986), and Yates and Curley (1985) for in-depth analyses and further references. A large number of unsolved problems in verification and measurement of calibrated forecasts remain, particularly where there are multiple factors or forecasts relevant to the outcomes and the decision being made.

PROBLEMS

8.1   Confirm the validity of the last paragraph in Section 8.2.2 (page 307).

8.2   For a two-category forecast show that
   a. The decision probabilities satisfy the inequality $p_0 < p_1$ if and only if the forecast likelihoods satisfy $f_0 < f_1$.
   b. When the forecast is well calibrated, $f_1 = p_1$ and $f_0 = p_0$.

8.3   For the four-category forecast of minimum temperature in Table 3.5 (p. 99), verify the probabilities found in Table 8.3 (p. 314).

8.4   Repeat the calculations leading to Table 8.3 (p. 314) for the calibrated version of the forecast in the following table. Comparing your results with those for the original forecast, comment on whether or not you think this forecast is more or less discriminating.

Forecasted minimum temperature		Actual minimum temperature category j				
Category i	Interval	1	2	3	4	Total days
1	−5° or below	65	20	5	0	90
2	−4° – 0°	20	230	30	2	282
3	1° – 5°	3	20	350	30	403
4	6° or above	2	12	18	193	225
Total days		90	282	403	225	1,000

8.5   Plot the cdf of $F|X = 1$ versus the cdf of $F|X = 0$ for the data in Table 3.4 (p. 98) and Table 8.4 (p. 322). These are known as the ROC curves. On the basis of these two curves can you decide which forecast discriminates better than the other?

8.6   Let the scale of two random variables F and X be changed to give random variables F' and X', where $F' = aF + b$ and $X' = cX + d$, where a, b, c, and d are given constants. Show that
   a. $Var[F'] = a^2 Var[F]$ and $Var[X'] = c^2 Var[X]$,
   b. $Cov[F', X'] = ac Cov[F, X]$,
   c. $\rho_{F'X'} = \rho_{FX}$.

8.7   Find the correlation coefficient for
   a. The probability forecast of rain using the data in Table 3.4 (p. 98),
   b. The categorical forecast of temperature using the data in Table 3.5 (p. 99).

8.8   Confirm the values of the correlation coefficients for the two calibrated probability forecasts for crop protection on page 323 in Section 8.7.

8.9   Based on the numeric examples for the crop protection problem in Section 8.7, comment on the properties of discrimination, calibration, and discrimination as important attributes of a forecast to help a decision maker.

8.10   Verify Equation (8.22) (p. 331), and find the amount by which the minimum expected return to the farmer is reduced by using this forecast rather than the one with the optimal cutoff level of 0.625.

8.11 Find $\mathbf{p}_F$, $\mathbf{F}^{(c)}$ and $\mathbf{P}^{(c)}$ for the remaining possible threshold values in Table 3.4 (p. 98), namely, between

   a. 1 and 0.75,

   b. 0.25 and 0.

8.12 In the aggregation of the four-category minimum temperature example on page 333, recall that $G = 1$ if $F = 1$ or $F = 2$. Use this to argue that $\mathbf{P}^{(A)}$ can be found directly from $\mathbf{P}^{(T)}$ using

$$p_{Y|G}(1|1) = \frac{p_{Y|F}(1|1)\,p_F(1) + p_{Y|F}(1|2)\,p_F(2)}{p_Y(1) + p_Y(1)}.$$

8.13 In Table 3.3 (p. 94) suppose that the category "Minimum temperature in the interval $(-4°, 0°)$" is of particular interest, and let $X = 1$ if this statement holds and $X = 0$ if it does not. Also let $F = 1$ if the statement is forecast to hold, and $F = 0$ if it is not. Find $Pr\{X = 1|F = 1\}$ and $Pr\{X = 1|F = 0\}$.

8.14 For the four-category forecast crop protection problem in Section 8.10,

   a. Find the decision policy that minimizes expected loss $l_{C4}{}^*$ as a function of r_2/r_1,

   b. Verify the equations given for $l_{C4}{}^*$ on page 336,

   c. Plot $l_{C4}{}^*$ in Figure 8-14 and compare with $l_{C2}{}^*$.

8.15*Assume that the joint pmf $p_{X,Y}(x,y)$ of two random variables X and Y is given by $p_{X,Y}(\sqrt{2}, 1) = p_{X,Y}(-\sqrt{2}, 1) = 1/4$ and $p_{X,Y}(0, -1) = 1/2$.

   a. Are X and Y independent? Explain.

   b. Find the correlation coefficient of X and Y. Are they uncorrelated?

You are asked to design a "good" point predictor of uncertain Y knowing $X = x$. To do so you formulate a decision model in which you define forecast error, e, as the difference between your point predictor f of Y, given $X = x$ and the actual outcome, which is $Y = y$. Your predictor (optimal decision) is the one that minimizes the expected value of squared forecast error.

   c. Carefully formulate the prediction problem as a decision problem, draw the influence diagram, and find the minimum expected squared error predictor of Y given X.

   d. Now, find the *linear* least-squared error predictor (regression line) of Y given $X = x$. Express your result in terms of the correlation coefficient.

   e. Find the minimum expected squared error predictor of X given $Y = y$. Does it differ from the answer in (c)?

   f. What is the conditional variance of the error using the predictor in (c)? Of the predictor in (d)?

   g. Find the unconditional expected squared error before Y is observed when you use (e).

CHAPTER 9

ADVANCED CONCEPTS

9.1 INTRODUCTION

For a diagram or picture to be useful in helping a decision maker understand a decision problem it must reveal the structure and interrelationships of the key variables in the problem. When clients explain a problem they usually make statements about the existence of relevant relationships between uncertain quantities, what is known when decisions are taken, and which quantities directly influence future decisions. It is important to capture this structure before probabilities are estimated because the conditioning structure is critical to assessment. The more delicate task of eliciting probabilities and payoffs can be undertaken once the influences of key variables are understood. We have already seen that influence diagrams and decision trees are of considerable help in problem formulation and in illustrating structure.

We know that decision trees enumerate unique paths with sequences of decisions and random events that eventually lead to a return, loss, or utility function at the end of the path. The dependence or independence of events along different paths of a decision tree are not always apparent, and it becomes increasingly important to have an organized procedure for enumerating paths, probabilistic dependencies, and returns. Decision tree representations of even the simplest real-life problems quickly become large and bushy; explicit drawing of the tree quickly becomes impractical for even modest-size problems. A decision problem with only ten variables, each able to take on three possible values, leads to a decision tree with $3^{10} = 59{,}049$ terminal nodes. Drawing such a tree is clearly impractical, and even the most ardent analyst would require that we first state such a problem in a way that allows one to conceptualize the tree or state the relevant data and relationships so that computations can be organized in a meaningful way. In this book we do not enter into the analysis of numerically efficient computing procedures, but, in the interest of completeness, we discuss the main features of several methods that not only are promising from the point of view of numeric efficiency but also allow us to make sensitivity analyses and gain insight into the complex structure and solutions of large decision problems.

Several concepts related to the drawing and simplification of influence diagrams were introduced in Chapters 3 and 4; for example, arc reversal and Bayes' rule in Section 3.4.6, cycles and no-forgetting arcs in Section 4.2, and irrelevant decision nodes in Sections 4.7 and 4.8. These were demonstrated by examples without formal proofs, because the proofs require a more complex mathematical notation than has been necessary to this point. The first objective of this chapter is to introduce the reader to a deeper understanding of chance influence diagrams or belief nets. The mathematical notation and proofs are covered in Sections 9.2 and 9.3 together with illustrative examples. We show several examples of influence diagrams in which it is possible to directly reduce and simplify the influence diagram before the analyst undertakes the difficult and costly assessments of probabilities and values or attempts any numerical calculations. In Section 9.3 we offer a proof that, by addition or deletion of appropriate directed arcs, equivalent chance influence diagrams can be obtained under arc reversal and barren node removal operations. We also show how implicit conditional independence statements can be deduced directly from an influence diagram and how the graph helps in understanding the propagation of information.

The second objective of this chapter is to extend the rollback algorithm of Section 4.4. The first step is to design a node-labeling scheme in decision trees using the identical naming and numbering system used for nodes in influence diagrams. The second is to recognize the data processing required to generate path-dependent costs, branch probabilities and the sets associated with node outcomes or decisions. The third is to modify the dynamic programming procedures to incorporate complex path histories and nonlinear utilities.

Section 9.4 includes a compact way to represent decision problems and offers the general design guidelines for algorithms that efficiently solve large decision problems. In Section 9.5 we discuss multiattribute problems that exploit the labeling procedures developed in Section 9.4 and see how the rollback algorithms can be extended and modified to include trade-offs between multiple attributes. A decision problem in nuclear power generation is used as an illustration.

The third and final objective, addressed in Section 9.6, is to show how rollback can be performed directly on an influence diagram without explicitly drawing a decision tree.

9.2 CLASSIFYING INFLUENCE DIAGRAMS

In order to pursue a more in-depth analysis of influence diagrams we need to introduce some additional concepts and definitions. An influence diagram can be represented as a directed graph without cycles. Let $\mathcal{G} = (\mathcal{N}, \mathcal{A})$ denote the graph with node set $\mathcal{N}$ and directed arc set $\mathcal{A}$. If i and j are two nodes in $\mathcal{N}$ joined by a directed arc from i to j, this arc is designated (i, j) and is a member of $\mathcal{A}$. The convention on a directed arc or path is that one must traverse it in a direction that agrees with the arrowhead, as in following a one-way street. The number of elements in $\mathcal{N}$ is the total number of nodes in the influence diagram. If the chance node set is denoted by $\mathcal{C}$, the decision node set by $\mathcal{D}$ and the value node by $\mathcal{V}$, then[1]

$$\mathcal{N} = \mathcal{C} \cup \mathcal{D} \cup \mathcal{V}.$$

We also adopt the convention that the value node is always the final node and either a chance or decision node is the origin node; the latter is equivalent to the root or terminal node in the associated decision trees.

Two node sets are particularly useful when characterizing influence diagrams. For a given node i the set of nodes connected directly to i with incoming arcs is called the *direct predecessor set*; similarly, the set of nodes connected directly to i with outgoing arcs is called the *direct successor set*. Formally we define for any node i in the influence diagram,

1. The set of nodes connected to i by directed arcs leading into i is called the *direct predecessor* set of i and denoted by $\mathcal{P}(i)$. Mathematically,

$$\mathcal{P}(i) = \{k \in \mathcal{N} \mid (k, i) \in \mathcal{A}\}, \tag{9.1}$$

2. The set of nodes connected to i by directed arcs leading out of i form the *direct successor* set of i and is denoted by $\mathcal{S}(i)$.

$$\mathcal{S}(i) = \{k \in \mathcal{N} \mid (i, k) \in \mathcal{A}\}. \tag{9.2}$$

The set of variables whose nodes are connected (not necessarily directly) to node i by a directed path into i are called *predecessors* (sometimes called *ancestors*) of i; the set of variables whose nodes are connected (not necessarily directly) to i by a directed path out of i are called *successors* (sometimes called *descendants*). Alternatively, we can think of a predecessor as either a direct predecessor of the node or a direct predecessor of another predecessor. The analogous recursive definition holds for a successor.

Because we do not allow cycles, $\mathcal{V}$ can never be included as a predecessor node in (9.1), whereas it is always a direct successor of one or more nodes in (9.2). It is always possible to trace one or more paths from origin to value node, that is, we do not use disconnected graphs to represent influence diagrams.

It is useful to repeat what is meant by a *no-forgetting arc*. We assume that (1) there is a complete ordering of decision nodes such that directed arcs lead from each decision node to all later decision nodes, and (2) if an outcome at a chance node is known when a given decision D_i is made, then that information is also known when each later decision D_j is made ($j > i$ means that decision D_j occurs after D_i). Thus, for some decision D_i, (1) there is a directed arc into D_i from each chance node whose outcome is known to

---

[1] In earlier chapters we used result and value node interchangeably. In this chapter we use the term *value* to denote the name of the node and result to denote the mathematical result or payoff function associated with the value node.

the decision maker before decision D_i is made, and (2) there is a directed arc from each of the same chance nodes into each decision node D_j that follows D_i.

In order to study the simplification and reduction of influence diagrams that represent well-posed and well-formulated decision problems it is helpful to develop the concept of a proper influence diagram (PID) and an extensive form influence diagram (EFID). The former is associated with the idea that certain simple rules must be observed if an influence diagram is to represent a well-posed decision problem, and the latter is associated with the additional structure that a proper influence diagram needs if it is to offer a valid decision tree representation of the problem.

9.2.1 A Proper Influence Diagram

A PID is a connected directed graph

1. With a single-origin node and a single-value node,
2. Without cycles,
3. Whose origin node has no predecessors and whose value node has no successors,
4. In which the "no-forgetting" principle is applied to all nodes.

The inclusion of "no-forgetting" arcs guarantees that any information or decisions known to the decision maker when he or she makes a decision are known at subsequent decisions. Figure 9-1 is a PID for the aircraft part problem formulated in the influence diagram of Figure 5-5 (p. 198). The difference between it and Figure 9-1 is that the nodes F and X are renamed T and C to denote Test and Condition of Part. Note that the first decision (test or not test) is remembered when the second decision (replace or rework) is made. The condition of the part, C, is not known when the replacement decision is made, but the influence diagram shows that the probability distribution assigned to the test outcome T, which precedes the decision, is conditioned on knowing C. Thus, although the influence diagram in Figure 9-1 is a PID, it does not capture the correct timing of real events.

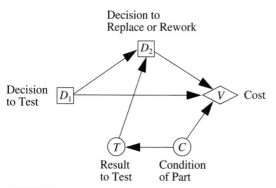

FIGURE 9-1
Proper influence diagram for aircraft part problem.

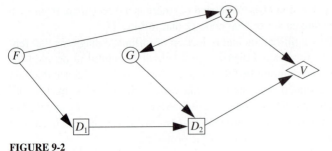

FIGURE 9-2
Example of an improper influence diagram.

In Figure 9-2 we consider another example that includes two forecasts, F and G, of the uncertain quantity X. The forecast F is known to the decision maker when the decision D_1 is made but is forgotten at decision D_2. Thus, we have an example of an improper influence diagram. It can be converted into a PID by adding a directed arc from F to D_2. Note also that without this arc, the only direct successor of node D_1 is the decision node D_2 so that the first decision does not affect or influence the conditional probability of G or the payoff at the value node. Although this feature would not disqualify it from being a proper influence diagram, it will have an important effect on the relevance of the first decision, a subject we discuss in some detail later in this chapter.

Another improper influence diagram is shown in Figure 9-3. In this diagram the first decision affects the outcomes of the probability distribution of G known to the decision maker at the time the second decision is made. The diagram does not indicate that the first decision is remembered at the time the second decision is made. Clearly the information contained in knowing the decision alternative selected at D_1, as well as the outcome of G, may be very different from just knowing the outcome of G; this additional information can affect the solution of the problem. Figure 9-3 can be converted to a PID by adding an arc from D_1 to D_2 and from F to D_2 at which point it can also be represented by a decision tree where the ordering of nodes coincides with the real timing of events and decisions.

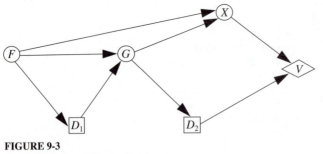

FIGURE 9-3
A second improper influence diagram.

9.2.2   Influence Diagrams in Extensive Form

We say that an influence diagram is in extensive form (EFID) if it is a proper influence diagram and the ordering of nodes corresponds to the real timing of actual events and decisions in the associated decision tree. Thus, an EFID corresponds to a proper influence diagram in which no arc reversal operations using Bayes' rule are required. The concept of an EFID plays an important role in discussing the relationship between a solution obtained from a decision tree and the corresponding influence diagram.

An influence diagram is said to be in extensive form when, in addition to being a proper influence diagram (i.e. assumptions (1) through (4) in Section 9.2.1), it is also true that

5. a chance node X_i that is a predecessor of any decision node D_j must be a direct predecessor of D_j. That is to say, there must be a directed arc (X_i, D_j) if X_i lies on *any directed path* leading to D_j.

Figure 9-1 is a PID but not an EFID. For the direct predecessor requirement of (5) to hold there would have to be a directed arc from node C to the second decision node. This would imply that we have perfect information about the condition of the part, and we would then conclude that the test node T was irrelevant to the decision problem.[2] An EFID for the same problem is shown in Figure 9-4 where the typical scenario in the decision tree is (1) decide to test or not, (2) observe the outcome of the test, (3) decide to install or rework, (4) observe the condition of the part, and (5) evaluate the cost. Note that node T is now a direct predecessor of the second decision node but C is not. In this example we have explicitly assumed the independence of test outcomes and part condition on the decision to test or not; thus, there is

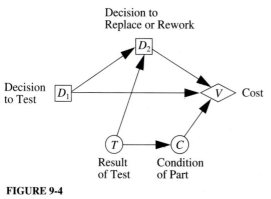

FIGURE 9-4
EFID for aircraft part problem.

---

[2] We discuss irrelevant chance and decision nodes in Section 9.2.3.

no need for a directed arc from the Test decision (D_1) node to the condition of the part (C).

The influence diagram shown in Figure 9-5 is a replica of the one in Figure 4-1 (p. 139) for the engine maintenance example. It is not only a PID but also an EFID because the time ordering of decisions and events allows us to formulate the problem as a decision tree. Compare the numbering of the nodes in Figure 9-5 with those in Figure 9-6, which is a decision tree that can be used to solve the maintenance problem. We see that the node numbers in the decision tree coincide with those used in the influence diagram. Whereas a node number occurs exactly once in an EFID, in the decision tree the same node number may be repeated many times.

Another example of an EFID is shown in Figure 9-7, which is similar to the two-sensor problem studied earlier. All no-forgetting arcs leading into decision nodes are included. One can draw a decision tree based on this PID whose ordering of nodes correctly reflects the real timing of decisions, events, and the availability of information. Note that there is a directed path from F to G to D_2 and that there is a directed arc connecting F to D_2.

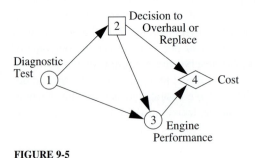

FIGURE 9-5
Engine maintenance problem.

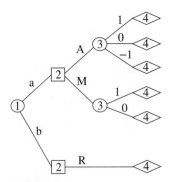

FIGURE 9-6
Associated decision tree with ID node numbers.

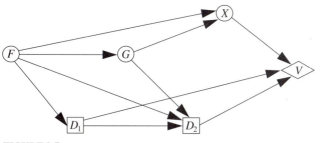

FIGURE 9-7
An EFID with two forecasts F and G.

9.2.3   Irrelevant Decision and Chance Nodes

Although we have used the term previously, we need to understand more fully the concepts of irrelevant chance and decision nodes.[3] The common meaning of the word *irrelevant* is that of no effect or influence. When applied to influence diagrams it means that a chance or decision node of interest is not relevant to, does not influence, or has no effect on the solution of the problem. Irrelevant nodes have a more formal definition. We say that

> A decision node in a PID is irrelevant if its only direct successors are decision nodes.
> A chance node in an EFID is irrelevant if its only direct successors are decision nodes.

In Figure 9-8(a) the chance node Y is irrelevant but X is not; knowing the distribution of Y or a particular value, say $Y = y$, does not affect the decision or the solution of the decision problem. Note that Y influences neither the distribution of X nor the payoff at the value node. The arc from X to V shows that the solution to the problem is affected by the distribution assigned to X. However, if there were an arc leading from Y to V, it would then be relevant to the decision D. This might occur, for example, if a particular outcome of Y affected the decision choices at D and, therefore, the possible payoff or result at the value node.

In Figure 9-8(b) the decision node D_1 is irrelevant because the decision D_2 is affected only by the distribution of X and the payoff, neither of which are affected by D_1. The choice of D_1 does not in any way affect the actions taken at D_2 because D_2 is based solely on an uncertain future outcome. In Figure 9-8(c) the decision D_1 is relevant because it directly influences the distribution of X, which we already know has an effect on our choices at D_2. In Figure 9-8(d) Y is relevant to the decision because it also influences the conditional probability of X.

An irrelevant chance or decision node is one whose only direct successor node is a decision node. Its direct successors do not include a value node or any other chance nodes. Because an irrelevant node cannot affect future policies, future outcomes, or op-

---

[3] See, for example, Section 4.7 (p. 179), where we describe an irrelevant decision node.

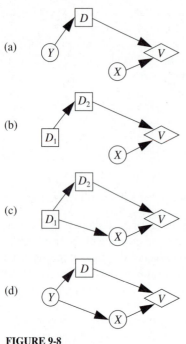

FIGURE 9-8
Relevant and irrelevant nodes.

timal policies, it can be removed from the influence diagram without having any effect on the optimal solution. By recognizing and removing irrelevant nodes an analyst or decision maker can usually simplify problem formulation and clarify the roles of the remaining chance and decision variables.

9.3   CHANCE INFLUENCE DIAGRAMS

In this section we discuss some basic properties of chance influence diagrams or belief nets, first mentioned in Section 3.4.6 (p. 103). The reader who has followed the material to this point is by now familiar with the basic concepts and properties of influence diagrams and their contribution to model formulation for decision problems. As we have seen, they contain three distinct types of nodes: decision, chance, and value. In this section we concern ourselves exclusively with influence diagrams (or subparts of influence diagrams) that include only chance nodes. We refer to these as chance influence diagrams or belief nets. They are of interest because their structure is essential to understanding and being able to calculate conditional probabilities of interest to the decision maker. In order to develop computer algorithms for calculating conditional probabilities, it is necessary to gain a deeper understanding of their structural properties.

A chance influence diagram is a directed acyclic graph in which all nodes in the graph denote random quantities. Associated with each node is a conditional probability function. Directed arcs joining nodes denote possible statistical dependence. The absence of a directed arc between two nodes denotes the very strong statement that the two nodes are conditionally or unconditionally independent.

When we deal with large numbers of random variables and many conditional independencies, it is useful to have a formal way of listing the random quantities and accounting for the directed arcs that does not require a different letter for each random quantity. Assume that $\mathbf{X}$ is a vector of n random variables $X_1, X_2, \ldots, X_n$ with joint probability density or mass function $p_{\mathbf{X}}(\mathbf{x}) = p_{\mathbf{X}}(x_1, x_2, \ldots, x_n)$. Using the rules of probability $p_{\mathbf{X}}(\mathbf{x})$ can be expressed as the product of conditional probability functions

$$p_{\mathbf{X}}(\mathbf{x}) = \prod_{i \in C} p(x_i | x_1, x_2, \ldots, x_{i-1}) \, p(x_1), \tag{9.3}$$

where $x_k \in C(k)$, $k \in C$.

Recall that C is our notation for the set of chance nodes in an influence diagram, so in this example $C = \{1, 2, \ldots, n\}$; $C(i)$ is the set of possible outcomes of the random variable X_i. In the remainder of this chapter we often use the simpler notation $p(\mathbf{x})$ in lieu of $p_{\mathbf{X}}(\mathbf{x})$ because the meaning is usually obvious from the context.

Rather than write down explicitly the conditional independence statements implicit in the factorization of $p(\mathbf{x})$ we can instead represent the sets of statements in the form of a directed graph. The right-hand side requires the conditioning that is consistent with the dependencies illustrated by the directed arcs in the influence diagram. Draw a node to represent each of the random variables $X_1, X_2, \ldots, X_n$, number these nodes $i = 1$, $2, \ldots, n$, and draw a directed arc from node i to j whenever j is conditionally dependent on i. Of course, X_j may be conditionally dependent on more than one node so that there may be many directed arcs leading into node j.

A chance influence diagram (often referred to as a belief net or knowledge map), represents conditional independence statements in a directed acyclic graph (i.e., it has no cycles or connected loops with all arcs going in the same clockwise or counterclockwise direction). All nodes that directly influence node j are connected to node j. Given a directed acyclic graph together with conditional probabilities specified at each node, there exists a unique joint probability distribution of the random quantities represented by the nodes of the graph. This specification is possible because a directed graph is acyclic if and only if there exists a list of the nodes such that any successor of a node i in the graph follows node i in the ordered list of nodes. The sparser the graph of a chance influence diagram, the larger the number of conditional independence statements embedded within it.

9.3.1 Directed Graphs and Predecessor and Successor Sets

Earlier in Section 9.2.1 we defined *direct predecessor* and *direct successor* sets of a given node i, that is, those nodes connected directly to i with incoming and outgoing arcs, respectively.

A chance influence diagram is also represented as a directed graph without cycles. Let $G = (C, \mathcal{A})$ denote the graph with node set C and directed arc set $\mathcal{A}$. As before, the *direct predecessor* and *direct successor* nodes of a given node i are those nodes connected directly to i with incoming and outgoing arcs, respectively. The set of *direct predecessors* of node i is

$$\mathcal{P}(i) = \{k \in C \,|(k, i) \in \mathcal{A}\},\tag{9.4}$$

and the set of *direct successors* of node i is

$$S(i) = \{k \in C \,|(i, k) \in \mathcal{A}\}.\tag{9.5}$$

If i and j are any two distinct nodes (connected or not), it may be the case that both i and j have a set of common direct predecessor nodes. Denote this set by

$$\mathcal{W}(i,j) = \mathcal{P}(i) \cap \mathcal{P}(j) = \{k \in C \,|\, (k, i) \in \mathcal{A}, (k,j) \in \mathcal{A}\}.\tag{9.6}$$

The set of direct predecessors of either node i or j is $\mathcal{P}(i) \cup \mathcal{P}(j)$, and the sets of nodes that are exclusive to only node i or to node j are given by

$$\mathcal{U}(i,j) = \mathcal{P}(i)\backslash\mathcal{W}(i,j) \quad \text{and} \quad \mathcal{U}(j,i) = \mathcal{P}(j)\backslash\mathcal{W}(i,j),\tag{9.7}$$

respectively.[4] In Equation (9.7) we can think of the set $\mathcal{U}(i, j)$ as the nodes having arcs directed into i but *not* into j and $\mathcal{U}(j,i)$ as the nodes having arcs directed into j but *not* into i. Note that if arc (i, j) is in $\mathcal{A}$, then $\mathcal{U}(j,i)$ must include node i.

The conditional probability of X_j is a function of its direct predecessors, which means that we can write

$$p(x_j \,|\, x_1, \ldots, x_{j-1}, x_{j+1}, \ldots, x_n) = p(x_j \,|\, \mathbf{x}_{\mathcal{P}(j)}),$$

with the understanding that $\mathbf{x}_{\mathcal{P}(j)}$ is the vector list of all x_k such that $k \in \mathcal{P}(j)$. In Figure 9-9, for example, $\mathcal{P}(4) = \{2, 3\}$, and $\mathbf{x}_{\mathcal{P}(4)} = (x_2, x_3)$ is the vector list of the random variables indexed by the set $\mathcal{P}(4)$. The conditional probability can therefore be written as $p(x_4|x_1, x_2, x_3) = p(x_4|\mathbf{x}_{\mathcal{P}(4)}) = p(x_4|x_2, x_3)$, because the directed graph explicitly states the conditional independence of x_4 and x_1 given x_2 and x_3.

A directed path from node i to node j is a sequence of connected, ordered pairs of nodes, each pair of nodes corresponding to a directed arc. The directed path from node

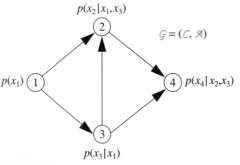

FIGURE 9-9
A chance influence diagram with four nodes.

---

1 to 4 via nodes 3 and 2 in Figure 9-9 can be written as the connected ordered sequence of arcs $\{(1,3),(3,2),(2,4)\}$. In this figure with four nodes and five arcs there are a total of three directed paths leading from node 1 to 4, two from node 1 to 2 but only one from node 1 to 3.

In terms of the defined nodes and directed arcs we have the predecessor sets: $\mathcal{P}(1) = \emptyset$, $\mathcal{P}(2) = \{1,3\}$; $\mathcal{P}(3) = \{1\}$, $\mathcal{P}(4) = \{2,3\}$; $\mathcal{W}(2,3) = \{1\}$, $\mathcal{W}(2,4) = \{3\}$, $\mathcal{W}(3,4) = \emptyset$; $\mathcal{U}(1,2) = \emptyset$, $\mathcal{U}(2,4) = \{1\}$, $\mathcal{U}(4,2) = \{2\}$. As node 2 is a direct predecessor of node 4, the conditional joint probability of X_2 and X_4 can be written as

$$p(x_2, x_4 | \mathbf{x}_{\mathcal{P}(2)}, \mathbf{x}_{\mathcal{P}(4)\backslash 2}) = p(x_4 | \mathbf{x}_{\mathcal{P}(4)}) p(x_2 | \mathbf{x}_{\mathcal{P}(2)})$$

$$= p(x_4 | x_2, x_3) p(x_2 | x_1, x_3) = p(x_2, x_4 | x_1, x_3). \tag{9.8}$$

The first factor in the bottom equation can be written as $p(x_4 | x_2, x_3)$ because x_4 and x_1 are conditionally independent.

9.3.2 Equivalent Chance Influence Diagrams

Two chance influence diagrams $\mathcal{G} = (\mathcal{C}, \mathcal{A})$ and $\mathcal{G}' = (\mathcal{C}, \mathcal{A}')$ are said to be *equivalent* if the joint probability distribution of all random quantities identified by the nodes of the directed graph $\mathcal{G}$ equals the joint probability distribution of the nodes on $\mathcal{G}'$ or one is the marginal of the other. Note that the node sets of $\mathcal{G}$ and $\mathcal{G}'$ are identical, but the set of directed arcs $\mathcal{A}'$ in general differs from $\mathcal{A}$. Compare the diagrams in Figures 9-9 and 9-10, each with four nodes numbered 1 through 4. The first graph, $\mathcal{G}$, includes five directed arcs, with arc set given by

$$\mathcal{A} = \{(1,2),(1,3),(3,2),(3,4),(2,4)\}.$$

The second, $\mathcal{G}'$, includes six directed arcs, with arc set given by

$$\mathcal{A}' = \{(1,2),(1,3),(1,4),(3,2),(3,4),(4,2)\}.$$

In $\mathcal{G}'$ arc $(1,4)$ has been added, and arc $(4,2)$ is the reverse of $(2,4)$ in the original (unprimed) graph. The chance influence diagrams are said to be equivalent if $p_{\mathcal{G}}(x_1, x_2, x_3, x_4) = p_{\mathcal{G}'}(x_1, x_2, x_3, x_4)$ with subscript $\mathcal{G}$ for the original graph on which the joint probability is defined and $\mathcal{G}'$ for the second graph. Obviously, the direct predecessors of nodes 2 and 4 are different in $\mathcal{G}$ and $\mathcal{G}'$.

In the general case with n nodes, two chance influence diagrams are said to be equivalent if the joint probability distributions are the same, that is, if

$$p_{\mathcal{G}}(x_1, x_2, \dots, x_n) = p_{\mathcal{G}'}(x_1, x_2, \dots, x_n) \text{ for all } x_i \in C_i, i \in C. \tag{9.9}$$

To show this equivalence for Figures 9-9 and 9-10 we must show that the joint probability of x_2 and x_4 conditional on x_1 and x_3 in $\mathcal{G}$,

$$p(x_2, x_4 | x_1, x_3) = p(x_4 | x_2, x_3) \, p(x_2 | x_1, x_3), \tag{9.10}$$

is equal to the joint probability of x_2 and x_4 on $\mathcal{G}'$,

$$p(x_2, x_4 | x_1, x_3) = p(x_2 | x_1, x_3, x_4) \, p(x_4 | x_1, x_3). \tag{9.11}$$

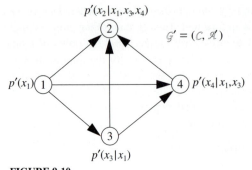

FIGURE 9-10
An equivalent chance ID with four nodes and six arcs.

If the marginal probability at node 1, $p(x_1)$, and the conditional probability at node 3, $p(x_3|x_1)$, are equal on both $\mathcal{G}$ and $\mathcal{G}'$, we conclude that the joint probabilities are equal and therefore the two chance influence diagrams are equivalent. We anticipate a later discussion and result on barren nodes by noting, in passing, that node 2 has no successors. Thus, in the decomposition of the right-hand side of Equation (9.9) into conditional probabilities no factor other than $p(x_2|x_1, x_3, x_4)$ depends on X_2.

9.3.3   Bayes' Rule and Arc Reversal

Arc reversal is used to making inferences and calculate posterior probabilities from likelihoods. It corresponds to the use of Bayes' rule for the two random variables represented by the nodes at either end of the directed arc being reversed. We saw in Section 3.4.6 that arc reversal in chance influence diagrams corresponds to node reversal in event trees. Arc (2,4) in Figure 9-9 and (4,2) in Figure 9-10 are an example of arc reversal.

Consider arc $(i, j) \in \mathcal{A}$ of the directed graph $\mathcal{G}$, and assume that there is no other directed path from i to j. As already seen, there are many circumstances when we want to reverse an arc (i, j). To do this we may have to add a new set of directed arcs resulting in a new directed graph $\mathcal{G}'$ with new conditional node probabilities in $\mathcal{G}'$. When the conditional probabilities on the new graph are properly calculated, $\mathcal{G}$ and $\mathcal{G}'$ are equivalent chance influence diagrams.

Focus attention on the conditional probabilities associated with the set of direct predecessors of nodes i and j in $\mathcal{G}$ and $\mathcal{G}'$; we do not need to consider the conditional probability distributions of other nodes. We can show equivalence if the joint probability of all chance nodes before reversal of arc (i, j) is equal to the joint probability after reversal, that is,

$$p_{\mathcal{G}}(x_1, x_2, \ldots, x_n) = p(x_1) \cdots p(x_i, x_j | \mathbf{x}_{\mathcal{P}(i)}, \mathbf{x}_{\mathcal{P}(j) \setminus i}) \cdots p(x_n | \mathbf{x}_{\mathcal{P}(n)})$$

$$= p'(x_1) \cdots p'(x_i, x_j | \mathbf{x}_{\mathcal{P}'(i) \setminus j}, \mathbf{x}_{\mathcal{P}'(j)}) \cdots p'(x_n | \mathbf{x}_{\mathcal{P}'(n)})$$

$$= p_{\mathcal{G}'}(x_1, x_2, \ldots, x_n). \tag{9.12}$$

If the conditional probabilities and direct predecessors on all nodes other than nodes i and j where the arc reversal occurs are unchanged, then the corresponding (primed) probabilities equal the original (unprimed) values.[5] To demonstrate that the graphs are equivalent we then only have to show that the inner two terms for the joint probability of x_i and x_j are equal, that is,

$$p'(x_i, x_j \mid \mathbf{x}_{\mathcal{P}'(i) \backslash j}, \mathbf{x}_{\mathcal{P}'(j)}) = p(x_i, x_j \mid \mathbf{x}_{\mathcal{P}(i)}, \mathbf{x}_{\mathcal{P}(j) \backslash i}). \tag{9.13}$$

This is a surprising result because it is possible to obtain the equality by only making changes in the membership of the *direct* predecessor sets and conditional probabilities for nodes i and j; in other words we only have to concern ourselves with *local* changes because all other conditional probabilities and direct predecessor sets are unchanged from their values in $\mathcal{G}$.

To ensure that the correct dependencies are taken into account, it is easy to state a general rule for construction of the direct predecessor sets of nodes i and j after arc reversal. The rule is that the new direct predecessors of node j (after arc reversal) includes the direct predecessors of node i and j but not i itself whereas the new direct predecessors of i include all the direct predecessors of i and j but not j. To see why this is so we compare the sets of direct predecessors for nodes i and j in $\mathcal{G}$ and $\mathcal{G}'$.

The primed index sets $\mathcal{P}'(i)$, $\mathcal{P}'(j)$, $\mathcal{W}'(i, j)$ are the new direct predecessors of nodes i and j in the directed graph $\mathcal{G}'$ following arc reversal of arc (i, j). Membership of the primed predecessor sets is defined as follows:

$$\mathcal{W}'(i, j) = \{ \mathcal{U}(i, j) \cup \mathcal{W}(i, j) \cup \mathcal{U}(j, i) \backslash i \}, \tag{9.14}$$

$$\mathcal{P}'(j) = \mathcal{W}'(i, j), \tag{9.15}$$

$$\mathcal{P}'(i) = \{ j \cup \mathcal{W}'(i, j) \}, \tag{9.16}$$

$$\mathcal{P}'(k) = \mathcal{P}(k) \qquad k \neq i, j. \tag{9.17}$$

The definition of these new sets means that $\mathcal{W}'(i, j)$ in $\mathcal{G}'$ includes all direct predecessors of both nodes i or j in $\mathcal{G}$ but excludes node i itself. It follows that node j is the only node in $\mathcal{G}'$ that is a direct predecessor of node i not in $\mathcal{W}'(i, j)$. Obviously, following arc reversal, node j must be a direct predecessor of node i. It follows that $\mathcal{U}'(i, j) = \{ j \}$ and $\mathcal{U}'(j, i) = \emptyset$.

Notice that if Figures 9-9 and 9-10 are equivalent, then, by this construction, the direct predecessor sets of nodes 2 and 4 before and after arc reversal are

$$\mathcal{P}(2) = \{1, 3\}, \qquad \mathcal{P}'(2) = \{1, 3, 4\},$$

$$\mathcal{P}(4) = \{2, 3\}, \qquad \mathcal{P}'(4) = \{1, 3\}.$$

We also define the new conditional (primed) distributions on $\mathcal{G}'$ as follows:

---

[5] For emphasis we add primes to the new (transformed) conditional probabilities in this section even though our notation for $p(x_i \mid \mathbf{x}_{\mathcal{P}(i)})$ does not require it.

$$p'(x_k|\mathbf{x}_{\mathcal{P}'(k)}) = p'(x_k|\mathbf{x}_{\mathcal{P}(k)}) = p(x_k|\mathbf{x}_{\mathcal{P}(k)}), \qquad k \neq i, j, \tag{9.18}$$

$$p'(x_j|\mathbf{x}_{\mathcal{P}'(j)}) = p'(x_j|\mathbf{x}_{\mathcal{W}'(i,j)}) = p'(x_j|\mathbf{x}_{\mathcal{U}(i,j)}, \mathbf{x}_{\mathcal{W}(i,j)}, \mathbf{x}_{\mathcal{U}(j,i)\backslash i})$$

$$= \int p(x_j|\mathbf{x}_{\mathcal{P}(j)}) p(x_i|\mathbf{x}_{\mathcal{P}(i)}) dx_i > 0, \tag{9.19}$$

$$p'(x_i|\mathbf{x}_{\mathcal{P}'(i)}) = \frac{p(x_i|\mathbf{x}_{\mathcal{P}(i)}) \, p(x_j|\mathbf{x}_{\mathcal{P}(j)})}{\int p(x_j|\mathbf{x}_{\mathcal{P}(j)}) \, p(x_i|\mathbf{x}_{\mathcal{P}(i)}) \, dx_i}. \tag{9.20}$$

In these expressions the left-hand sides depend only on primed quantities (after arc reversal) whereas the right-hand side depends only on unprimed probabilities defined on $\mathcal{G}$ (e.g., before arc reversal). Also, $\mathcal{P}'(k) = \mathcal{P}(k)$ and $p'(x_k|\mathbf{x}_{\mathcal{P}'(k)}) = p(x_k|\mathbf{x}_{\mathcal{P}(k)})$ for any $k \neq i$, j in $\mathcal{G}'$. Thus, we only have to show the equality of the factor that represents the joint probability of nodes i and j, that is,

$$p(x_i, x_j|\mathbf{x}_{\mathcal{P}(i)}, \mathbf{x}_{\mathcal{P}(j)\backslash i}) = p'(x_i, x_j|\mathbf{x}_{\mathcal{P}'(i)\backslash j}, \mathbf{x}_{\mathcal{P}'(j)}).$$

To do this we multiply the defining expressions for $p'(x_j|\mathbf{x}_{\mathcal{P}'(j)})$ and $p'(x_i|\mathbf{x}_{\mathcal{P}'(i)})$ to obtain

$$p'(x_j|\mathbf{x}_{\mathcal{P}'(j)}) p'(x_i|\mathbf{x}_{\mathcal{P}'(i)}) = p'(x_j, x_i|\mathbf{x}_{\mathcal{P}'(j)}, \mathbf{x}_{\mathcal{P}'(i)\backslash j})$$

$$= p(x_i|\mathbf{x}_{\mathcal{P}(i)}) p(x_j|\mathbf{x}_{\mathcal{P}(j)}) = p(x_i, x_j|\mathbf{x}_{\mathcal{P}(i)}, \mathbf{x}_{\mathcal{P}(j)\backslash i}),$$

which shows that the chance influence diagrams are equivalent after arc reversal. Bayes' rule shows us that the conditional probability of X_i, given its direct predecessor set in $\mathcal{G}'$, is just equal to the joint probability $p'(x_i, x_j|\mathbf{x}_{\mathcal{P}'(i)\backslash j}, \mathbf{x}_{\mathcal{P}'(j)})$ divided by $p'(x_j|\mathbf{x}_{\mathcal{P}'(j)})$:

$$p'(x_i|\mathbf{x}_{\mathcal{P}'(i)}) = \frac{p'(x_i, x_j|\mathbf{x}_{\mathcal{P}'(i)\backslash j}, \mathbf{x}_{P'(j)})}{p'(x_j|\mathbf{x}_{\mathcal{P}'(j)})} = \frac{p(x_i|\mathbf{x}_{\mathcal{P}(i)}) \, p(x_j|\mathbf{x}_{\mathcal{P}(j)})}{\int p(x_j|\mathbf{x}_{\mathcal{P}(j)}) \, p(x_i|\mathbf{x}_{\mathcal{P}(i)}) \, dx_i}.$$

It is this result that motivated the choice of the primed conditional probabilities on $\mathcal{G}'$ and ensures that all probability calculations are coherent.

9.3.4   Barren Nodes

It may happen that either in the original formulation or following arc reversal we obtain a chance influence diagram where all directed arcs connected to a particular chance node j lead into the node but none lead from it. When this occurs, the direct successor set is empty so that $\mathcal{S}(j) = \emptyset$. We can therefore write

$$p(\mathbf{x}) = p(x_1, x_2, \dots, x_n) = p(x_j|\mathbf{x}_{\mathcal{P}(j)}) \prod_{i \in C \backslash j} p(x_i|\mathbf{x}_{\mathcal{P}(i)}), \tag{9.21}$$

where it is understood that any or all nodes of the graph may be in the direct predecessor set of j, but node j is explicitly excluded as a direct predecessor of any other node. Thus, no factor in the product term depends on x_j. If we need to calculate the marginal distribution of all other chance variables, excluding node j from the list, we can integrate both sides of this equation. But the integral (summation) of the barren node term involving only node j on the right-hand side of the equation equals 1, so that the joint distribution of the remaining nodes is

$$\int p(x_1, x_2, \ldots, x_n) dx_j = \prod_{i \in C \setminus j} p(x_i \mid \mathbf{x}_{\mathcal{P}(i)}).$$

We can implement this mathematical operation in a mechanical way by deleting the barren node from the directed graph and removing all the directed arcs that lead into it. Any probability calculations that affect or are affected only by the remaining nodes in the influence diagram will therefore be unaffected by the disappearance of the barren node.

Suppose, for example, that the payoff function in a decision problem depends only on the random quantities at nodes 3 and 4 in Figure 9-10. Clearly, the outcome of node 2 would not have any effect on the payoff, and thus, from the point of view of the decision maker, it can be disregarded. If we have a result or payoff function $R(\mathbf{x}')$ that depends on some subset $\mathbf{x}'$ of the chance vector $\mathbf{x}$ but excludes node j, then the conditional expectation of payoff at the value node does not depend on the barren node; moreover, removal of the barren node will not affect optimal decisions. The next two sections contain examples that illustrate the use of these results in simplifying chance influence diagrams and calculating posterior probabilities.

9.3.5   Cancer Diagnosis

We offer an example of a medical diagnosis problem to illustrate the use of arc reversal and barren node removal. In a medical report on diagnostic medicine it is stated that

> Metastatic cancer (M) is a possible cause of a brain tumor (T) and is also an explanation for increased total serum calcium (S). In turn, either of these could explain a patient falling into a coma (C). Severe headache (H) is also possibly associated with a brain tumor. However, in patients with confirmed brain tumors, knowledge about a coma provides no information about severe headaches."[6]

We draw a chance influence diagram to visualize the conditional probabilities that must be assessed so that one can then calculate the joint probability of the five random variables $C, H, M, S,$ and T. The random variables take on the small letters c, h, m, s, and t, respectively, when a condition is present; the absence of each condition is indicated by the (barred) letters $\bar{c}, \bar{h}, \bar{m}, \bar{s},$ and $\bar{t}$. Conditional probabilities are described in the usual way. For example, the probability a patient has a headache is $p(h) = Pr\{H = h\}$, the probability a patient has no tumor is $p(\bar{t}) = Pr\{T = \bar{t}\}$ and that the patient has a headache given the absence of a tumor is $p(h \mid \bar{t}) = Pr\{H = h \mid T = \bar{t}\}$.

The purpose of our analysis in this problem is to obtain an expression for the conditional probability that a patient will lapse into a coma given that only a brain tumor has been verified. In other words we are interested in calculating

$$p(c \mid t) = Pr\{\text{coma present} \mid \text{tumor present}\}.$$

---

[6] The original problem appeared in Spiegelhalter (1986).

We shall see that calculation of this conditional probability involves the reversal of an arc and the deletion of a barren node in the chance influence diagram.[7]

The conditional dependencies among the factors and symptoms is shown in Figure 9-11. The presence or absence of serum calcium and tumors depends on the presence or absence of cancer; either or both of these influence the probability of presence or absence of a coma. The presence or absence of a headache is influenced by the presence or absence of a tumor, but it influences no other variable. The important conditional independency is that coma is independent of the presence or absence of cancer given presence or absence of tumor and serum calcium.

We observe that headaches do not enter into the calculations because H is a barren node in Figure 9-11; it has no influence on any of the other variables and can be removed. Thus, we are left with a simpler diagram having only nodes M, S, T, and C shown in Figure 9-12.

The second observation is that the presence of a tumor ($T = t$), affects the conditional probability of having or not having cancer. In evaluating the conditional probabil-

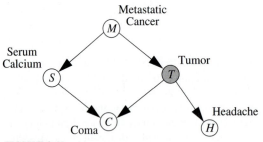

FIGURE 9-11
Chance influence diagram for cancer problem.

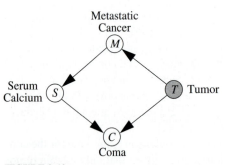

FIGURE 9-12
Modified chance influence diagram for cancer problem.

---

[7] In the interests of simplicity in this section we do not use primes to denote posterior probabilities calculated following arc reversal. The reader should understand that henceforth each arc reversal yields posterior (primed) probabilities.

ity $p(c|t)$, that is, the probability of a coma given that *only* a brain tumor has been verified, one is unable to observe whether there is or is not increased serum calcium or whether the patient has or does not have metastatic cancer. Because none of these conditions can be observed, one must allow for all possibilities, m and $\overline{m}$ as well as s and $\overline{s}$ in evaluating $p(c|t)$. Using Bayes' rule we can write the conditional probability of metastatic cancer given a tumor as

$$p(m|t) = \frac{p(t|m)\,p(m)}{p(t)} = \frac{p(t|m)\,p(m)}{p(t|m)\,p(m) + p(t|\overline{m})\,p(\overline{m})}. \tag{9.22}$$

Because M has no direct predecessors and T has only M as its direct predecessor, Equations (9.14), (9.15), and (9.16) yield

$$\mathcal{P}(T) = \{M\}, \; \mathcal{P}(M) = \emptyset \quad \text{and} \quad \mathcal{P}'(T) = \emptyset, \; \mathcal{P}'(M) = \{T\}.$$

Thus, there is no need to add additional arcs leading into M or T following arc reversal. Of course, the probability of not having cancer, given a brain tumor present, $p(\overline{m}|t)$, is just 1 minus the preceding expression. We also know that

$$p(c|t) = \sum_{x \in \mathcal{M}} \sum_{y \in \mathcal{S}} p(c \mid t, y)\,p(y \mid x)\,p(x \mid t), \tag{9.23}$$

where summation over the sets $\mathcal{M}$ and $\mathcal{S}$ confirms that we include both presence and absence of metastatic cancer and increased serum calcium.

Equation (9.23) is important to discuss in the context of Figure 9-12 because we can identify two different paths of influence that confirmation of a tumor has on the presence or absence of a coma. One path leads directly from T to C, the other from T to M to S to C. The T node is shaded to emphasize that its condition is known, not uncertain, whereas S, M, and C are uncertain. From the original problem formulation it is clear that the presence of a tumor directly affects the probability of a coma. What is not so obvious is how a tumor indirectly affects the probability of a coma when neither serum calcium nor cancer can be observed. The summations in Equation (9.23) have the effect of marginalizing over the random nodes M and S. The result is a conditional probability for a two-node chance influence diagram with one directed arc leading from T to C. Although there are two different paths of influence that lead from T to C in this problem, in more complex situations there may be many more than two paths of influence.

This is a good place to illustrate the use of conditional expectation to remove a chance node and calculate marginal distributions. Consider the summation over S in Equation (9.23). The conditional expectation,

$$E_{S|M}[p(c|t, S)|\, M = m] = \sum_{y \in S} p(c|\, t, y)\,p(y|\, M = m), \tag{9.24}$$

removes the unobserved S node from the chance influence diagram as illustrated in Figure 9-13(a). Similarly, by taking expectations with respect to the metastatic cancer condition, we obtain Figure 9-13(b) that shows us that the conditional probability of a coma can be assessed in terms of the presence or absence of a tumor when neither S nor M are observed.

Consider these operations when we are given the following numeric data:

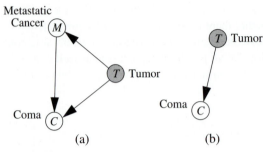

FIGURE 9-13
Removal of nodes S and M with conditional expectation.

1. Severe headaches are common but are more likely if a tumor is present; $p(h| t) = 0.8$, and $p(h| \bar{t}) = 0.2$.
2. Coma is rare but likely if either tumor or serum calcium is present; $p(c| t,s) = 0.8$, $p(c| \bar{t},s) = 0.8$, $p(c| t,\bar{s}) = 0.8$, $p(c| \bar{t},\bar{s}) = 0.05$.
3. Serum Calcium is unlikely but much more likely if metastatic cancer is present; $p(s| m) = 0.8$, $p(s| \bar{m}) = 0.2$.
4. Brain tumor is rare but somewhat more likely if metastatic cancer is present; $p(t| m) = 0.2$, $p(t| \bar{m}) = 0.05$.
5. In the population of interest one in five has metastatic cancer, that is, $p(m) = 0.2$.

These probability assessments are consistent with the influence diagram in Figure 9-11. As noted, because H is a barren node, the information in (1) is not required to determine $p(c| t)$ or $p(\bar{c}| t)$. We use Equation (9.22) to calculate $p(m| t) = 0.5$, so that our data are now in the form required by Figure 9-12. Finally, we use Equation (9.23), which has four terms, to find the result in Figure 9.13(b)

$$p(c| t) = p(c| t,s)p(s| m)p(m| t) + p(c| t,\bar{s})p(\bar{s}| m)p(m| t)$$
$$+ p(c| t,s)p(s| \bar{m})p(\bar{m}| t) + p(c| t,\bar{s})p(\bar{s}| \bar{m})p(\bar{m}| t) = 0.8.$$

9.3.6   Arc Reversal and Barren Node Removal

Consider a slightly more complicated inference problem related to the chance influence diagram shown in Figure 9-9 (p. 352). Here our objective is to determine the conditional probability $p(x_1| x_4)$. A possible sequence of arc reversals and barren node removals is shown in Figure 9-14, with the desired result in Figure 9-14(e). Note that the first of these coincides with Figure 9-10 (p. 354).

We want to eliminate nodes 2 and 3 from the graph by appropriate probabilistic and graphical manipulations. The first operation we consider is reversal of arc (2,4). Had we attempted to reverse either arc (1,2) or arc (3,4), we would have created directed cycles that are not allowed; thus, neither of these operations is performed. After the reversal of arc (2,4), x_4 has conditional probability

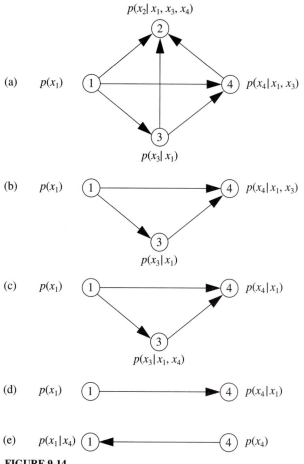

FIGURE 9-14
A sequence of equivalent chance influence diagrams.

$$p'(x_4|\, x_3,\, x_1) = \int p(x_4|\, x_3,\, x_2)\, p(x_2|\, x_3,\, x_1)dx_2, \tag{9.25}$$

and node 2 now has a conditional probability $p(x_2|\, x_1,\, x_3,\, x_4)$, calculated using Bayes' rule. After arc reversal, arrows from nodes 1 and 3 lead to 2 and also to 4. This introduces possible additional dependency relations. The new predecessor sets are: $\mathcal{P}'(2) = \{1, 3, 4\}$; $\mathcal{P}'(4) = \{1, 3\}$; $\mathcal{W}'(2, 4) = \{1, 3\}$ so that $\mathcal{U}'(2, 4) = \{4\}$, $\mathcal{U}'(4, 2) = \emptyset$.
The conditional probability $p(x_2|\, x_1,\, x_3,\, x_4)$ is

$$p'\left(x_2\middle|x_4,\, x_{\mathcal{W}'(2,4)}\right) = p'\left(x_2\middle|x_1,\, x_3,\, x_4\right) = \frac{p\left(x_2\middle|x_1,\, x_3\right) p\left(x_4\middle|x_2,\, x_3\right)}{p'\left(x_4\middle|x_1,\, x_3\right)}.$$

Node 2 in Figure 9-14(a) now has no successor nodes, carries no information on the relation between x_1 and x_4, is a barren node, and is therefore irrelevant to the calculation of $p(x_1|\, x_4)$. Hence, node 2, together with arcs leading into it, can be deleted from the graph. Figure 9-14(b) shows the graph at this stage with the same joint probability for (x_1, x_3, x_4) as would be obtained in Figure 9-14(a). Continuing in this way, we compute $p(x_1| x_4)$ by reversing arc (1,3) (Figure 9-14(c)), eliminating barren node 3, and finally reversing the arc (1,4). This final arc reversal yields the desired result, $p(x_1| x_4)$, in Figure 9-14(e).[8]

9.4   PATH HISTORY AND ROLLBACK COMPUTATIONS

An important requirement of any good procedure for solving a decision analysis problem is the organization and assembly of the appropriate data required to solve the problem. The data can be represented in formulation tables or spreadsheets associated with the nodes and outcome sets of an influence diagram or by labeling nodes and branches of a decision tree; the latter may not have to be represented graphically in the usual sense of drawing a picture. This construct and the assessment of costs and probabilities are as important and usually much more difficult than are the dynamic programming or rollback calculations once the data have been organized, assembled, and displayed in graphical or tabular form. There are two essential phases of any procedure to assemble relevant data and then solve a decision problem: the first consists of identifying costs and other variables associated with branch payoffs and probabilities that may be path- or scenario-dependent. In this phase the computational procedure computes and moves from left to right (forward in time) because the probabilities of outcomes and costs incurred in traversing branches depend on what has happened in the past *not upon what may happen in the future*. The second phase concentrates on a recursive rollback algorithm using the data and other intermediate results that have been assembled and calculated in the first phase. As pointed out, the second phase consists of calculations based on the principle of optimality that proceed from right to left (toward the root or backward in time) on the decision tree.

The tree in Figure 9-6 (p. 348) shows how events and decisions unfold over time in the replacement decision whose extensive form influence diagram was given in Figure 9-5. It has three chance nodes, two decision nodes, six value nodes, and ten branches. In the influence diagram there are only two chance nodes, one decision node, one value node, and five directed arcs. Unlike in earlier chapters, decision trees now use the same node-numbering convention as the influence diagram. In discussing rollback algorithms heretofore each chance, decision, or value node was given a unique name or number, whereas in Figure 9-6 the numbering of nodes in the decision tree coincides with the numbering system used in the influence diagram. It should be noted that although the decision node is used on both branches leaving node 1, the choices differ; there are two choices at the upper node 2 and only one at the lower node 2. Node 3 outcomes need not be considered in the bottom branch whereas there are three in the top (A) and two in the bottom (M) branch. Value node 4 appears in each of the six distinct paths.

---

[8] This problem was selected from Barlow (1988).

Although there are great advantages to using the new node-numbering convention, it has the unfortunate feature that the sequence of node numbers on a path *does not* uniquely define a path. For example, there are five paths with the node sequence {1, 2, 3, 4}; in the design of rollback algorithms that find optimal decisions one must find a way to label and identify the probabilities and costs or returns associated with different paths even though they may use the same sequence of numbered nodes. Path labels must recognize both the node and outcomes sets in a properly formulated decision tree.

With distinct node numbers on a decision tree, say, i and j, it was possible to identify the branch from node i to j by the convention (i,j); this is no longer feasible as there may be several different branches between the same (i,j) node pair; moreover, these would conflict with the completely different meaning of the directed arc (i,j) in the influence diagram. From now on the jth branch leading out of node i in a decision tree will denote the jth outcome. The path that traverses node i in the decision tree and then traverses branch j leads to the next node, k.[9] Although this modified notation seems simple enough, it has been our experience that there is often confusion of outcomes or alternatives (branches) in decision trees with flow of information or influences (directed arcs) in influence diagrams. They must not be confused.

An algorithm that makes use of the node-numbering convention in Figure 9-6 must recognize that node numbers in the decision tree are not unique; it must therefore use *branch* identifiers from the $C(i)$ and $\mathcal{D}(i)$ sets to enumerate outcomes at chance nodes and alternatives at decision nodes and thereby enumerate distinct paths from origin to terminal nodes. Node labels should compactly represent the additional information that influences costs or probabilities on the path leading to the node of interest; we therefore construct a path history vector that carries the information required to provide a dynamic programming rollback calculation. For the rollback computation to be successful all data must be in place on terminal nodes and branches, including branch probabilities and costs that occur along a particular path.

9.4.1   A History Vector Algorithm

Let $\mathbf{h}_i$ be a *state or history vector*[10] that contains a list of nodes (chance or decision) and branches (outcomes or alternatives) that uniquely identify a path leading from the starting node to node i.[11] Because there may be many paths to a given node we associate with node i a set $\mathcal{H}_i = \{\mathbf{h}_i\}$. Assume that we reach node i via history $\mathbf{h}_i$ and then follow branch j (a random outcome or a decision alternative) that leads us to the next node k. Then the state vector at node k is given by $\mathbf{h}_k = [\mathbf{h}_i,(i,j)]$. Here, each element of the history vector is a pair of numbers in which the first element is a node number and the second element

---

[9] In the literature the next node, k, is sometimes denoted by the function n_{ij}.

[10] Our history vectors $\mathbf{h}$ differ from the state s notation used by Kirkwood (1993). In this section we only require information on nodes and branches, no payoffs or costs incurred in traversing the path. However, see Section 9.5.5 (p. 376).

[11] The node i refers to the node number in the influence diagram of which there may be many copies in the tree.

is a branch number that represents either an outcome or an alternative connecting node i to k; in other words, branch j is a member of either $C(i | \mathbf{h}_i)$ or $D(i | \mathbf{h}_i)$. It is important for the reader to understand the meaning of each element of $\mathbf{h}_i$, and how our notation differs from that used in the basic rollback algorithm on page 168 in Section 4.4.4.

The following procedure generates all history vectors. The number of paths in the tree is equal to the count of members of the disjoint sets $\mathcal{H}_i$, $i \in \mathcal{V}$. Assume that the starting node is numbered 1.

1. Define a set $\mathcal{H}_i$ for each node $i \in \mathcal{N}$, and set $\mathcal{H}_i = \emptyset$ (the empty set) for all i.
2. Set $\mathbf{h}_1 = \emptyset$ (the empty vector) and $\mathcal{H}_1 = \{\mathbf{h}_1\}$.
3. Consider the next higher numbered node. If at node i and $i \in D$, each branch j with $j \in D(i | \mathbf{h}_i)$ leads to a node k. Set $\mathbf{h}_k = [\mathbf{h}_i, (i, j)]$, and add it to $\mathcal{H}_k$.
4. If at node i and $i \in C$, each branch j with $j \in C(i | \mathbf{h}_i)$ leads to a node k. Set $\mathbf{h}_k = [\mathbf{h}_i, (i, j)]$, and add it to $\mathcal{H}_k$.
5. Repeat steps 2 and 3 for all chance and decision nodes to determine the disjoint sets $\mathcal{H}_i$, $i \in \mathcal{V}$.

At this point the procedure is terminated. The union of $\mathcal{H}_i$ sets with $i \in \mathcal{V}$, contains all the paths in the tree.

The history vector algorithm allows great flexibility in constructing path-dependent outcome and alternative sets, conditional probabilities, payoffs, and cost functions. Although this formulation of the history vector calculation provides only that information required to identify the nodes and branches traversed in following a particular path from origin to terminal node, the path history can easily include functions assigned to node variables and branches that are used to construct utilities at terminal nodes.

9.4.2 Engine Maintenance

We demonstrate the procedure using the engine maintenance problem EFID in Figure 9-5 (p. 348). The reader is encouraged to follow the procedure without referring to the decision tree in Figure 9-6. Greater appreciation will be gained for the difficulty of designing a computer algorithm to find every path through the tree; the computer cannot see!

1. Set $\mathbf{h}_1 = \emptyset$ and $\mathcal{H}_1 = \{\mathbf{h}_1\}$.
2. $C(1) = \{a, b\}$ so that

$$\mathcal{H}_2 = \{[(1, a)], [(1, b)]\}.$$

3a. $D(2 | [(1, a)]) = \{A, M\}$ so that

$$\mathcal{H}_3 = \{[(1, a), (2, A)], [(1, a), (2, M)]\}.$$

b. $D(2 | [(1, b)]) = \{R\}$ so that

$$\mathcal{H}_4 = \{[(1, b), (2, R)]\}.$$

4.

$$C(3|[(1, a), (2, A)]) = \{1, 0, -1\} \text{ and}$$

$$C(3|[(1, a), (2, M)]) = \{1, 0\}, \text{ so that}$$

$$\mathcal{H}_4 = \{[(1, a), (2, A), (3, 1)], [(1, a), (2, A), (3, 0)],$$

$$[(1,a), (2, A), (3, -1)], [(1, a), (2, M), (3,1)],$$

$$[(1, a), (2, M), (3, 0)], [(1, b), (2, R)]\}.$$

Because $\mathcal{V} = \{4\}$ has only one element, the set $\mathcal{H}_4$ contains all six paths from the origin node to the single-value node. Although it may look overly complex for hand calculations, this scheme can be easily implemented on a computer to quickly generate all paths in a decision tree.

Having constructed the sets containing all paths, we are now interested in finding the best decisions and alternatives. Recall that in the earlier rollback algorithm on page 168 in Section 4.4.4 we used v_i to label node i on the optimal path leading backward from a terminal node. The only difference here is that v_i labels are explicit functions of node and branch numbers contained in the history vector.

9.4.3 Rollback Using Path History Vectors

Assume that we are given the payoff $r_j(\mathbf{h}_i)$ for every branch j leaving node i in the decision tree.[12] Because every terminal node is at the end of a single branch we assume that any payoff associated with a terminal node is assigned to that single branch. In what follows we also assume that we are interested in maximizing expected value, v_1, at the origin node. The following algorithm finds the optimal paths, decisions, and returns:

1. Label all terminal nodes at the end of every path with 0; this means that we set $v_i(\mathbf{h}_i) = 0$ for every $\mathbf{h}_i \in \mathcal{H}_i$, $i \in \mathcal{V}$.
2. Find an unlabeled node i where all later-occurring nodes k connected to it are labeled.
   a. If $i \in \mathcal{D}$, set

$$v_i(\mathbf{h}_i) = \underset{j \in \mathcal{D}(i|\mathbf{h}_i)}{Max} \{r_j(\mathbf{h}_i) + v_k(\mathbf{h}_k)\} \qquad (9.26)$$

   and d_i^* equal to the j that yields this maximum.
   b. If $i \in C$, set

$$v_i(\mathbf{h}_i) = \sum_{j \in C(i|\mathbf{h}_i)} p_j(\mathbf{h}_i)[r_j(\mathbf{h}_i) + v_k(\mathbf{h}_k)]. \qquad (9.27)$$

Step 2 is repeated until the starting node is labeled. This starting node label gives the maximum expected value for the problem; moreover, the v_i's at each node give the max-

---

[12] This notation explicitly allows branch costs or payoffs to depend on history.

imum expected value from that point forward and the d_i*'s computed for each decision node give the optimal decisions. At each decision node $i \in \mathcal{D}$ one may also want to attach a second label that represents the optimal decision, d_i*, to be taken at that node.

The recursion described in this section for calculating optimal expected values is identical to the rollback algorithm in Chapter 4 that used a unique node-numbering system, except where computations reveal explicitly how path history influences branch probabilities and branch-dependent payoffs or costs. We have labeled costs and returns on the branches of the tree to emphasize the location and timing of their occurrence; alternatively, we could have carried all costs and payoffs forward to terminal nodes. As formulated, the algorithm can be used to calculate optimal expected values when the decision maker is risk neutral.

9.4.4 A Nuclear Reactor Decision Example

We illustrate the algorithms of Section 9.4.3 in a decision problem on whether to build a conventional fission or advanced fusion reactor to provide future energy needs. The economic gains, costs, and operating characteristics of the conventional reactor are well understood but are much more uncertain in the case of the fusion reactor. If an experimental test of the fusion reactor yields promising results, there is potential for great savings. On the other hand there is the possibility that the newer fusion reactor design will yield much worse operating conditions and greater costs—in other words it is riskier. The objective is to maximize the expected lifetime net payoff. We first construct an influence diagram, and using the data from the problem construct a formulation table, then use a spreadsheet to perform the rollback calculations. As the reader will see, this example is small enough to be solved by hand but is complex enough to demonstrate the need for and details of a computer-based algorithm that records path histories and makes use of data contained in a formulation table.

There are two decisions to be made; first whether to test (T) or not test (N) the advanced reactor design, second whether to build a conventional (C) or advanced (A) reactor. If it is decided to test, the outcome of the test cannot be predicted with certainty. We assume that the result of the test will be either positive (p) or negative (n). If a conventional reactor is to be built, its future performance, although well understood and not dependent on the outcome of the test of the advanced design, is still uncertain. It will be either a success (s) or a failure (f). If an advanced reactor is to be built, its future performance is more uncertain. It may be a success (s), lead to a minor accident (m), or lead to a large accident (l). If a decision is made to test, the decision as to which reactor to build cannot be made until the results of the test are known.

The statements in the preceding paragraph imply that $\mathcal{D}$ has two elements, C has three, and $\mathcal{V}$ has one, so that the influence diagram has a total of six nodes. The numbering system shown yields the sets $\mathcal{D} = \{1,3\}$, $C = \{2,4,5\}$, and $\mathcal{V} = \{6\}$. The probabilistic dependencies are seen in the diagram in Figure 9-15.

Knowing the sets $\mathcal{D}$, C, and $\mathcal{R}$, the alternatives and chance outcomes given above and the structure shown in the extensive form influence diagram of Figure 9-15, we can use the History Vector Algorithm to construct the decision tree. The construction of the history vectors in Table 9.1 can be easily made; the results are shown graphically in the decision tree in Figure 9-16.

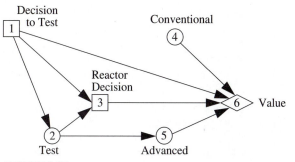

FIGURE 9-15
Influence diagram for reactor decision problem.

TABLE 9.1
Path history vectors for the reactor problem in Figure 9-15

Node i	Node type: description	$C(i\|h_i)$ or $\mathcal{D}(i\|h_i)$	Next node, k	History vectors, h_i in $\mathcal{H}_i$
1	Decision: test or not	N: No test	3	∅
		T: Test	2	
2	Chance: test outcome	p: positive	3	[(1, T)]
		n: negative	3	
3	Decision: reactor type	C: Conv	4	[(1, N)]
		C: Conv	4	[(1, T), (2, p)]
		A: Advanced	5	
		C: Conv	4	[(1, T), (2, n)]
		A: Advanced	5	
4	Chance: conventional reactor outcome	s: success	6	[(1, N), (3, C)]
		f: failure	6	
		s: success	6	[(1, T), (2, p), (3, C)]
		f: failure	6	
		s: success	6	[(1, T), (2, n), (3, C)]
		f: failure	6	
5	Chance: advanced reactor outcome	s: success	6	[(1, T), (2, p), (3, A)]
		m: minor	6	
		l: large	6	
		s: success	6	[(1, T), (2, n), (3, A)]
		m: minor	6	
		l: large	6	
6	Result: total energy value			[(1, N), (3, C), (4, s)]
				[(1, N), (3, C), (4, f)]
				[(1, T), (2, p), (3, C), (4, s)]
				[(1, T), (2, p), (3, C), (4, f)]
				[(1, T), (2, p), (3, A), (5, s)]
				[(1, T), (2, p), (3, A), (5, m)]
				[(1, T), (2, p), (3, A), (5, l)]
				[(1, T), (2, n), (3, C), (4,s)]
				[(1, T), (2, n), (3, C), (4,f)]
				[(1, T), (2, n), (3, A)), (5, s)]
				[(1, T), (2, n), (3, A), (5, m)]
				[(1, T), (2, n), (3, A), (5, l)]

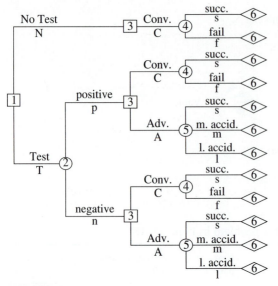

FIGURE 9-16
The reactor problem decision tree.

The reader should be able to trace all paths in the tree. Note that at the value node (6) there are a total of twelve different paths with different history vectors. These contain different sequences of nodes and events because of different chance outcomes or different choices of decision alternatives on each of the twelve paths through the tree.

The information required to solve the decision tree is shown in the first four columns of Table 9.1 and columns 3 and 4 of Table 9.2. These latter contain the branch probabilities and branch payoffs (in $ billions) corresponding to cost of test, construction, cleanup costs, and so on. Recall that Table 9.1 identifies the node type, name, number, outcomes, and alternatives as well as the next node in the history vector that leads from node 1 to node 6, the value node.

Starting at terminal node 6 the algorithm assigns values of 0 to all terminal or value nodes. Node 5 is a chance node, and we therefore use step (2b) of the algorithm to calculate expected payoffs for each branch. Moving to node 4, another chance node, we perform similar calculations. Node 3 is a decision node, and we therefore use step (2a) of the algorithm; the expected payoffs are shown in column 5 and the optimal decision in column 6. We calculate expected values at chance node 2 and finally find the optimal decision and maximum expected payoff at decision node 1. The reader is urged to work through the algorithm to produce Tables 9.1 and 9.2 and provide appropriate node labels in Figure 9-16. To summarize the results, the optimal policy is:

1. Proceed with the test of the advanced reactor.
2a. If the test is positive, build the advanced reactor.
2b. If the test is negative, build the conventional reactor.

TABLE 9.2
Rollback solutions for the reactor problem

Node	Next node	Branch probability $p_j(\mathbf{h}_i)$	Branch value, $r_j(\mathbf{h}_i)$	Expected node payoff	Optimal policy decision
1	3	—	0		
	2	—	−1	8.13	Test
2	3	0.60	—	9.13	
	3	0.40	—		
3	4	—	−4	7.76	Conventional
	4	—	−2		
	5	—	−4	10.04	Advanced
	4	—	−2	7.76	Conventional
	5	—	−2		
4	6	0.98	10		
	6	0.02	−2	9.76	
	6	0.98	10		
	6	0.02	−2	9.76	
	6	0.98	10		
	6	0.02	−2	9.76	
5	6	0.40	16		
	6	0.46	−2		
	6	0.14	−6	14.04	
	6	0.90	16		
	6	0.06	−2		
	6	0.04	−6	4.64	
6	none	—	0	0	

Columns 5 and 6 in Table 9.2 display the optimal expected returns and policies obtained from the rollback calculations. The expected payoff with this policy is $8.13 billion. Sensitivity analyses can be performed easily by changing some or all the data in columns 3 and 4. The influence diagram, in conjunction with a formulation table containing the outcome and alternative sets and branch probabilities and arc costs, can be assembled in such a way that the implementation of the history path algorithms is invisible to the user. A modern spreadsheet program can be used to perform the calculations without ever having to explicitly draw the tree from the influence diagram.

In a decision as momentous as whether or not to proceed with an advanced fusion reactor there are many factors to be considered and many ways to measure the outcome of the decision. We therefore revisit this problem in the next section and demonstrate how different measures of performance can lead to conflicting optimal decisions. As discussed in Chapter 7, this possibility forces us to consider trade-offs among different attributes and to see how the optimal policies and overall objectives change as different trade-off weights are considered.

9.5 MULTIATTRIBUTE ROLLBACK WITH AND WITHOUT TRADE-OFFS

Thus far we have described rollback algorithms that can be used to solve decision trees whose objectives are defined in terms of a single attribute. Before we incorporate the notion of trade-offs and conflict resolution to solve objectives that contain more than one attribute, it is useful to see how the labeling algorithm proceeds when we perform rollback calculations separately for each attribute as if they were independent of one another. For example, one might be interested in finding optimal decision sequences if minimum cost policies were sought as well as optimal sequences that maximized reliability or minimized risk.

9.5.1 Calculating Noninferior Points with Two Attributes

The rollback algorithm just described can be extended without difficulty to multiple attributes. By making history path and rollback calculations separately for each attribute, we can then focus on those decisions and chance events where conflicts arise. If the number of conflicts is small, the decision maker can use informal comparisons to identify preferences and subjectively select a preferred choice. At a given node all attributes may yield a dominant solution in which case there is no policy conflict. At other nodes, however, the optimal policy to follow from that point on is attribute-dependent; to resolve the conflicts, one must implicitly or explicitly indicate a preference for one solution vector over another.

If we are comparing expected values, the rollback recursion is similar to the procedures discussed in earlier sections. Consider the case of two attributes. Let $r_j(\mathbf{h}_i)$ denote the payoff (costs are negative) in traversing branch j from node i for the first attribute, and let $s_j(\mathbf{h}_i)$ denote the payoff for the second. Also, let (v_i, u_i) denote the optimal expected return at node i for attributes 1 and 2, respectively.[13] At decision nodes we have

$$[v_i(\mathbf{h}_i), u_i(\mathbf{h}_i)] = Max_j\{r_j(\mathbf{h}_i) + v_k(\mathbf{h}_k); s_j(\mathbf{h}_i) + u_k(\mathbf{h}_k)\}, \qquad i \in \mathcal{D}, j \in \mathcal{D}(i|\mathbf{h}_i), \quad (9.28)$$

where the meaning of the vector maximization in (9.28) is that at the decision node i the best alternative j is selected for each attribute. When there is a conflict, the optimal path will lead to different "next" nodes k. At chance nodes on the decision tree we have

$$[v_i(\mathbf{h}_i), u_i(\mathbf{h}_i)] = \sum_j p_j(\mathbf{h}_i)[r_j(\mathbf{h}_i) + v_k(\mathbf{h}_k), s_j(\mathbf{h}_i) + u_k(\mathbf{h}_k)], \qquad i \in \mathcal{C}, j \in C(i|\mathbf{h}_i). \quad (9.29)$$

As before, the path history vector at the next node is obtained from the recursion defined in Section 9.4.1 (p. 363).

Proceeding in this way we can calculate all node labels that denote the optimal expected value from that node to the terminal or value node. It is then possible to

---

[13] In this section we depart from our usual convention and use u and v to denote node labels for two different attributes. In all other sections of the book u is reserved for a utility and v is either a node label or a value function. U is reserved for the multiattribute utility function and V is the name of the value node.

reject inferior (dominated) solutions and only retain those which identify a conflict and the possibility of a trade-off between attributes. By graphing (in two dimensions for two attribute) or listing those solutions that cannot be rejected, it may be possible for the decision maker to identify a subset of feasible solutions and then make a preferred choice even when one is unable to specify a formal or numeric trade-off between attributes.

9.5.2   Rollback with Linear Trade-offs

In a two-attribute problem, assume that we use a trade-off weight of unity for the first attribute and a trade-off weight w for the second attribute. Define $v_i(\mathbf{h}_i, w)$ as the optimal return at node i conditional on the path history $\mathbf{h}_i$ and the scalar trade-off weight w. At decision and chance nodes we can write the general recursions for the value function as we did before: use of the optimality principle provides a dynamic programming equation that now includes w. Obviously there will only be one optimal return label at each node that is measured in the same units as those used for the payoffs $r_j(\mathbf{h}_i)$. With w fixed

$$v_i(\mathbf{h}_i, w) = Max_j[(r_j(\mathbf{h}_i) + ws_j(\mathbf{h}_i)) + v_k(\mathbf{h}_k, w)], \qquad j \in \mathcal{D}(i|\mathbf{h}_i), i \in \mathcal{D} \qquad (9.30)$$

and

$$v_i(\mathbf{h}_i, w) = \sum_j p_j(\mathbf{h}_i)[(r_j(\mathbf{h}_i) + ws_j(\mathbf{h}_i)) + v_k(\mathbf{h}_k, w)], \qquad j \in C(i|\mathbf{h}_i), i \in C. \qquad (9.31)$$

Whatever procedure we developed previously for the recursion can be used with one small modification: along branches that represent random outcomes or decision alternatives, we simply substitute $(r_j(\mathbf{h}_i) + ws_j(\mathbf{h}_i))$ for the quantity $r_j(\mathbf{h}_i)$ used with one attribute. Thus, the rollback procedures described in Section 9.4 extend easily to the multiattribute case.

The inherent simplicity of the algorithm for the additive linear case can be seen by defining the mixture of node-branch payoffs

$$r_j(\mathbf{h}_i | \mathbf{w}) = \sum_l w_l r_{jl}(\mathbf{h}_i), \qquad (9.32)$$

where $r_{jl}(\mathbf{h}_i)$ is the payoff of the lth attribute on branch j and the summation is over all attributes. The vector $\mathbf{w}$ has n elements corresponding to the number of attributes. The rollback equations then lead to the optimal expected value node labels

$$v_i(\mathbf{h}_i, \mathbf{w}) = Max_j[r_j(\mathbf{h}_i | \mathbf{w}) + v_k(\mathbf{h}_k, \mathbf{w})], \qquad j \in \mathcal{D}(i|\mathbf{h}_i), i \in \mathcal{D} \qquad (9.33)$$

and

$$v_i(\mathbf{h}_i, \mathbf{w}) = \sum_j p_j(\mathbf{h}_i)[(r_j(\mathbf{h}_i | \mathbf{w}) + v_k(\mathbf{h}_k, \mathbf{w})] \qquad j \in C(i|\mathbf{h}_i), i \in C. \qquad (9.34)$$

We emphasize that $\mathbf{w}$ is an n-dimensional trade-off vector but that the node labels are scalar path-dependent expected value functions. The rollback algorithm can obviously be used to make sensitivity analyses on trade-off weights, conditional probabilities, optimal return functions, and attribute costs to reveal policy regions and boundaries that are important to a decision maker.

9.5.3   Reactor Decision Revisited

We return to the reactor problem described in Section 9.4.4 on whether to build a conventional or advanced fusion reactor. The basic formulation is identical in all respects to the original problem except that we now measure the results with two criteria: (1) economic return, and (2) safety. Costs arise in the production and operation of the reactor, revenue is generated from the sale of electric power, and safety and reliability issues arise because there may be loss of life if reactor accidents occur. The economic returns and safety characteristics of the conventional reactor are well understood, but we are much less certain about each of these in the case of the advanced reactor. If an experimental test of the advanced reactor yields promising results, there is great potential for savings and increased safety (expressed in terms of reduced numbers of lives lost). On the other hand the advanced reactor design may yield much riskier operating conditions at greater cost should the test prove unsuccessful. Dependencies among events and decisions are illustrated in the influence diagram in Figure 9-15 (p. 367). Clearly, the reactor decision must be made before the uncertain outcomes are known. Using loss of life as the attribute by which to measure safety, our objective is to choose those decisions that both maximize the expected net economic return and minimize the expected loss of life over the lifetime of the reactor. It is not clear that both of these can be attained with the same decisions.

In the case of the conventional reactor there are only two possible outcomes: successful and unsuccessful operation; with the advanced reactor there are three possible outcomes: success, a minor accident, and a large accident reflecting the likelihood that with the advanced design there are great potential benefits, but there is also the risk that the decision may lead to a large accident. Conditional probabilities for the advanced reactor design are identical to those shown in Table 9.1 and Table 9.2. The costs, economic returns (both in $ billions), and estimated loss of life (in thousands) are given in Table 9.3. The operating characteristics of the conventional reactor are independent of any decisions or test results. The test decision is also expensive. We recall that the two possible outcomes with the building of a conventional reactor are independent of other outcomes or decisions or characteristics of the advanced reactor; on the other hand the outcomes for the advanced reactor are influenced by whether we have positive or negative test results.

All branches in the decision tree in Figure 9-17 have a two-element vector associated with them, the data for which are obtained directly from Table 9.3. The first ele-

TABLE 9.3
Economic and loss of life data for reactor problem

Reactor type	Construction cost	Accident cleanup cost		Accident loss of lives		Economic return
		Minor	Large	Minor	Large	
Conventional	2		2		4	10
Advanced	4	2	6	1	3	16

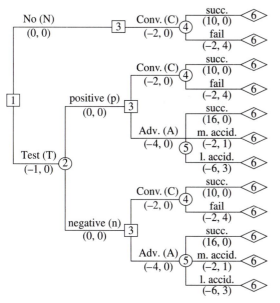

FIGURE 9-17
The reactor decision problem with two attributes.

ment shows the net economic return, and the second element the lives lost. Net economic return is obtained by subtracting costs from revenues and is seen to be negative on some of the branches.

Using the rollback algorithm we denote the expected value of the optimal policy at the ith node by v_i. We begin by setting the economic return and loss of life at node 6 in Figure 9-17 equal to zero (i.e., $v_6 = 0$). In the calculations that follow multiplications and additions are described on two-element vectors, one element for each of the attributes. Recall that we keep track of paths by a paired (node, branch) notation; for example, the path history $[(1,T),(2,p),(3,A)]$ denotes the decision to test at node 1, the outcome of a positive report at node 2, and the selection of the advanced reactor at decision node 3.

Conditional expectations at chance nodes are calculated as the sum of products of the form $p_j(\mathbf{h}_i)[r_j(\mathbf{h}_i) + v_k(\mathbf{h}_k)]$, where j is a branch corresponding to a particular outcome (with $j \in C(i|\mathbf{h}_i)$) and k is the next node on branch j identified in the history vector $\mathbf{h}_k$. Recall that $\mathbf{h}_i$ denotes the path history from the origin to node i.

At the nodes 5, the expected outcomes following testing and both positive and negative reports are

$$v_5([(1,T), (2,p), (3,A)]) = 0.9(16,0) + 0.06(-2,1) + 0.04(-6,3)$$

$$= (14.04, 0.18),$$

$$v_5([(1,T), (2,n), (3,A)]) = 0.4(16,0) + 0.46(-2,1) + 0.14(-6,3)$$

$$= (4.64, 0.88).$$

The expected outcome from a conventional reactor is independent of the path taken to get to node 4 so that we have

$$v_4([(1,N), (3,C)]) = v_4([(1,T), (2,p), (3,C)]$$

$$= v_4([(1,T), (2,n), (3,C)]$$

$$= 0.98(10,0) + 0.02(-2,4)$$

$$= (9.76, 0.08).$$

At decision nodes we use a different calculation. To illustrate we calculate the best alternatives at node 3 for each of the three different paths leading to a node 3 decision. The No Test branch at the top of the tree yields

$$v_3([(1,N)]) = (9.76 - 2, 0.08 + 0)$$

$$= (7.76, 0.08),$$

whereas in the bottom branch that corresponds to the Test decision followed by a negative test report we have

$$v_3([(1,T), (2,n)]) = (Max\{9.76 - 2, 4.64 - 4\}, Min\{0.08 + 0, 0.88 + 0\})$$

$$= (7.76, 0.08).$$

The conventional reactor yields both the largest expected economic return and at the same time the lowest expected number of lives lost; thus, we are able to reject a strictly *inferior* solution (i.e., worse in *both* attributes) that corresponds to selecting an advanced reactor when the test yields a negative report.

The remaining decision alternatives to be considered at node 3 occur when the test results are positive. The maximization at decision node 3 yields

$$v_3([(1,T),(2,p)]) = (Max\{9.76 - 2, 14.04 - 4\}, Min\{0.08 + 0, 0.18 + 0\})$$

$$= (10.04, 0.08). \tag{9.35}$$

In the first attribute we find that we should make the decision to construct the advanced reactor but with the second attribute (lives lost) we should construct the conventional reactor. Thus, the optimal path is attribute-dependent and poses a policy conflict; resolving this conflict is at the heart of the decision problem. Because we cannot resolve it at this time, we keep both vectors in the label. The node label with two-row vectors emphasizes that the expected economic return is maximized by building an advanced reactor, but the expected number of lives lost is minimized by building a conventional reactor.

The optimal decision alternative is unclear at node 3 along the [(1,T),(2,p)] path; this conflict will *continue* to propagate through all chance and decision nodes that occur earlier in time. Unless we find a way to resolve the conflict at this decision node, the calculations of expected economic benefit at earlier decision points must include the possibility that later decisions will not be able to reject strictly inferior solutions. At node 2, for example, we have

$$v_2([(1,T)]) = 0.6\{(7.76, 0.08); (10.04, 0.18)\}$$
$$+ 0.4\{(7.76, 0.08); (7.76, 0.08)\}$$
$$= \{(7.76, 0.08); (9.128, 0.14)\}. \tag{9.36}$$

At node 1 along the test branch the expected economic return is reduced by 1, the cost of testing; thus, the optimal policy is a choice between the conventional and advanced reactor,

$$v_1(\emptyset) = \{(7.76, 0.08); (8.128, 0.14)\}. \tag{9.37}$$

The choice cannot be resolved because the former is safer in terms of lives lost and the latter produces greater expected economic return. If we use attribute 1, we should not pursue the testing alternative; on the other hand if we use the second attribute, we should pursue the results of the test. Thus, our inability to resolve the reactor choice at node 3 has now propagated to the first decision node where we must decide whether to test or not.

9.5.4 Economic Value per Life Saved

In the nuclear reactor decision, assume that there is a dollar trade-off per life saved that we denote by w. Although the idea of such a trade-off may disturb some people who think that no monetary amount can be equated to the value of an individual's life, such a person might be willing to spend an amount of money, w, on safety devices to reduce risk and increase the probability of saving a life.[14] Using the notation for multiattribute problems developed in Chapter 7, for the economic return attribute, $\bar{r} > \underline{r}$ and so $\Delta > 0$. For the loss of life attribute, $\bar{r} < \underline{r}$, and so $\Delta < 0$. Because we wish to maximize expected economic return and minimize expected loss of life, using the sign convention shown in Table 7.2 (p. 266), we combine net economic return with loss of life using a negative weight w and maximize the equivalent expected net economic return. This weight w can be thought of as the cost you are prepared to incur in millions of dollars to prevent one loss of life. If a is the net economic return produced by power generation and b is the number of lives lost, then an "equivalent" economic return is given by $(a + wb)$. To understand better the connection between the trade-off weight and the uncertainty of the test report, assume that p is the probability of a positive test report, and that $1 - p$ is the probability of a negative report. We now have two unknown parameters, w and p, and want to investigate the policy regions in terms of these parameters. Are there different policy regions in the (p,w) space? If there are, how do we calculate the boundaries between them?

At decision node 3 along the $[(1,T),(2,p)]$ path the label is seen to be $\{(7.76, 0.08); (10.04, 0.18)\}$. We are indifferent to the advanced and conventional reactor decisions when

---

[14] See Chickering (1994).

$$7.76 + 0.08w = 10.04 + 0.18w.$$

Solving for w yields the value $\overline{w} = -\$22.8$ million/life lost. Notice that this indifference value at node 3 in Figure 9-17 is independent of the probability of a positive test report. When $w < \overline{w}$, we have placed an even higher value on lives lost than $\overline{w}$, in which case it is optimal to construct the conventional reactor. In this case, the label on node 2 is $(7.76, 0.08)$ and on node 1 is

$$v_1(\{\emptyset\}) = (Max\{7.76 - 1, 7.76\}, Min\{-0.08, 0.08\})$$

$$= (7.76, -0.08).$$

The optimal initial decision is not to test, *independent of p.*

For the case where $w > \overline{w}$, the reader can show that the optimal decision is to test and, based on the report, decide whether or not to build an advanced reactor. At node 3 on the $[(1,T),(2,p)]$ path the decision to build an advanced or conventional reactor is independent of the probability p; the decision rests only on the size of w relative to $\overline{w}$. At node 1, on the other hand, the decision whether to test or not is dependent on both p and w. It is not difficult to show that at node 1 the decision to test is optimal when p satisfies

$$p > \frac{1}{2.28 + 0.1w}. \tag{9.38}$$

The results for both nodes 1 and 3 are shown graphically in Figure 9-18. The curve from Equation (9.38) in the northwest corner of (a) subdivides the (p,w) plane into the Test and Do Not Test subregions at node 1. The vertical line at $\overline{w}$ divides the rectangle in (b) into two subregions for the decision at node 3.

9.5.5   History and Rollback with Nonlinear Utilities

In the computational procedures that we have described the first calculation is the path history, the second is a calculation that, starting at the terminal nodes, uses rollback and the inclusion of new cost data on branches traversed along different paths. These are appropriate when we are dealing with expected value or risk-neutral utility; the calculations have the appealing feature that one can construct audit trails that indicate where and when new costs and benefits arise. Costs that have occurred in the past can be treated as fixed sunk costs that do not enter into the calculations of optimal future decisions. This is appealing to the accounting profession and to decision makers whose primary concern is to focus their attention on marginal costs above and beyond those that have already been committed. This formulation and solution procedure does not take into account how risk aversion must assess asset position, which is determined not only by future uncertain outcomes but also by the costs and benefits of past events and decisions. For example, historical commitment to a major allocation of resources may make moot the possibility of further allocations to a highly desirable new alternative. To analyze preferred alternatives while recognizing risk aversion requires that one construct the sequence of costs, returns, and payoffs on each path and then assess and compare the global utility function on each path.

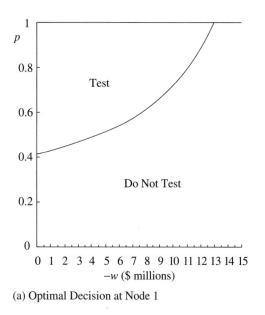

(a) Optimal Decision at Node 1

Advanced	Conventional

$0 \qquad\qquad -\overline{w} = \$22.8 \text{ millions} \qquad -w$

(b) Optimal Decision at Node 3 if Test Result Is Positive

FIGURE 9-18
Optimal policy regions for the reactor problem.

We are interested in risk attitudes and utility functions such as those described in Chapter 6. Rather than incorporate utility calculations in the rollback procedure, one can easily incorporate branch data in the state or history path vector and then compute the utility of each path whenever a terminal node is reached. Recall that the jth branch payoff or cost from node i is denoted by $r_j(h_i)$, where $j \in \mathcal{D}(i|h_i)$ if i is a decision node and $j \in \mathcal{C}(i|h_i)$ if i is a chance node. In general, the utility depends not only on the nodes and branches traversed on the path reaching each terminal node but also on the branch costs and payoffs. In order to calculate utilities at terminal nodes, the history vector should also include branch data. If k is the next downstream node after traversing branch j from node i, we can define a path history or state at node k as

$$\mathbf{h}_k = [\mathbf{h}_i, (i, j, r_j(h_i))],$$

which means that the label on node k is a triple rather than a pair of numbers as described in Section 9.4.1 (p. 363). For example, the history vector associated with the bottom path in Table 9.1 (alternatively, the bottom path in Figure 9-16) is now given by

$$h_6 = [(1, T, -1), (2, n, 0), (3, A, -4), (5, 1, -6)],$$

and the utility can be written as a function of the branch data in that path. If the utility is simply the total payoff of that path, we obtain a value equal to $u_6((1, T, -1), (2, n, 0), (3, A, -4), (5, 1, -6)) = -(1+4+6) = -11$. Similarly, the top path in Figure 9-16 would yield $u_6((1, N, 0), (3, C, -2), (4, s, 10)) = 0 - 2 + 10 = 8$.

In general, the utility function at the terminal node is not simply a total cost (or payoff) equal to the sum of branch and node costs (payoffs). In many applications, particularly if there are long periods elapsing between decisions or commitment of large financial resources, it may not be appropriate to decompose the utility for several different cost expenditures into the sum of the utilities of each separate cost component when we use the nonlinear risk-averse utility functions described in Chapter 6. The utility of a particular path leading to a terminal or value node is represented by $u_i(\mathbf{h}_i)$, which signifies that the utility is a (possibly algebraic) function of the path history taken to get to the terminal node i. This utility function in general depends on the entire history of nodes visited and branches traversed.

The following rollback algorithm finds the optimal solutions:

1. Calculate the path-dependent utility at *all* terminal nodes and set the labels at these nodes to $v_i = v_i(\mathbf{h}_i) = u_i(\mathbf{h}_i)$ for every $\mathbf{h}_i \in \mathcal{H}_i$, $i \in \mathcal{V}$. In general, the utility can depend on all nodes visited and branches traversed in a given path.
2. Find an unlabeled nonterminal node i where all later-occurring nodes k connected to it are labeled.
   a. If $i \in \mathcal{D}$, set

$$v_i(\mathbf{h}_i) = \underset{j \in \mathcal{D}(i|\mathbf{h}_i)}{Max}\{v_k(\mathbf{h}_k)\} \qquad (9.39)$$

   and d_i^* equal to the j that yields this maximum.
   b. If $i \in \mathcal{C}$, set

$$v_i(\mathbf{h}_i) = \sum_{j \in \mathcal{C}(i|\mathbf{h}_i)} p_j(\mathbf{h}_i) v_k(\mathbf{h}_k) \qquad (9.40)$$

As before, branch j is traversed in going from node i to the next node k. Step 2 is repeated until the origin node is labeled, at which point the label on the origin node gives the maximum expected utility solution for the problem.

The recursion mimics the rollback algorithm described in Section 9.4.3, except that the node labels are calculated solely in terms of the utility at terminal nodes.[15] Note that the initial labels on terminal nodes differ from the algorithm in Section 9.4.3 (p. 365) and are, in general, non-zero. The terminal node labels are deterministic utility functions; however, upstream labels have the interpretation of expected utility because an expectation is computed as soon as the first upstream chance node is encountered. In

---

[15] The counterparts for Equations 9.39 and 9.40 are Equations 9.26 and 9.27.

the current procedure utility is calculated once and for all at each terminal node; rollback computations select the best expected utility to be used at successive upstream nodes.

The functional dependence on a state or path history allows us great flexibility in problem formulation because it is possible to incorporate completely general utilities with complex path dependencies at terminal nodes. The v_i's at each node give the maximum expected utility and the d_i^*'s at each decision node yield the optimal decisions from that point forward. Although historical costs may represent fixed costs at an interior node, this construction correctly assumes that the path utility depends on the asset position and total level of resources available at each decision.

The rollback procedure is easily extended to multiattribute decision problems by substituting the appropriate multiattribute utility function $U(\mathbf{r})$ described in Section 7.10 (p. 291). The notation becomes slightly more complicated because one must keep track of the vector of attributes as well as the path history. If we denote path dependence in the multiattribute case by $U(\mathbf{r}; \mathbf{h}_i)$, the assignment of terminal node labels is modified slightly to:

1'. Calculate the path dependent multiattribute utility at all terminal nodes, and assign labels $v_i = v_i(\mathbf{r}; \mathbf{h}_i) = U_i(\mathbf{r}; \mathbf{h}_i)$ at all terminal nodes, that is for every $\mathbf{h}_i \in \mathcal{H}_i, i \in \mathcal{V}$.

All calculations of expected utility at upstream nodes remains unchanged.

Theoretically, subsequent numeric calculations should be able to make use of algebraic dependencies at interior nodes or the storage of intermediate numbers that are used repeatedly. The general experience with available computer technology and the largest decision tree problems currently being solved seems to be that the additional effort to memorize, store, and recall previously computed values is more time-consuming than is the brute-force approach of recomputing values as needed. One should expect to see further efficiencies and improvements in computer algorithms that recognize special structure in the models or new developments in the technology of parallel computation.

9.5.6 Nonlinear Utilities in the Nuclear Reactor Problem

We again return to the nuclear reactor problem where we now use an exponential utility function over the payoff interval $(-11, 11)$ expressed in billions of dollars. We compare the optimal solutions in Table 9.4 for two different levels of risk, $\beta = 2$ and $\beta = 10$, the latter corresponding to the case where the measure of willingness to accept risk is ten times as great at the low end as at the high end. We use the exponential utility function described in Section 6.4.1 (p. 242) with the best result being $r = 11$ and the worst being $r = -11$; thus, an algebraic expression for the utility is

$$u(r) = \frac{\beta}{\beta - 1}\left(1 - \beta^{-\left(\frac{r + 11}{22}\right)}\right). \tag{9.41}$$

The beta parameter is a measure of relative risk aversion for the largest payoff relative to the smallest payoff. The two columns corresponding to $\beta = 1$ correspond to the expected value solution we obtained earlier in Table 9.2. The next nodes, the branch

TABLE 9.4
Rollback solutions for the reactor problem

Node	Next node	Expected utility $\beta = 1$	Optimal policy decision	Expected utility $\beta = 2$	Optimal policy decision	Expected utility $\beta = 10$	Optimal policy decision
1	3	0.852		0.891	Do Not Test	0.951	Do Not
	2	0.869	Test	0.890		0.934	Test
2	3	0.911		0.914		0.934	
	3	0.807		0.855		0.934	
3	4	0.853	Conv.	0.891	Conv.	0.951	Conv.
	4	0.807		0.855		0.934	Conv.
	5	0.911	Adv.	0.914	Adv.	0.923	
	4	0.807	Conv.	0.855	Conv.	0.934	Conv.
	5	0.484		0.509		0.574	
4	6	0.864		0.901		0.959	
	6	0.318		0.396		0.577	
	6	0.818		0.866		0.942	
	6	0.273		0.344		0.518	
	6	0.818		0.866		0.942	
	6	0.273		0.344		0.518	
5	6	1.0		1.0		1.0	
	6	0.182		0.237		0.38	
	6	0		0		0	
	6	1.0		1.0		1.0	
	6	.182		0.237		0.38	
	6	0		0		0	
6	None						

costs and returns, and the branch probabilities are identical with those used in Table 9.2 (p. 369). The only difference between Tables 9.2 and 9.4 is that the numbers in the former give the utilities at the beginning of each branch leading out of the node designated in column 1 to the next nodes specified in column 2, whereas in the latter the optimal utility label at each node is the utility from the origin node to the terminal node via the path that traverses the given node. Because Table 9.2 was calculated using expected value functions that added branch returns and costs as they occurred, the $\beta = 1$ calculations include fixed or "sunk" costs whereas the labels in Table 9.2 do not.

The reader will see that the expected utilities and the optimal policies change as one goes from $\beta = 1$ to $\beta = 2$ to $\beta = 10$. It is interesting that at $\beta = 2$, the decision maker is almost indifferent between the No Test decision and the Test decision at the first node. In the first two cases where risk aversion is low one is willing to invest in a test and then select the advanced reactor if the test report is positive. On the other hand when the decision maker is much more risk averse, the decision to select the conventional reactor always dominates; the best policy is to not make the test because the added cost of testing will not affect the final construction decision. Notice that the optimal policy at the first node has changed because the optimal policy, given a favorable test report, is now the conventional nuclear reactor design.

9.6 REDUCING INFLUENCE DIAGRAMS

An important step in model building is the ability to transform and reduce a given influence diagram into an equivalent influence diagram. We are particularly interested in reduction operations that recursively yield equivalent EFIDs. In this section we examine operations that can be used to reduce, simplify, and eventually solve EFIDs; we spell out in some detail the mathematical structure of two reduction operations, based on the principle of optimality, that can be successively applied to an EFID to yield a smaller but equivalent EFID.

In thinking about influence diagram reduction algorithms it is helpful to understand the effect that different operations have on the ensuing size and complexity of the graph. As seen in Section 9.3.3, arc reversal often adds many arcs but does not alter the number of nodes in the influence diagram. The deletion of a barren node results in the removal of at least one node and all the arcs leading into it, which may, in some circumstances, be a very large number. The removal of an irrelevant chance or decision node always reduces the number of nodes by one; simultaneously it may remove one or more arcs. Common sense suggests that a "good" algorithm is one that removes as many barren and irrelevant nodes as early as possible, delaying the arc reversal operation as late as possible.

We again emphasize that we distinguish between the name of the terminal node, which we refer to as V (for value node) and the result or payoff function denoted by R (or L if we refer to a loss function).

Even though there is general agreement on the importance of and need for direct solutions of influence diagrams, slight differences have developed in the definitions of terms and the details of the procedures for the reduction operations. In this chapter we concern ourselves exclusively with the reduction of influence diagrams in extensive form, sometimes referred to as a "decision-tree network." With large influence diagrams a substantial amount of computational effort may have to be devoted to transformations (arc reversal, barren node removal) that convert the influence diagram in the original formulation to one in extensive form.

Various algorithms have been developed to directly reduce influence diagrams, but few require that the influence diagram be in extensive form until the final set of calculations. However, it seems desirable that one should simplify an influence diagram by removing as many irrelevant and barren nodes as early as possible in the sequence of calculations and then transform the influence diagram into an equivalent one in extensive form. At this stage the influence diagram is subjected to an alternating sequence of conditional expectation and maximization (minimization) reduction operations that invoke the principle of optimality.

9.6.1 Equivalent Influence Diagrams

We extend the notion of equivalent chance influence diagrams used in Section 9.3.2 to equivalent influence diagrams that include decision nodes and a value node. Consider two separate graphs $\mathcal{G}$ and $\mathcal{G}'$ that may have different sets of chance or decision nodes with directed arcs connecting these nodes. As before, we denote the vector of chance nodes in G by $\mathbf{X} = (X_1, X_2, \ldots, X_n)$, the vector of values of these random variables by $\mathbf{x}$, the vector of decision nodes by $\mathbf{D} = (D_1, D_2, \ldots, D_m)$, and the decision vector by $\mathbf{d}$. The direct predecessor and successor nodes of a node i are $\mathcal{P}(i)$ and $S(i)$, respectively. For a given decision rule, the joint distribution of the random quantities is $p(\mathbf{x}|\mathbf{d})$. In general, the number of chance nodes, n, differs from the number of decision nodes, m, and there is only one value node.

Two influence diagrams G and G' are said to be equivalent if:

(1) $p(\mathbf{x}'|\mathbf{d}')$ and $p(\mathbf{x}|\mathbf{d})$ are coherent, and
(2) G' and G have the same optimal expected value.

Equivalent influence diagrams do not necessarily have the same optimal policies even though their expected values are equal.

In (1) we mean that the joint probability distribution on G and G' are the same or that one is the marginal of the other. In the reduction operations described in this and Section 9.6.3 G and G' always differ by one chance or decision node whose only successor in one of the diagrams is the value node. The direct predecessors and direct successors, as well as the new conditional probabilities on the reduced graph will differ from those in the original, but they must be coherent. In (2) we mean that we get the same optimal expected result by solving the problem on G' as we do when we solve it on G.

A simple example may help us understand the concept of an equivalent influence diagram. Consider the EFID G and a typical path in the associated decision tree in Figure 9-19. The value node V has a payoff function $R(X, Y, d)$ where both X and Y are random and $D = d$ is a decision taken before either X and Y have occurred. As before, we use the notation $x \in X, y \in \mathcal{Y}, d \in \mathcal{D}$. For fixed x and d we define a new payoff function,

$$R'(x, d) = E[R(x, Y, d)| X = x, D = d], \tag{9.42}$$

associated with the value node of the reduced influence diagram, denoted by G' in Figure 9-20. We emphasize that the expectation is over Y conditional on known x and d. A typical decision tree path for the new influence diagram is shown in the bottom half

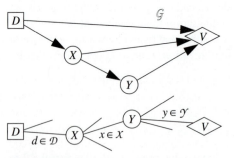

FIGURE 9-19
A four node EFID, G, and a path in the tree.

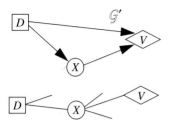

FIGURE 9-20
The reduced EFID $\mathcal{G}'$ with payoff R'.

of the figure. Note that the reduced influence diagram in Figure 9-20 is also in extensive form, the only difference between the two influence diagrams being that chance node Y has been removed by using the expectation operator.

The influence diagram in Figure 9-20 is equivalent to the one in Figure 9-19 when

$$p'(x|\,d) = \sum_{y\,\in\,\mathcal{Y}} p(x,y|\,d) = \sum_{y\,\in\,\mathcal{Y}} p(x|y,\,d)p(y|\,d) \tag{9.43}$$

and $Max_d\,E_{X|D}[R'(X,d\,)|\,D = d] = Max_d\,E_{X|D}[E_{Y|X}\,[R(X,Y,d)|\,X]|\,D = d].$ (9.44)

Equation (9.43) states that the marginal distribution of the chance node X in Figure 9-20 is coherent with the joint distribution of X and Y in Figure 9-19, and Equation (9.44) states that the optimal expected value for the original and reduced EFIDs are equal.

We could equally well have given an example of an EFID with chance nodes preceding a decision node, the latter being the last node on a decision tree path before the value node. Of course, this means that the decision maker has perfect information on all chance nodes that precede the last decision node. Define the new payoff function in the reduced influence diagram (after removing the decision node) as

$$R'(x,y) = Max_d\,[R(X,\,Y,\,d)|\,X = x,\,Y = y]. \tag{9.45}$$

The joint probability of X and Y would not be affected by the removal of the decision node so that we would have an equivalent influence diagram for the reduced EFID as

$$E[R'(X,Y)] = E[Max_d\,R(X,Y,d)|\,D = d]. \tag{9.46}$$

Another example is given in Figure 9-4 (p. 347) for the aircraft part replacement problem. The original influence diagram has two decision nodes, two chance nodes, and one value node; the rightmost one in Figure 9-28 has only one chance node and one value node. However, as we shall show, they have the same optimal policies, the same expected costs, and the probability distribution for the test node in Figure 9-28 is coherent with the probability distributions in Figures 9-4 and 9-27. Thus, the influence diagrams are equivalent.

9.6.2 An Example of EFID Reduction

Suppose that instead of using the rollback procedure to move from right to left on each path of the decision tree in Figure 9-6 (p. 348), we only made the calculations for the terminal node ($V = 4$) and labeled all nodes numbered 3 before making any other roll-back calculations.[16] By doing this in the engine maintenance problem we would take expectations over possible outcomes and thereby substitute a numeric label at each node 3; the label would represent the conditional expectation of the uncertain performance outcomes of the engine. We thereby remove node 3 from the decision tree in Figure 9-6 and obtain the smaller tree shown in Figure 9-21(a). The value node is still numbered $V = 4$, but its payoff is now an expectation.

The smaller decision tree can be associated with the three-node influence diagram shown in Figure 9-21(b). Thus, by using the appropriate conditional expectations and removing node 3, we have obtained not only a smaller and less bushy decision tree in Figure 9-21(a) but also a reduced and equivalent influence diagram in Figure 9-21(b). What should now be clear from the previous section is that the payoff function on terminal node 4 differs from the original payoff function of the terminal node in Figure 9-6. The payoff function on node 4 of Figure 9-21(a) is now given by the conditional expectation calculated from Equation (9.42).

When we initially analyzed the problem it was clear that there were two uncertainties, one the uncertainty due to the outcome of the test, the other the uncertainty for the eventual performance of the repaired or replaced engine. Once node 3 has been removed, the decision maker and model builder can focus on the only remaining uncertainty, which is how the outcomes of the diagnostic test affect optimal decisions and solutions. We want to stress to readers that what is useful and revealing about Figure 9-21 is not

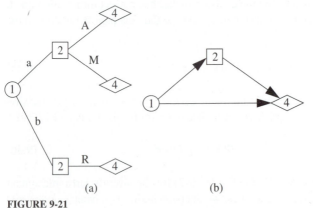

(a)   (b)

FIGURE 9-21
Reducing the influence diagram in Figure 9-5.

---

[16] Such a procedure is not unlike shortest path algorithms in which the label for the shortest distance is placed on a parent node only after shortest path labels to all children of that parent have been calculated.

the uninteresting details of the numeric procedure that might be used but, rather, the insights and revelations that come from an equivalent problem that has one rather than two uncertainties.

9.6.3 Chance Node Removal through Expectation

When an influence diagram is in extensive form and there are no barren or irrelevant nodes, only two reduction operations need to be considered: chance node removal by expectation and decision node removal by maximization (minimization).[17] In this section we consider the reduction operation that is possible when the value node is the only successor of a chance node in an EFID, as is the case illustrated in Figure 9-22. Thus, the chance node is the last node before the value node in paths of the associated decision tree.

Because the value node is the only direct successor of this chance node, it is possible to remove the chance node using conditional expectation. This chance node has the highest index in the list of all chance nodes, and we can therefore number it node n. Because only a chance node is being removed, decisions are not affected, so that $\mathbf{d'} = \mathbf{d}$. The joint probability distribution of the nodes in the reduced graph is given by

$$p(\mathbf{x'}|\mathbf{d}) = p(x_1 \ldots, x_{n-1}|d_1 \ldots, d_m) = \sum_{x_n \in X_n} p(x_1 \ldots, x_n|d_1 \ldots, d_m)$$

$$= \sum_{x_n \in X_n} p(x_n|x_1 \ldots, x_{n-1}, d_1 \ldots, d_m)\, p(x_1 \ldots, x_{n-1}|d_1 \ldots, d_m). \qquad (9.47)$$

Using our notation for direct predecessors, note that the first term inside the summation of Equation (9.47) could have been written as $\sum p(x_n|\, \mathcal{P}(X_n))$ and is a conditional probability distribution that sums to 1.

After removal of the node X_n, the direct predecessors of the new value node are $\mathcal{P'}(V) = \mathcal{P}(V) \cup \mathcal{P}(X_n)\backslash X_n$ where the direct predecessors of X_n may be chance and/or decision nodes. A necessary condition for this reduction operation to be allowed is that the

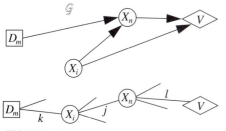

FIGURE 9-22
Chance node with value node as direct successor.

---

[17] Covaliu (1993) defines a *Strict* EFID as an EFID that has been stripped of all barren and irrelevant nodes.

value node be the *only* direct successor of X_n. As a result, the value node in the reduced influence diagram inherits all the direct predecessors of the removed chance node.

The equality of optimal expected return and therefore the equivalence of the reduced influence diagram can be shown because

$$R'(P'(V)) = \sum_{x_n \in X_n} R(P(V)) p(x_n|P(X_n)) \qquad (9.48)$$

is the payoff function on the reduced graph. Of course, the payoff function in the original graph is a function of the direct predecessors of V. Unless X_n is a barren node, the payoff function depends on X_n. The new (primed) payoff function is itself a conditional expectation. Note that the dependence on **d** in Equation (9.48) may be due to the fact that D_m is a direct predecessor of V or that the new value node inherits a dependence on D_m because the latter was originally a direct predecessor of X_n.

The influence diagrams are equivalent when optimal expected return in the original EFID is equal to the optimal expected return in the reduced EFID with the new (primed) payoff function given in (9.48):

$$\underset{d \in \mathcal{D}}{Max}\{E_{\mathbf{X}}[R(P(V))]\} = \underset{d \in \mathcal{D}}{Max}\{\sum_{\mathbf{x'} \in \mathcal{X'}} R'(P'(V)) p(\mathbf{x'}|\mathbf{d})\}. \qquad (9.49)$$

In Figure 9-22 the influence diagram shows that the only direct successor of chance node X_n is the value node; the direct predecessors of the value node are two chance nodes $P(V) = \{X_n, X_i\}$, and the conditional probability of X_n is $p(x_n| d_m, x_i)$. D_m is not a direct predecessor of V.

The direct predecessors of X_n are $P(X_n) = \{D_m, X_i\}$ so that the direct predecessors of V after X_n is removed by expectation are

$$P'(V) = P(V) \cup P(X_n)\backslash X_n = \{X_n, X_i, D_m\}\backslash X_n = \{D_m, X_i\}.$$

The reduced influence diagram is shown in Figure 9-23. In the remainder of this section we denote branch outcomes of chance nodes by j and l and *alternatives at decision nodes by the index k*. Reduction of chance nodes can be continued in the manner just described until the value node has no chance node with the property that its only successor is the value node. When this occurs, the value node *must* either have a decision node with this property or the decision problem has been solved; in the former case reduction can take

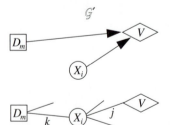

FIGURE 9-23
Reduced EFID after chance node removal.

place by maximization over alternatives available at a decision node. We now consider this possibility in the procedure for reducing an EFID.

9.6.4 Node Removal through Maximization

Consider the operation that arises when a decision node is the last node before the value node in a path leading from origin to value node. In such cases there is perfect information about all chance nodes because they precede the decision node. Clearly, this decision node can be assigned the highest index among all ordered nodes other than the value node; let us assume that this number is m. If the only successor of decision node D_m is the value node and if the direct predecessors of the value node are a subset of the direct predecessors of the decision node, then the decision node can be removed by maximizing (or minimizing, depending on the objective) over available alternatives. An example of such an influence diagram and a typical path in the associated decision tree are shown in Figure 9-24.

The jth chance outcome, $j \in C(X_n)$, is followed by the kth decision alternative, $k \in \mathcal{D}(D_m)$. At node D_m there is perfect information on all chance nodes in the entire tree because the EFID is a proper influence diagram. Thus, this final decision is nothing more than a problem in which there are no uncertainties and the best of a list of deterministic alternatives is to be selected. In this particular example the removal of D_m would be followed by the removal of X_n, but in general one might find yet another decision node.[18] The removal of this decision node does not affect the number, the joint distribution, or the outcomes of any chance nodes, so that in the reduced graph we have $\mathbf{x}' = \mathbf{x}$. The joint probability of all chance variables in the reduced graph is identical to what it was before the removal of D_m.

Because the decision node is last in the list of ordered nodes before the value node, the conditional probability in the reduced graph is

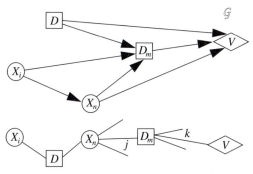

FIGURE 9-24
Decision node with value node as direct successor.

---

[18] With the understanding, of course, that if the direct predecessor of D_m is a decision node, it must not be irrelevant.

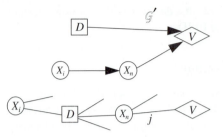

FIGURE 9-25
Reduced EFID following removal of decision node.

$$p(\mathbf{x}' \mid \mathbf{d}') = p(x_1, x_2, \ldots, x_n \mid d_1, d_2, \ldots, d_{m-1}) = p(\mathbf{x} \mid \mathbf{d}'). \tag{9.50}$$

The new payoff function for the reduced graph is given by

$$R'(\mathbf{x}, \mathbf{d}') = R'(P'(V)) = \underset{d \in \mathcal{D}_m}{Max} \{R(P(V))\}. \tag{9.51}$$

Thus, we see that the value node in the reduced graph does not inherit any additional direct predecessors of the decision node that were not already direct predecessors of the value node in the original. In Figure 9-24 the direct predecessors of the value node are $\mathcal{P}(V) = \{D, D_m, X_n\}$, whereas in the reduced diagram in Figure 9-25 we have

$$\mathcal{P}'(V) = \mathcal{P}(V) \backslash D_m = \{D, D_m, X_n\} \backslash D_m = \{D, X_n\}. \tag{9.52}$$

It is easy to show that the original and reduced EFIDs are equivalent. Notice that the optimal policy for decision node m is at most a function of all X_i observed and decisions D_j made before D_m is taken.

Once an influence diagram is represented as an EFID, one can proceed "backward" from the value node and, as shown in the flowchart of Figure 9-26, remove either a chance or decision node by use of the rules just described. Each reduction operation results in a new EFID smaller by one node and, usually, many directed arcs. The associated decision tree becomes progressively less bushy with each reduction operation. By induction, one can continue the reduction operations until only one node remains and the decision problem is solved. If we start with an EFID and follow the rules described, at each step we obtain a new equivalent EFID. We encourage the reader to experiment with his or her own influence diagram formulations as well as problems given at the end of the chapter.

9.6.5   Revisiting the Aircraft Part Problem

We return to the aircraft part replacement problem already discussed in Chapters 3 and 5. In order to illustrate the use of reduction operations on EFIDs, we begin with the client's formulation of the influence diagram shown in Figure 9-4 (p. 347). Although the sequence of reduction operations used in this example is unique, there are generally many possibilities. Note that chance node C has $\mathcal{S}(C) = \{V\}$. Thus, we can take expec-

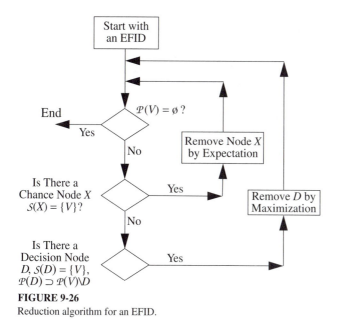

FIGURE 9-26
Reduction algorithm for an EFID.

tations over all outcomes of the C node to obtain Figure 9-27. The direct predecessors of the value node are now given by $\mathcal{P}(V) = \{D_1, D_2, T\}$ and a loss function L is associated with the value node. The conditional expectation for any test outcome $T = t$ and decision alternatives is

$$E[L|\, D_1 = d_1, D_2 = d_2, T = t] = E_{C|T}[E[L|\, D_1 = d_1, D_2 = d_2, C]|\, T = t]$$

$$= \Sigma_c\, E[L|\, D_1 = d_1, D_2 = d_2, C = c]\, p(C = c|\, T = t). \tag{9.53}$$

Note that $E[L|\, d_1, d_2, c]$ is a deterministic function of (d_1, d_2, c) but not of t. Obviously, the sets of alternatives for decision nodes remain unchanged. In Figure 9-27 there are no chance nodes with only the value node as a direct successor; thus a decision node must precede the value node in the EFID.

We now turn our attention to the D_2 node whose only direct successor is the value node and whose direct predecessors are a subset of the direct predecessors of the value

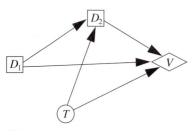

FIGURE 9-27
Removal of the part condition node.

node. A reduction operation using minimization results in one less decision node and three fewer directed arcs after removing those from D_2 to V, from D_1 to D_2, and from T to D_2.

At this point we have reduced the original influence diagram to a much simpler one with one node of each type: the value node V now has a loss function L', a decision node D_1, and a test node T. The resulting EFID in Figure 9-28(a) is identical in structure to the newsboy problems discussed in Chapter 2. It is useful to note that node D_1 cannot be removed at this stage; the direct successor of node T is V and so the next step is to remove the test node by taking expected values:

$$E[L|\, D_1 = d_1] = p(T = \text{``d''})E[L|\, T = \text{``d,''}\ d_1] + p(T = \text{``n''})E[L|\, T = \text{``n,''}\ d_1].\ [19]$$

Thus node T is removed as is the arc from T to V; the result is the EFID in Figure 9-28(b). The final step is to remove the decision node D_1 by minimization:

$$E[L] = Min_d\{E[L|\, D_1 = d]\}$$

$$= Min\ [r/5,\ r/8,\ c + 3r/22] = Min\ [r/8,\ c + 3r/22].$$

As the reader is already aware from earlier discussions of this problem, the optimal solution depends on whether c is larger or smaller than the value of perfect information. At each stage in the sequence of reductions, a simpler EFID is obtained whose optimal decisions and solutions are embedded within the optimal solutions of the original formulation.

9.7   SUMMARY AND INSIGHTS

The chapter focuses on some of the issues that arise when formulating large-scale decision problems with influence diagrams and decision trees. As the size of a problem grows, it is increasingly important to understand the interrelationships among its random quantities, the number of choices and assessments that are essential to the formulation, and data relevant to the decision problem. It also becomes increasingly important while building a model to make simple checks of the relevance, consistency, and meaning of nodes and, directed arcs and the timing of information. Establishing the relevance of in-

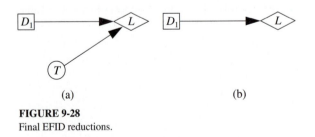

(a)   (b)

FIGURE 9-28
Final EFID reductions.

---

[19] "d" for forecast of defective; "n" for non-defective.

formation and decisions is greatly enhanced by following some simple rules that can be used to check the formulation and design of influence diagrams as well as some simple techniques for simplifying and reducing them to more tractable and intuitive structures.

Influence diagrams that provide meaningful representations of decision problems conform to a simple set of guidelines for the construction of proper influence diagrams (PIDs). The use of an extensive form influence diagram (EFID) helps us understand how the principle of optimality can be simultaneously invoked on decision trees and influence diagrams in order to calculate optimal expected returns. We note the useful role of graphical reduction techniques and the important insights and interpretations that can be gleaned from direct reduction before one undertakes the estimation of cost and the difficult assessment of conditional probabilities.

In an EFID the removal of a chance node by expectation causes the value node in the reduced influence diagram to inherit all the direct predecessors of the original value node and the chance node. The payoff function for the value node in the reduced graph is a conditional expectation that can be expressed as a function of the inherited predecessor chance and decision nodes. The removal of a decision node by maximization (minimization) causes the value node in the reduced influence diagram to inherit only the direct predecessors of the original value node. The new payoff function for the value node in the reduced graph is therefore only a function of these inherited predecessor chance and/or decision nodes.

Decision trees for realistic problems often become very large, and it may become impractical to draw or even visualize them; to a great extent this may be due to the inexperience of a model builder who does not fully understand the structure of the decision problem, or how it can be represented in a more compact form suitable for solution. One can make effective use of reduction operations on influence diagrams to quickly reduce the dimensionality of the influence diagram and the "bushy mess" of very large decision trees. These simplifying operations are made easier by the early recognition and removal of irrelevant and barren nodes. By reducing the influence diagram and the associated decision trees early in the model-building process one can focus on the essential rather than the irrelevant and can often spot inconsistencies or difficulties that are not apparent to a casual observation and understanding of the decision problem. These concerns for simplicity may appear to be a trivial issue to the mathematician but to the model builder they are often the difference between success and failure in solving realistic decision problems.

A reduction operation that removes one (nonvalue) node in an extensive form influence diagram is equivalent to using a one-step rollback procedure at each terminal node of the corresponding decision tree and then replacing the original chance or decision node with a terminal node having an equivalent payoff function. If at any stage the procedure results in an influence diagram whose conditional probability structure does not make sense or leads to an unreasonable result, one should reconsider the original formulation before gathering data, estimating costs, or undertaking the difficult assessment of conditional probabilities. Perhaps the most important aspects of reduction operations are the interpretation and insights derived from the reduced influence diagrams.

A coherent notation for the solution of decision problems on influence diagrams and decision trees must take into account the sequence of path-dependent outcomes and decisions. Any algorithm for solution of large decision problems must therefore include a state or path labeling procedure calculated prior to rollback. The first part of a two-step procedure calculates states resulting from paths traversed in the left-to-right direction, the second calculates optimal decisions in the right-to-left direction. Whether the path-dependent utilities are calculated at the end of each path (terminal nodes) or during the rollback calculations depends on whether the utility function is additive and/or linear, the particular application and the form in which data and relevant forecasts are made available.

The first explicit use of chains of influence among conditionally independent random quantities appeared in a paper by Good (1961). To our knowledge the first explicit formulation of an influence diagram linking decision variables and random quantities with conditional independence assumptions was reported by Miller, Merkhofer, Howard, Matheson and Rice (1976). One of the first reported uses of an influence diagram in formulation of a decision problem was by North and Stengel (1982).

The notion of an influence diagram that could be associated with a decision tree was first used by Howard and Matheson (1984) and called a decision tree network. The notion of a proper influence diagram and the use of influence diagrams in extensive form was first made by Smith (1989). The decision tree network and the extensive form influence diagram have essentially the same interpretation. The definition and first use of barren nodes was introduced by Shachter (1986) and the use of irrelevant nodes was first introduced by Covaliu (1993). The use of sequential diagrams that capture both the asymmetric and sequential characteristics of decision problems has been introduced by Covaliu (1993) and Covaliu and Oliver (1995).

The study of chance influence diagrams and belief nets has, by now, a very extensive literature (see, for example, Shenoy and Shafer (1986), Shachter (1988), Smith (1989), Spiegelhalter (1986), (1989), Lauritzen and Spiegelhalter (1988), and Spiegelhalter and Lauritzen (1990)). A fundamental requirement is that probability calculations are coherent and observe the laws of probability; the basic tool for inference is the arc reversal operation which corresponds to the use of Bayes' rule.

The origins of path-dependent histories in rollback calculations is difficult to trace. Apparently, the first algorithm for a path-dependent history was described by Rousseau and Matheson (1967) in analyses of the flight of the *Voyager* spacecraft. The first presentations of a general purpose algorithm in the open literature were by Cheung and Kirkwood (1989), Call and Miller (1990), and Kirkwood (1991, 1993), although numerous undocumented versions were claimed to have been used in proprietary software products. Depending on the application, relevant information may be carried in a data structure as a list or array; the particular choice depends to a great extent upon the computer language being used and the extent to which repeated calculations are required or may be stored efficiently for later recall.

Kirkwood (1991) uses a node-numbering convention for decision trees that is identical with the influence diagram; these ideas have been expanded and exploited by Covaliu (1993) and Covaliu and Oliver (1995). General mathematical structures to reflect more complex path dependencies, asymmetry in the outcome sets of the chance variables and decisions, and nonlinear utilities can be incorporated in node and branch labels. Articles by Shachter (1986), Smith (1989), and more recently the thesis by Covaliu (1993) suggest algorithms that can be used to reduce influence diagrams directly. Few require that the influence diagram be in extensive form until the final step.

It is possible to represent data for very large decision trees compactly in lookup tables or spreadsheets. Such structure has been exploited in several algorithms, Call and Miller (1990), Kirkwood (1991, 1994), Covaliu (1993), and Covaliu and Oliver (1995), with apparent success in reducing computation times. General algorithms for computing and retaining noninferior points with many attributes in large decision trees are discussed by Haimes, Li, and Tulsiani (1990).

The transformation and reduction of influence diagrams has been studied by many authors, and the list of proposed procedures and algorithms is growing rapidly. To name a few, these include Olmsted (1983), Howard and Matheson (1984), Shachter (1986), Matheson (1989), Smith (1989), Call and Miller (1990), Tatman and Shachter (1990), and more recently Covaliu (1993). Smith (1989) and Covaliu (1993) have shown that during the reduction of EFID's by removal of chance nodes either the value node eventually has a decision node as its only direct predecessor or the decision problem has been solved. An overview of solution methods for decision analysis is contained in Kirkwood (1992c).

PROBLEMS

9.1 Assemble a spreadsheet with data for the aircraft part replacement problem described in Section 5.3.1, clearly indicating branch costs and probabilities. Calculate the optimal policies using the spreadsheet data and a tree whose node numbers coincide with those of the influence diagram.

9.2 Calculate the optimal policies for the decision on the conventional/advanced reactor Section 9.4.4 (p. 366) when we allow the possibility of selecting the advanced nuclear reactor in the no-test case.

9.3 In the metastatic cancer problem described in Section 9.3.5, make a list of the conditional probabilities that must be assessed for you to obtain the joint probability of the five factors stated for the problem. In terms of these probabilities, obtain expressions for the conditional probability

  a. That increased serum calcium is present when it is not possible to observe the presence or absence of tumors.

  b. That a patient has no cancer given no severe headache.

  c. That a patient has cancer given a severe headache and the presence of increased serum calcium.

9.4 As in the text in Section 9.3.5 let m denote Metastatic Cancer, s denote Increased Serum Calcium, t denote the presence of a Tumor, c denote a Coma, and h denote a Severe Headache. Assume that $p(m) = 0.2$, $p(s|m) = 0.8$, $p(s|\overline{m}) = 0.2$, $p(t|m) = 0.2$, $p(t|\overline{m}) = 0.05$. If a tumor is confirmed, show that the conditional probability of having cancer is increased two and a half times over the value it would have when there is no tumor.

9.5 In the same problem express $p(m,s,t|h)$ in terms of $p(t|h)$, and show that if the probability of a brain tumor given a severe headache is found to be 0.104, then the joint probability of increased calcium, tumor, and metastatic cancer has increased about 30% given evidence of severe headaches.

9.6 Construct an EFID of your choice where the typical path in the decision tree is (X, D_1, Y, D_2, V). Reduce the influence diagram by removing D_2, Y, and D_1. Write out the explicit form for the payoff function at each step, and show the equivalence of all three influence diagrams (the original plus the two reductions).

REFERENCES

Aitchison, J., and I. R. Dunsmore. (1975). *Statistical Prediction Analysis.* Cambridge University Press, Cambridge.

Baird, B. F. (1989). *Managerial Decisions under Uncertainty.* Wiley, New York.

Barlow, R. E. (1988) "Using Influence Diagrams." Proceedings of the conference on *Accelerated Life Testing and Experts' Opinions in Reliability,* Edited by C. A. Clarotti and D. V. Lindley, North Holland Physics Publishing, 145–157.

Behn, R. D., and J. W. Vaupel. (1982). *Quick Analysis for Busy Decision Makers.* Basic Books, New York.

Bellman, R. E., and S. E. Dreyfus. (1962). *Applied Dynamic Programming,* Princeton University Press, Princeton, NJ.

Belobaba, P. (1987). "Airline Yield Management. An Overview of Seat Management Control." *Transportation Science,* 21, 63–73.

———. (1989). "Application of a Probabilistic Decision Model to Airline Seat Management Control." *Operations Research,* 37, 183–197.

Belton, V., and T. Gear. (1983). "On a Shortcoming of Saaty's Method of Analytic Hierarchies." *Omega,* 11, 228–230.

Brier, G. W. (1950). "Verification of Forecasts Expressed in Terms of Probability." *Monthly Weather Review,* 78, 1–3.

Bunn, D. W. (1978). *The Synthesis of Forecasting Models in Decision Analysis.* Series in Interdisciplinary Systems Research, No. 56. Birkhauser, Basel, Switzerland.

———. (1984). *Applied Decision Analysis.* McGraw-Hill, New York.

Call, H. J., and W. A. Miller. (1990). "A Comparison of Approaches and Implementations for Automating Decision Analysis." *Reliability Engineering and Systems Safety,* 30, 115–162.

Chankong, V., and Y. Y. Haimes. (1983). *Multiobjective Decision Making.* North-Holland, New York.

Cheung, D. C., and C. W. Kirkwood. (1989). *EXPRESSION TREE: A Decision Tree Analyzer with Variables and Expressions.* Report No. DIS-88/89-9. Department of Decision and Information Systems, Arizona State University, Tempe.

Chickering, A. L. (1994). "How Much Is a Life Worth." *San Francisco Chronicle Newspaper,* Editorial Page, Feb. 9.

Clemen, R. (1990). *Making Hard Decisions: An Introduction to Decision Analysis.* PWS-Kent, Boston.

Corner, J. L., and C. W. Kirkwood. (1991). Decision Analysis Applications in the Operations Research Literature, 1970–1989. *Operations Research*, 39, 206–219.

Covaliu, Z. (1993). "Representation and Solution of Decision Problems Using Decision Diagrams." Ph.D. Thesis. Department of Industrial Engineering and Operations Research, University of California, Berkeley.

———, and R. M. Oliver. (1995). Representation and Solution of Decision Problems Using Sequential Decision Diagrams. *Management Science* (to appear).

Crook, J. N., and D. B. Edelman, Editors (1992). *Journal of Mathematics Applied in Business and Industry,* Oxford University Press, 4(1), 1–123.

DeGroot, M. H. (1970). *Optimal Statistical Decisions.* McGraw-Hill, New York.

French, S. (1989). *Readings in Decision Analysis.* Chapman & Hall, New York.

———. (1993). *Decision Theory: An Introduction to the Mathematics of Rationality.* Ellis Horwood Limited, Chichester, England.

Gillman, L. (1991). "The Car and Goats Fiasco." *Focus.* (The American Mathematical Monthly Newsletter), 11(June), 8.

———. (1992). "The Car and the Goats." *The American Mathematical Monthly,* 99(1), 3–4.

Good, I. J. (1961). "A Causal Calculus." *British Journal of Philosophy of Science,* 11, 305—318.

Granger, C. W. J., and P. Newbold. (1975). Economic Forecasting: An Atheists Viewpoint. In G. A. Renton, (Ed.), *Modeling the Economy.* Heinemann, London.

Haimes, Y. Y., D. Li, and V. Tulsiani. (1990). "Multiobjective Decision Tree Analysis." *Risk Analysis,* 10(1), 111–129.

Hand, D. J. (1981). *Discrimination and Classification.* Wiley, New York.

Hijn, J. M., and C. R. Johnson. (1988). "Evaluation Techniques for Paired Ratio-Comparison Matrices in a Hierarchical Decision Model." In W. Eichhorn (Ed). *Measurement in Economics,* p. 269–288, Physica-Verlag, Heidelberg.

Hillier, F., and G. Lieberman. (1994). *Operations Research.* McGraw-Hill, New York.

Hopper, M. A., and E. M. Lewis. (1991). "Development and Use of Credit Profit Measures for Account Management." *IMA Journal of Mathematics Applied in Business and Industry,* 4(1), 3–17.

Howard, R. A. (1988). "Decision Analysis: Practice and Promise." *Management Science* 34(6), 679–695

———. (1989). *From Influence to Relevance to Knowledge.* In Oliver and Smith (1990), 3–23.

———, and J. E. Matheson. (1984). *Readings in the Principles and Applications of Decision Analysis,* vols. I, II. Strategic Decision Group, Menlo Park, CA.

———, ———, and D. W. North. (1972). "The Decision to Seed Hurricanes." *Science,* 176, 1191–1202.

Katz, R. W., and A. H. Murphy. (1987). "Quality-Value Relationships for Imperfect Information in the Umbrella Problem." *The American Statistician,* 41(3), 187–189.

———, and ———. (1990). "Quality-Value Relationships for Imperfect Weather Forecasts in a Prototype Multistage Decision-Making Model." *Journal of Forecasting,* 9, 75–86.

Keeney, R. L. (1980). *Siting Energy Facilities.* Academic Press, New York.

———, and H. Raiffa. (1976). *Decisions with Multiple Objectives: Preferences and Value Trade-offs.* Wiley, New York.

Kirkwood, C. W. (1991), *ADAM2: An Algebraic Decision Analysis Modelling System for Research.* Technical Report No. DIS 91/92-9. Department of Decision and Information Systems, Arizona State University, Tempe.

———. (1992a). *Notes on Applied Multiattribute Preference Theory.* Technical Report No. DIS 91/92-12. Arizona State University, Tempe.

———. (1992b). *Notes on Attitude toward Risk Taking and the Exponential Utility Function.* Technical Report No. DIS 91/92-8. Department of Decision and Information Systems, Arizona State University, Tempe.

———. (1992c). "An Overview of Methods for Applied Decision Analysis." *Interfaces,* 22(6), 28–39.

———. (1993). "An Algebraic Approach to Formulating and Solving Large Models for Sequential Decision Under Uncertainty." *Management Science,* 39(7), 900–913.

———. (1994). "Implementing an Algorithm to Solve Large Sequential Decision Analysis Models." *IEEE Transactions on Systems, Man, and Cybernetics,* 24, 1425–1432.

Klein, I. (1993). "Comments on 'Let's Make a Deal: The Player's Dilemma'." *The American Statistician,* 45, 284-287, 47, 82–83.

Kmietowicz, Z. W., and H. Ding. (1993). "Statistical Analysis of Income Distribution in the Jiangsu Province of China." *The Statistician*, 42(2), 107–121.

Larson, H. J. (1982). *Introduction to Probability Theory and Statistical Inference,* 3rd ed. Wiley, New York.

Lauritzen, S. L., and D. J. Spiegelhalter. (1988). "Local Computations with Probabilities on Graphical Structures and Their Application to Expert Systems." *Journal of the Royal Statistical Society Series B,* 50, 157–224.

Lee, P. M. (1989). *Bayesian Statistics: An introduction.* Oxford University Press, London.

Lewis, E. M. (1991). *An Introduction to Credit Scoring.* The Athena Press, San Rafael, CA.

Lindley, D. V. (1982). "The Improvement of Probability Judgements." *Journal of the Royal Statistical Society, Series A,* 145, Part 1, 117–126.

———. (1985). *Making Decisions,* 2d ed. Wiley, London

Luce, R. D., and H. Raiffa (1957), *Games and Decisions.* Wiley, New York.

Matheson, J. E. (1989). "Using Influence Diagrams to Value Information and Control." In Oliver and Smith (1990). pp. 25–48.

McLachlan, G. J. (1992). *Discriminant Analysis and Statistical Pattern Recognition.* Wiley, New York.

Mendenhall, W., D. D. Wackerly, and R. L. Scheaffer. (1990). *Mathematical Statistics with Applications,* 4th ed., PWS-Kent, Boston.

Miller, A. C., M. W. Merkhofer, R. A. Howard, J. E. Matheson, and T. R. Rice. (1976). "Development of Automated Aids for Decision Analysis." Stanford Research Institute, Menlo Park, CA.

Miller, M. E., C. D. Langefeld, W. M. Tierney, S. L. Hui, and C. J. McDonald. (1993). "Validation of Probabilistic Predictions." *Medical Decision Making,* 13, 49–58.

Morse, P. M., and G. E. Kimball. (1951). *Methods of Operations Research.* MIT Technology Press, Cambridge, MA.

Murphy, A. H. (1991). "Forecast Verification: Its Complexity and Dimensionality." *Monthly Weather Review,* 119(7), 1590–1601.

———, and R. L. Winkler. (1987). "A General Framework for Forecast Verification." *Monthly Weather Review,* 115(7), 1330–1338.

———, and ———. (1992). "Diagnostic Verification of Probability Forecasts," *International Journal of Forecasting,* 7, 435–455.

———, B. G. Brown, and Y. Chen. (1989). "Diagnostic Verification of Temperature Forecasts." *Weather and Forecasting,* 4(4), 485–501.

Nelson, T. O. (1986). "ROC Curves and Measures of Discrimination Accuracy: A Reply to Swets," *Psychological Bulletin,* 100, 128–132.

North, D. W., and D. N. Stengel. (1982). "Decision Analysis of Program Choices in Magnetic Fusion Energy Development," *Management Science,* 28, 276–288.

O'Hagan, A. (1988). *Probability: Methods and Measurement.* Chapman and Hall, London.

Oliver, R. M. (1992). "Comments on 'Quality/Value Relationships for Imperfect Weather Forecasts' by Katz and Murphy." *Journal of Forecasting,* 11(1), 81–86.

———, and J. Q. Smith. (1990). "Influence Diagrams, Belief Nets and Decision Analysis." Proceedings of a Conference on Influence Diagrams for Inference, Prediction and Decision Making held at Berkeley, CA, May 9–11, 1988. Wiley, New York.

Olmsted, S. M. (1983). "On Representing and Solving Decision Problems." Ph.D. Thesis. Engineering-Economic Systems Department, Stanford University, Stanford, CA.

Phillips, R. L. (1991). *Yield Management: What Is It? When Can It Help?* Technical Report No. 8300-178. Decision Focus, Mountain View, CA.

———. (1993a). *A Dynamic Algorithm for Determining Optimal Airline Passenger Mix.* Technical Report No. 8085-11. Decision Focus, Mountain View, CA.

———. (1993b). *A Marginal Value Approach to Airline Origin and Destination Revenue Management.* Technical Report. Decision Focus, Mountain View, CA.

———, D. W. Boyd, and T. A. Grossman. (1991). "An Algorithm for Calculating Consistent Itinerary Flows." *Transportation Science,* 25, 225–239.

Pratt, J. W. (1964). "Risk Aversion in the Small and in the Large." *Econometrica,* 32, 122–136.

———, H. Raiffa, and R. Schlaifer. (1964). "The Foundations of Decision under Uncertainty: An Elementary Exposition." *Journal of the American Statistical Association,* 59, 353–375.

Raiffa, H. (1968). *Decision Analysis.* Addison-Wesley, Reading, MA.

Rao, U. L. G., and A. Y. P. Tam. (1987). "An Empirical Study of Selection and Estimation of Alternative Models of the Lorenz Curve." *Journal of Applied Statistics,* 14, 275–280.

Ross, S. (1990). *A First Course in Probability.* 3rd ed. Wiley, New York.

Rousseau, R. D., and J. E. Matheson. (1967). *Computer Programs for Decision Analysis: Tree Generation and Analysis.* SRI Report No. 188531-168. Stanford Research Institute, Menlo Park, CA.

Saaty, T. L. (1990). "How to Make a Decision: The Analytic Hierarchy Process." *European Journal of Operations Research*, 48, 9–26.

Shachter, R. D. (1986). "Evaluating Influence Diagrams." *Operations Research*, 34(6), 871–882.

———. (1987). *DAVID Influence Diagram Processing System for the Macintosh, User's Manual.* Center for Academic Computing. Duke University, Durham, NC.

———. (1988). "Probabilistic Inference and Influence Diagrams." *Operations Research*, 36(4), 589–604.

Shenoy, P. P., and G. Shafer. (1986). "Propagating Belief Functions with Local Computations," *IEEE Expert*, 1, 43–52.

Smith, J. Q. (1988a), *Decision Analysis, A Bayesian Approach.* Chapman & Hall, London.

———. (1988b). "Models, Optimal Decisions and Influence Diagrams." Bayesian Statistics III, Conference Proceedings, pp. 765–776. Oxford University Press, London.

———. (1989). "Influence Diagrams for Bayesian Decision Analysis." *European Journal of Operational Research*, 40, 363–376.

Spiegelhalter, D. J. (1986). "Probabilistic Reasoning in Predictive Expert Systems." In *Uncertainty in Artificial Intelligence,* L. M. Kanal and J. Lemmer (Eds.), pp. 357–370. North Holland, Amsterdam.

———. (1989). "Fast Algorithms for Probabilistic Reasoning in Influence Diagrams, with Applications in Genetics and Expert Systems." Proceedings of a Conference on Influence Diagrams for Inference, Prediction and Decision-Making held at Berkeley, CA, May 9–11, 1988. Wiley, New York.

———, and S. L. Lauritzen. (1990). "Sequential Updating of Conditional Probabilities on Directed Graphical Structures." *Networks*, 20, 579–605.

Swets, J. A. (1986). "Indices of Discrimination or Diagnostic Accuracy: Their ROCs and Implied Models." *Psychological Bulletin*, 99(1), 100–117.

———, R. M. Pickett, S. F. Whitehead et al. (1979). "Assessment of Diagnostic Technologies." *Science*, 205, 753–759.

Taha, H. A. (1987). "Operations Research: An Introduction," 4th ed. Macmillan, New York.

Tatman, J. A., and R. D. Shachter. (1990). "Dynamic Programming and Influence Diagrams." *IEEE Trans. on Systems, Man and Cybernetics.* 20, 2.

Thomas, L. C., J. N. Crook, and D. B. Edelman. (1992). *Credit Scoring and Credit Control.* Clarendon Press, Oxford, England.

Tierney, J. (1991). "Behind Monty Hall's Doors: Puzzle, Debate and Answer?" *The New York Times*, July 21.

Tukey, J. W. (1961), "Discussion Emphasizing the Connection between Analysis of Variance and Spectrum Analysis", *Technometrics*, vol. 3, pp. 191–219.

Urkowitz, H. (1983). *Signal Theory and Random Processes.* Artech House, Dedham, MA.

Vargas, L. G., and F. Zahedi. (1993). *Mathematical and Computer Modelling*, 17, (4–5).

von Neumann, J., and O. Morgenstern. (1953). *Theory of Games and Economic Behavior.* Princeton University Press, Princeton, NJ.

vos Savant, M. (1991). "Ask Marilyn," *Parade Magazine,* 17 February.

———. (1992), *Ask Marilyn,* pp. 199–212. St. Martin's Press, New York.

Weinstein, M. C., and H. V. Fineberg. (1980). *Clinical Decision Analysis.* Saunders, Philadelphia.

Wilkie, A. D. (1992). "Measures for Comparing Scoring Systems." Proceedings of a Conference on Credit Scoring and Credit Control, L. C. Thomas, J. N. Crook, and D. B. Edelman (Eds.), pp. 123–138. Oxford University Press, Oxford.

Winkler, R. L. (1972). *Introduction to Bayesian Inference and Decision.* Holt, Rinehart, and Winston, New York.

———. (1990). "Decision Modeling and Rational Choice: AHP and Utility Theory." editor of a set of papers by J. S. Dyer, T. L. Saaty, P. T. Harker, and L. G. Vargas in *Management Science*, 36, 247--275.

Yates, J. F., and S. P. Curley. (1985). "Conditional Distribution Analyses of Probabilistic Forecasts." *Journal of Forecasting,* 4, 61–73.

AUTHOR INDEX

SUBJECT INDEX